ENVIRONMENTAL SCIENCE, ENGINE

POTABLE WATER AND SANITATION

ENVIRONMENTAL SCIENCE, ENGINEERING AND TECHNOLOGY

Additional books in this series can be found on Nova's website under the Series tab.

Additional E-books in this series can be found on Nova's website under the E-books tab.

ENVIRONMENTAL SCIENCE, ENGINEERING AND TECHNOLOGY

POTABLE WATER AND SANITATION

JOEL M. MCMANN
EDITOR

Nova Science Publishers, Inc.
New York

For permission to use material from this book please contact us:
Telephone 631-231-7269; Fax 631-231-8175
Web Site: http://www.novapublishers.com

LIBRARY OF CONGRESS CATALOGING ... DATA

Potable water and sanitation / editor, Joel M. McMann.
p. cm.
Includes bibliographical references and index.
ISBN 978-1-61122-319-4 (hardcover)
1. Sanitation. 2. Drinking water--Purification. 3.
Water-supply--Developing countries. I. McMann, Joel M. II. Title.
TD155.P68 2010
363.6'1--dc22

2010038655

Published by Nova Science Publishers, Inc. † New York

CONTENTS

Preface		**vii**
Chapter 1	Sanitation of Fresh-cut Fruit and Vegetables: New Trends, Methods and Impacts *Francisco Artés, Perla Gómez Di Marco, Alejandro Tomás-Callejas and Francisco Artés-Hernández*	**1**
Chapter 2	Health Impact of Sanitation in India *Brijesh C. Purohit*	**37**
Chapter 3	Hydrogeology and Groundwater Resources of Ghana: A Review of the Hydrogeology and Hydrochemistry of Ghana *Bruce Banoeng-Yakubo, Sandow Mark Yidana, Joseph Owolabi Ajayi, Yvonne Loh and Daniel Asiedu*	**77**
Chapter 4	Characterization of the Quality of Groundwater from the Major Aquifers in Ghana *Sandow Mark Yidana*	**115**
Chapter 5	The Challenge of Sustainable Waste Management in Ghana – Practices, Problems and Possibilities *Service Opare*	**151**
Chapter 6	Sanitation Development and Roles of Japan *Yoshiaki Tsuzuki*	**179**
Chapter 7	Industry Structure and Development in the Water and Wastewater Industry *Malcolm Abbott and Bruce Cohen*	**203**
Chapter 8	Plant Derivatives as Natural Sanitizers for the Control of Fungal Invasion *Ahmed A. Tayel*	**215**
Index		**227**

PREFACE

Groundwater is the mainstay of the rural water economy of Ghana. The resource has been tapped at various levels in the hydrogeological basins to supply potable water resources to communities. Also discussed are the water resources development and sustainable management in the Republic of Sudan. This book reviews the structural issues that have confronted the reform of the potable water supply and wastewater industries, and examines them in light of the forces that are expected to impact on this issue across the globe.

Chapter 1 – The association between foodborne outbreaks and fresh-cut plant produces has noticeably increased throughout the last years and, consequently, consumers are concerned about the quality and safety of this kind of commodities. Sanitation is the most important target for keeping overall quality of fresh-cut or minimally processed plant produces. Washing-disinfection is the only step that effectively reduces microbial load all over their production chain. The most commonly sanitizer used in the fresh-cut industry is chlorine. However, it has been identified some problems related to its use, such us potentially dangerous by-products formation upon reaction with organic matter with adverse effects on human health, its pH dependence, its gas emission that may affect worker's safety, and its relative poor efficiency in a number of horticultural products. Also, some European countries have forbidden its use and this may lead to new regulatory restrictions in the near future. Several studies have reported the efficiency of chemical and physical alternatives to be used for replacing chlorine. These possible options, which are presently considered as promising treatments, include antimicrobial solutions, electrolyzed water, O_3, UV-C radiation, intense light pulses, superatmospheric O_2 packaging, N_2O and noble gases-enriched packaging atmospheres. Theses emerging alternatives can be applied alone or combined. In this chapter, recent studies accomplished with these eco-innovative sanitizers on fresh-cut plant commodities are reviewed.

Chapter 2 - Despite significant investments over the last 20 years, India still faces the most daunting sanitation challenge in the world. In comparison to other countries, its sanitation is rated as the second worst in the world after China. The current situation depicts that only 28 per cent of people in rural areas, who comprise two-thirds of India's billion plus people, had access to toilets. Though in view of Millennium Development Goals, it aims at ending the appalling state of sanitation to half by 2015 which might have put severe burden on community, health services and cause losses to productivity to the nation.

This paper aims at assessing the status of sanitation and its direct and indirect impact on health in India. It integrates impact of water and sanitation and the efficacy of health systems and emphasizes that even in the presence of a public health system, these factors play an important role in influencing health outcomes. The study is divided into three sections. An overview of evolution of water and sanitation sector programmes in the first section of the paper is followed by an assessment of its direct impact on incidence of selected diseases in major Indian states. Using sub-state level (or district level) information, section third provides case studies of three Indian states belonging to rich, middle income and poor category respectively. The results of stochastic frontier analysis for these states indicate that the differential impact of health systems in terms of health outcomes could be explained by regional disparities and sanitation factors both at the state and district level. Based on the author estimates, the final section provides policy measures which emphasize that more efficient health outcomes could be achieved by overcoming investment needs through different innovative measures. These may, depending upon regions and locations across the country, include better coordination between social sector and economic policies, higher subsidy to sanitation, improved management of existing programmes, community participation and better opportunities for women in health, education and other rights.

Chapter 3 - This paper surveys the aquifers and the groundwater resources landscape in Ghana to identify the critical resource issues constraining the development and management of groundwater throughout the country. The research is based on a collation of the recent data on the hydrogeological properties of the different geological terrains in the country. This study finds that there are five main hydrogeological provinces distinguished by their well yields, lithology, and groundwater quality. This latest hydrogeological zonation, which is an improvement over previous works consists of the Birimian Province, the Crystalline Basement Granitoid Complex Province, the Voltaian Province, the Pan African Province, and Coastal Sedimentary Province. The hydrogeological properties of each province are discussed. The most prolific aquifers in the country have been identified among the fractured and weathered zones within the Togo Series, Buem Formation, Birimian and Tarkwaian System rocks. Aquifers of these rock formations also offer groundwater of the best quality for most uses in the country.

Hydrochemical data gathered from wells drilled through all the hydrogeological terrains in the country for this study suggest that groundwater hydrochemistry is controlled principally by the weathering of silicate minerals and cation exchange activity. These processes are pervasive throughout the country. In isolated areas, groundwater contamination due to domestic waste discharge and agricultural activities has been noted. Four groundwater types have been identified based on Q-mode hierarchical cluster analysis and conventional graphical methods. In the coastal areas such as the Keta basin, seawater intrusion has been identified as one of the major influences on hydrochemistry of groundwater in the shallow unconfined aquifers. On the basis of the hydrochemistry, groundwater suitability for irrigation purposes is copiously discussed.

The aquifer management problems can be posed as follows. There is inadequate knowledge base of the aquifer geometry and safe yield characteristics of most of the aquifers in the country. There is also inadequate information to address the status and trends in groundwater quality for most areas of the country. The following emerging issues have been identified: anthropogenic effects on groundwater quality; salt-water intrusion in coastal aquifers, groundwater depletion or inadequate supplies from groundwater to meet human needs; uncertainties about recharge and sustainability due to increased abstraction; as well as lack of institutional capacity to respond adequately to sustainable groundwater use. Recommendations to ameliorate these problems include the strengthening of institutional capacity for comprehensive national evaluation of groundwater resources, including aquifer delineation and pro-active strategies to reduce groundwater abstraction to safe yield levels and adoption of aquifer protection strategies for overall protection of groundwater quality.

Chapter 4 - In Ghana, groundwater plays a very crucial role in the overall water delivery system, especially in the rural areas. This resource is being assessed for possible exploitation for irrigation projects since the over reliance on rain-fed agriculture does not auger well for national development in the face of dwindling and equally erratic rainfall patterns in the country. In this study, Q-mode hierarchical cluster analysis (HCA) and R-mode factor analysis were used jointly with Geographic Information Systems (GIS) to determine the spatial groundwater quality types in Ghana, and the spatial variation of the major sources of variation in the hydrochemistry of groundwater in the country. In addition to assisting in the understanding of the different groundwater types in the country, the study demonstrates the applicability of multivariate statistical methods in groundwater quality classification in Ghana. This study finds that two major groundwater types exist in the country: the Na-Cl groundwater types which dominate the coastal aquifers where groundwater salinity is high and attributed partly to seawater intrusion and Na-K-HCO_3 groundwater types which dominate aquifers farther inland. It has also been determined that the effects of seawater intrusion on groundwater quality in Ghana is significant along the coast where over extraction of the resource has led to seawater up-coning. Farther inland, groundwater hydrochemistry is controlled principally be mineral weathering processes, among which silicate mineral weathering has been noted to be the predominant. The effects of chemical fertilizers, organic manure, and domestic sewage are significant in areas where the aquifers are relatively unconfined and shallow or where the well head is poorly protected and where agricultural activities are high. High nitrate concentrations have been noted in groundwater sample from the northern parts of the country, some parts of the forested areas and the Keta strip of the Volta region of Ghana. The oxidation of organic carbon is a minor factor because its effects are relatively insignificant when compared to mineral weathering and anthropogenic contamination. Interpolation maps have been generated to highlight the intensity of each of the major processes in the country. Groundwater quality for irrigation purposes is discussed in relation to the major hydrogeological basins in the country.

Chapter 5 - Adequate sanitation is essential for environmental quality and improved health. Developing countries such as Ghana that experience high mortality and morbidity levels could attain qualitative health improvements through proper waste collection and disposal. However, data from nationally representative surveys conducted in Ghana between 1991/92 and 2005/06 reveal that sanitation infrastructure and services are below optimal levels, households utilize improper waste disposal practices which worsen sanitation conditions and implementation of a number of foreign donor funded water and sanitation schemes has failed to register marked improvements in sanitation coverage. This chapter undertakes an analysis of various sanitation conditions in Ghana and constraints that limit efforts to expand coverage. Based on the findings, strategies whose implementation could ensure sustainable waste management in Ghana such as composting, enforcement of punitive sanctions and increased resource support for acquisition of improved sanitation facilities by households are proposed.

Chapter 6 - Water and sanitation sector development is demanded in face of the Millennium Development Goals (MDGs) target year of 2015. In Japanese sanitation sector, a zoning planning or geographical division scheme of centralized and on-site municipal wastewater treatment systems have been developed in several decades. On the contrary, there are countries with mandatory on-site treatment systems even in the areas served with centralized wastewater treatment systems to prevent infectious diseases. In the Introduction section, recent development especially in sanitation sector after 1990, the completion year of the Water Supply and Sanitation Decade, was chronologically and briefly summarized focusing on resources and benefits of sanitation in the world. In the second section, resources or investments and profits of sanitation were discussed. In the third section, Japanese sanitation scheme was described including typical on-site treatment systems, *johkasou*, which treats both black and gray water. Not only technological aspects but also regulation and institutional aspects are important for on-site treatment system development because of necessity of operation and maintenance. In the fourth section, differences between zoning scheme and mixture situations of centralized and on-site treatment systems were discussed. In the fifth section, technological alternatives in consideration with pollutant discharge reduction function of wastewater treatment systems and investments were discussed. Japanese experiences in sanitation sector will contribute to sanitation sector development in the world together with a huge stock of world experiences. For sanitation perspective, the author summarized his personnal ones on the basis of the government documents and academic literature for the sector development and the roles of Japan which are conducting huge investments in the sector.

Chapter 7 - The purpose of this chapter is to review the structural issues that have confronted the reform of the water supply and wastewater industries in the past, and examine them in light of those forces that can be expected to impact on the issue of structural reform in the future. These factors including such things as; changing climate patterns; rising demands for improved access to clean water, the desirability of diversifying water supplies and the reduction of environmental impacts.

Chapter 8 - Plant kingdom always donate human with lots of precious compounds and materials. Most of plant derivatives are recognized as safe, eco-friendly and low-cost agents regarding synthetic and chemical substances.

The global trend has shifted towards the reduction of synthetic biocides application in general and in food-dealing fields in particular. Despite of the prevalence applications of semi-synthetic and synthetic compounds in sanitation for the control of microbial pathogens attack, natural products are still viewed as potential alternatives.

Two innovative systems were constructed for evaluating the sanitation potency of plant derivatives in closed and semi-closed places against many fungal pathogens; the first applied plant extracts aerosol as sanitizers, whereas the second employed fumigation with smoldered plant fumes. Both of the two techniques exhibited promising and effective sanitation competence toward many fungal pathogens that attack human and plants. *Candida albicans* and *Penicillium digitatum* strains could be completely inhibited using aerosolization with pomegranate peel extract, while growth of *Aspergillus flavus* and *Botrytis cinerea* could be successfully stopped after exposure to smoldered plant fumes either in vitro or on the infected crops using fumigation room. The biodegradable and inoffensive nature of sanitizers from plant origins recommends their application for the effective elimination of fungal pathogens.

It is an important step toward desirable substitution of synthetic and chemical sanitizers with other human-friendly agents from biological derivations.

In: Potable Water and Sanitation
Editor: Joel M. McMann

ISBN 978-1-61122-319-4

Chapter 1

SANITATION OF FRESH-CUT FRUIT AND VEGETABLES: NEW TRENDS, METHODS AND IMPACTS

Francisco Artés[1,2,*], Perla Gómez Di Marco[2], Alejandro Tomás-Callejas[1,2] and Francisco Artés-Hernández[1,2]

[1]Postharvest and Refrigeration Group. Dept. Food Engineering. Technical University of Cartagena. P. Alfonso XIII, 48. 30203, Cartagena, Murcia, Spain

[2]Institute of Plant Biotechnology. Technical University of Cartagena. Campus Muralla del Mar. 30202, Cartagena, Murcia, Spain

ABSTRACT

The association between foodborne outbreaks and fresh-cut plant produces has noticeably increased throughout the last years and, consequently, consumers are concerned about the quality and safety of this kind of commodities. Sanitation is the most important target for keeping overall quality of fresh-cut or minimally processed plant produces. Washing-disinfection is the only step that effectively reduces microbial load all over their production chain. The most commonly sanitizer used in the fresh-cut industry is chlorine. However, it has been identified some problems related to its use, such us potentially dangerous by-products formation upon reaction with organic matter with adverse effects on human health, its pH dependence, its gas emission that may affect worker's safety, and its relative poor efficiency in a number of horticultural products. Also, some European countries have forbidden its use and this may lead to new regulatory restrictions in the near future. Several studies have reported the efficiency of chemical and physical alternatives to be used for replacing chlorine. These possible options, which are presently considered as promising treatments, include antimicrobial solutions, electrolyzed water, O_3, UV-C radiation, intense light pulses, superatmospheric O_2 packaging, N_2O and noble gases-enriched packaging atmospheres. Theses emerging

* E-mail: fr.artes@upct.es Tel: +34-968-325510; Fax: +34-968-325433.

alternatives can be applied alone or combined. In this chapter, recent studies accomplished with these eco-innovative sanitizers on fresh-cut plant commodities are reviewed.

1. Introduction

Consumers are more and more conscious about the relationship between the food they eat and their overall health and well-being. This fact has increased the demand of convenient fresh food, keeping their living fresh state, free from additives, with high nutritional value and powerful antioxidant properties which fight reactive oxygen species. The industry satisfies this particular demand by offering minimally fresh processed or fresh-cut (FC) fruit and vegetables. FC produces are elaborated free from additives by using light combined methods such as washing, cutting, disinfecting, and packaged at chilling temperatures under polymeric films able to generate optimum *modified atmosphere packaging* (MAP) conditions (Artés and Allende, 2005). FC commodities usually need no further processing prior to consumption, offering important advantages for consumers since, in addition to convenience and functionality, they have high quality and any or low wastage at a reasonable price (Wiley, 1994; Beuchat, 2002; Bruhn, 2002; Artés, 2004).

Despite the benefits derived from eating raw fruits and vegetables, due to their specific methods of preparation, FC plant foods are highly perishable and must be elaborated following strict control procedures for reducing overall quality loss and assuring its safety to consumers. FC processing of plant produce stimulates microorganisms growth, which may be potentially harmful to human health. Due to this, safety is an issue of concern as these foods have long been known to be vehicles for transmitting infectious diseases (Leistner and Gould, 2002; Artés and Allende, 2005).

FC produces are alive even when processing destroys plant tissues increasing metabolic activity and inducing senescence and loss of resistance to microbial spoilage. Plant damage stimulates respiration rate (RR), C_2H_4 emission, enzymatic and non-enzymatic colour change and nutrient loss from cells. These changes lead to quality loss and reduced shelf-life when compared to that found for the intact product (Wiley, 1994; Artés et al., 2007). There are also pre-harvest factors like plant *cv*, growing conditions, harvest time and maturity/ripening stage at harvest that have a high influence on postharvest life. Processing conditions (precooling, trimming, cleaning, conditioning, cutting, peeling, coring, handling, washing, disinfecting, draining, rinsing, drying, packaging) and distribution conditions (temperature, relative humidity -RH- , atmosphere composition and duration) highly determine produce characteristics. For being successful, FC industry must be seen as a highly integrated system with all processing steps considered in combination with the others (Shewfelt and Prussia, 1993; Artés, 2004).

FC industry is really concerned about fresh-like quality, safety and high nutritional value. Moreover, it is more and more oriented to look for sustainable procedures, especially those related to sanitation. The main preservation technique to prevent or delay spoilage is chilling storage combined to active or passive MAP. Also some chemical coadjutants like antimicrobial solutions, acidulants and antioxidants are commonly used (Leistner and Gould, 2002, Artés and Allende, 2005).

This chapter intends to review the major sanitation approaches which can be used for keeping quality and safety of FC plant commodities within a sustainable frame represented by antimicrobial solutions, electrolyzed water (EW), O_3, ultraviolet (UV-C) light, intense light pulses, and innovative MAP under superatmospheric O_2 and noble gases.

2. Antimicrobial Agents Applied as Solutions

In many situations of the FC industry, chlorination of water is one of the primary elements of a properly managed postharvest sanitation program. This operation is considered as the most important to disinfect water and minimize the transmission of pathogens from infested plant produce, water or debris, to non-infested surfaces such as those mechanically injured during harvesting, transportation or processing, wounds, or the natural plant surface openings (Artés et al, 2009). Disinfectant capacity of NaClO is related with its very high oxidizing properties making it useful for sanitizing both products and equipment of the processing area (Nieuwenhuijsen et al., 2000). It is effective for most situations, relatively inexpensive, and useful in operations of any size (Suslow, 1997). Effectiveness of NaClO against microorganisms mainly depends on concentration since their efficacy increases when increasing concentration, but also on pH, temperature, organic matter present in the washing water and plant commodity, time of exposure, and initial microbial load of raw material (Boyette et al., 1993; Gómez et al., 2010). However, very high NaClO levels may cause product tainting (Adams et al., 1989) and sodium residue on the product and equipments (Ritenour and Crisosto, 1996).

NaClO in contact with water induces an increase in pH and generates hypochlorous acid (HOCl), which is the active antimicrobial compound. The acid dissociates readily to hypochlorite ions (OCl^-) at high pH, or chlorine gas (Cl_2) at low pH, thus the pH must be kept in the range of 6.5 to 7.5 for HOCl to be stable and efficient (Boyette et al., 1993; Suslow, 1997). It should be indicated that, even if NaClO is more efficient at low pH levels, values between 6 and 7.5 should be selected for reducing the risk of corrosion of metallic processing equipment (Beuchat, 2000).

An understanding of NaClO chemistry together with a definition of terms is necessary to describe the amount of chlorine in any form available for oxidative reaction and disinfection. Total chlorine refers to the sum of free and combined chlorine in solution. Free chlorine, free available and free residual chlorine refer to chlorine in the form of elemental chlorine (Cl_2), hypochlorous acid (HClO) and hypochlorite ion (OCl^-) (Simons and Sanguansri, 1997).

NaClO may partially oxidize food constituents originating unhealthy by-products, such as chloroform ($CHCl_3$), haloacetic acids or other trihalomethanes (THM) that have known or suspected carcinogenic or mutagenic potential effect, with proved toxicity to liver and kidney (Nieuwenhuijsen et al., 2000; Hrudey, 2009; Ölmez and Kretzschmar, 2009). At the same time, at pH over 7.5 NaClO reacts with organic N-based materials to produce chloramines (Suslow, 1997). These negative aspects related with chlorine have induced some European countries, like Germany, The Netherlands, Denmark, Switzerland and Belgium to forbid the use of chlorine for disinfection of FC produces (Betts and Everis, 2005; Carlin and Nguyen-the, 1999). An option to reduce THM levels is to change chlorine by chloramine disinfection because they do not react with organic matter in the water to form THM. Adding ammonia to a chlorination system switches NaClO to chloramine (Capece, 2001). However chloramines could be considered as a rather less acceptable alternative because its efficacy as a disinfection agent is lower and could be risky for workers organs like eyes and respiratory tract.

It has been demonstrated that effectiveness of NaClO is limited to some products (Beuchat and Brackett, 1990). Actually, chlorine compounds are helpful in reducing the aerobic microbial counts in many leafy vegetables, but not necessarily in root vegetables. At the same time, it was reported that NaClO is not very effective for inhibiting *L. monocytogenes* growth in shredded lettuce or Chinese cabbage (Ahvenainen, 1996).

According with previously mentioned concerns, considerable research is nowadays done looking for alternatives to NaClO. Among them, some organic acids formulations, such as peroxyacetic acid (PA) combined with citric and ascorbic acids, ClO_2 and H_2O_2, calcium lactate, EW, steamer jet injections and biological compounds have been tested with varying results as it will be described below.

2.1. Peroxyacetic Acid

PA is a combination of peracetic acid (CH_3CO_3H) and H_2O_2, and typically commercialized as a liquid. PA is an interesting alternative to chlorine since its break-down products, acetic acid, O_2, CO_2, and water makes its use completely sustainable and eco-friendly. Its first application was for cleaning surfaces in concentrations ranging from 85 to 300 ppm. The Food and Drug Administration of the USA has set a minimum of 85 ppm PA for cleaning hard surfaces where food is handled (FDA, 1997), although for cleaning food's surface 50 ppm is commonly enough (Stampi et al., 2001) while levels used in the environmental and medical fields normally ranged from 1,200 to 2,600 ppm.

Related to the mechanism of action of PA, it has been suggested to act primarily on lipoproteins in the cell membrane, and it may be possible that it is equally effective against outer membranes lipoproteins, facilitating its action against Gram-negative cells (Leaper, 1984).

PA tolerates several adverse factors like a wide range of temperature application, pH changes (from 1 to 8), water solutions with high carbonate concentration and soil contaminated. By these positive sides, its current main area of application is in fruit and vegetables processing (Artés et al., 2009). For plant surfaces treatment, recommended formulations combine H_2O_2 11% and CH_3CO_3H 15%, at 80 ppm, followed by rinsing with tap water (Suslow, 1997). It has been reported that it was effective for controlling *E. coli*, *Salmonella* spp., and *L. monocytogenes* in FC products (Park and Beuchat, 1999; Rodgers et al., 2004; Ruiz-Cruz et al., 2007). In *Enterobacter sakazakii* counts decreased 5 log units in lettuce with PA applications (Kim et al., 2006). Compared to 150 ppm NaClO, 80 ppm PA reduced the psychrotrophic counts by 2 log units and mesophilic counts by 1 log unit in FC Galia melon, resulting in the pieces having a shelf life of 10 d at 5°C. However, this sanitizer decreased the total vitamin C and the antioxidant activity (Silveira et al., 2010). Vandekinderen et al., (2009) studied the impact of a decontamination step on the shelf-life, sensory quality and nutrient content of grated carrots under MAP and stored at 7°C. 80 mg PA L^{-1} showed possibilities for extending shelf-life without pronounced effects on nutritional content.

2.2. Chlorine Dioxide

ClO_2 is a stable dissolved gas with a strong oxidative power and penetration, higher than that found for NaClO (FDA, 1998). ClO_2 is a strong bactericide and virucide at levels as low as 0.1 ppm having also a very high effectiveness against spores. It attacks planktonic and sessile bacteria and fungi, disinfects surfaces, and prevents and removes rapidly bio films, avoiding bacterial re-growth (EPA, 1999). ClO_2 does not need a long contact time for being highly effective against pathogenic organisms such as *Legionella*, amoebal cysts, *Giardia cysts*, *E. coli*, and *Cryptosporidium* (Xie, 2003). Different compounds of bacterial cells membrane react with ClO_2, causing the interruption of several cellular processes. Cell membrane has been identified as the primary target of ClO_2 on microbial cells but ClO_2 also reacts directly with amino acids and the RNA attacking the cell structure and/or the acids inside the cell. The production of proteins is prevented and the cell membrane affected by changing membrane proteins and lipids (EPA, 1999).

Viruses are eliminated in a different way than bacteria. ClO_2 reacts with peptone, a water-soluble substance that originates from hydrolysis of proteins to amino acids, and kills viruses by prevention of protein formation being more effective than NaClO or O_3 (EPA, 1999).

Differently to chlorine and bromine, ClO_2 does not ionize to form weak acids neither to form carcinogenic by-products, like THM. This allows ClO_2 to be effective over a wide pH range. However, differently than PA, which is not dangerous, ClO_2 and its disinfection byproducts (chlorite and chlorate) can create problems for dialysed patients (EPA, 1999).

ClO_2 can be used as both gas and aqueous type. As a gas it has more penetrability than the aqueous sanitizers (Han et al., 2001). Different factors such us gas concentration (0.1-0.5 mg L^{-1}), time (7-135 min), RH (55-95%) and temperature (5-25°C) affect the antimicrobial effect of ClO_2 gas on the inactivation of *E. coli* O157:H7 (Han et al., 2001; Gómez-López et al., 2009). One of the most important qualities of ClO_2 is its high water solubility, especially in cold water, which makes it strongly recommended for using in the FC industry. ClO_2 remains as dissolved gas in solution being approximately 10 times more soluble in water than NaClO.

Main ClO_2 drawback is that it must be generated on-site by reacting sodium chlorite and acid or NaClO (EPA, 1999). Today, new technology allows a more easy production, like systems where the reactants are packaged in pouches for reacting together. However, ClO_2 is unstable and can be explosive when concentrations reach 9.5% or more in air (Betts and Everis, 2005; Jin et al., 2009).

While for drinking water less than 1 ppm ClO_2 is allowed, for whole produce a maximum concentration of 3 ppm is permitted (EPA, 1999; CFR, 2007a,b). The effect of ClO_2 to inhibit the main foodborne pathogens related with the FC commodities has been reported previously. Concentrations of 4-5 mgL^{-1} with contact times of 6 – 30 min were effective to reduce *Salmonella* spp., *E. coli* O157:H7 and *L. monocytogenes* inoculated onto cabbage, carrot, lettuce, strawberry and melon (Sy et al., 2005; Mahmoud et al., 2007; Mahmoud and Linton 2008a,b).

A solution of 5 ppm ClO_2 has been effectively used for apple, lettuce, strawberry and cantaloupe for inhibiting inoculated *E. coli* and *L. monocytogenes* (Rodgers et al., 2004). In MFP faba bean (*Vicia faba* L) the immersion for 2 min in 4 ppm ClO_2 was as effective as 150 ppm NaClO + 1g L^{-1} citric acid in reducing enterobacteriacea and mesophilic counts after 8 d at 5°C, decreasing browning and improving sensory quality (Artés et al., 2007; 2009).

Although the lack of knowledge about toxicity of liquid CD disinfection by-products make further studies necessary to establish the suitability of this chemical for FC produce sanitation. López-Gálvez et al. (2010) demonstrated that the use of aqueous 3 mg L^{-1} ClO_2 was equally effective as 100 mg L^{-1} NaClO and it did not cause any detrimental effect on the sensory and nutritional quality of FC Iceberg lettuce without the potential formation of THMs. Tomás-Callejas et al. (2010) evaluated *Salmonella* spp. removal and cross-contamination potential during the washing-disinfection, rinsing, and de-watering steps in FC baby Red chard. Typical industry rates of 20 ppm NaClO (at pH 7.0) and 3 ppm ClO_2 substantially prevented cross contamination by *Salmonella* among co-mingled Red chard leaves but neither fully disinfected the pathogen from inoculated leaves.

2.3. Hydrogen Peroxide

H_2O_2 is a strong oxidising agent that is commonly used in industry and in the medical field. It is considered a powerful bactericide, being effective even against spores. It is unstable and decomposed upon standing, agitation, and exposure to light or heating, producing water and O_2. However, H_2O_2 is also able to generate other cytotoxic oxidising species like hydroxyl radicals (Khadre and Yousef, 2001). A efficacy of H_2O_2 washing, similar to that of NaClO, has been demonstrated in extending shelf-life and reducing native microbial and pathogen populations, including *E. coli*, in whole grape, prune, apple, orange, mushrooms, melon, tomato, red bell pepper and lettuce, and in FC cucumber, zucchini, bell peppers, and melons (Sapers, 2003; Artés et al., 2009). H_2O_2 is not currently approved as a sanitizing agent for fresh produce even when is permitted for other uses in food processing and packaging because it leaves no potentially harmful residues and it has been approved by FDA.

For disinfection of FC commodities, the use of dilute H_2O_2 solution has shown to be promising. For example, washing with 5% H_2O_2 was more effective than with 1000 ppm NaClO and Na_3PO_4 for reducing the microbial load on cantaloupe rinds, thus improving microbial quality and shelf-life (Sapers and Simmons, 1998). Residual H_2O_2 in treated fruits and vegetables might be eliminated passively by the action of endogenous catalase, given enough time for reaction, or actively by rinsing immediately after treatment to avoid reactions between H_2O_2 and food constituents that might affect product quality or safety. Ukuku (2004) studied the effect of H_2O_2 on microbial quality and appearance of whole and FC melons contaminated with *Salmonella* spp. H_2O_2 treatments of whole melon (2.5% and 5%) for 5 min caused a 3 log CFU cm^{-2} reduction of the indigenous surface microflora and 3 log CFU cm^{-2} reduction in *Salmonella* in all melon surfaces. However, browning of shredded lettuce increased after dipping in a H_2O_2 solution (Parish et al., 2003).

Pedahzur et al. (1997) noticed an increase in the bactericidal effect of H_2O_2 and silver on *E. coli* when applied together, suggesting that such increase in the bactericidal effect of these two disinfectants might be due to the synergism between the two chemicals. Gopal et al. (2010) examined the use of silver and H_2O_2 as possible alternative to NaClO. Combination of electrochemically generated silver (5 ppm) and H_2O_2 (0.4 ppm) caused a reduction in the total plate count (0.87 log), Pseudomonas (2.66 log), Enterobactericeae (1.61 log) and yeast and mould (1.60 log) immediately after washing in comparison to water-washed shredded lettuce. The combination of 1% H_2O_2 with 25 μg mL^{-1} nisin, 1% sodium lactate, and 0.5% citric acid (HPLNC) for reducing transfer of bacterial pathogens from whole melon surfaces to FC pieces has been tested. HPLNC reduced the number of *E. coli* O157:H7 and *L. monocytogenes* by 3 to 4 log CFU cm^{-2} on melon and the natural microflora on FC melons were also substantially reduced (Ukuku et al., 2005).

The use of H_2O_2 as vapour instead of water solutions also appears to reduce microbial counts, extending shelf life and maintaining quality of FC green bell pepper, cucumber, and zucchini (Sapers, 2003).

2.4. Citric Acid – Ascorbic Acid – Calcium

Organic acids and calcium have been mostly applied for controlling enzymatic and non-enzymatic browning in many fruit and vegetables (Sapers, 1993). They have also a positive effect for avoiding texture deterioration (Rosen and Kader, 1989) and microbial growth (Yildiz, 1994) at levels that did not adversely affect taste and flavour of plant commodities. Organic acids and calcium are more effective for bacteria than for moulds and yeast due to the low pH (between 2.1 and 2.7) at which they are applied.

Studies on FC Chinese cabbage showed that citric acid (10 g L^{-1}) repressed the petiole sprouting (black speck) development and prolonged their shelf life from 10 (control) to 14 days at 5°C. When stored at 0°C the shelf life was not extended by citric acid, ascorbic acid and $CaCl_2$ dips, but citric acid improved quality by reducing black speck. In this experiment, treatment with citric acid produced only a slight pH reduction without affecting taste. The $CaCl_2$ (10 g L^{-1}) was beneficial in reducing browning while no microbial spoilage occurred after 35 d at 0°C or 21 d at 5°C under any treatment (Kim and Klieber, 1997).

FC Amarillo melon dipped in 0.52 mM citric acid for 30 s before MAP reached a shelf life of 10 days at 5°C. This treatment kept microbial safety and avoided translucency and discoloration. Compared to 1.4 mM NaClO, citric acid increased lightness and improved visual appearance of melon pieces (Aguayo et al., 2003). Dipping green celery crescents in a 0.5 M ascorbic and 0.1 M citric acid solution was as effective as 100 mg L^{-1} NaClO for reducing microbial counts and improving consumers acceptability (Gómez and Artés, 2004). Ölmez and Temur (2010) investigated the effects of 0.25 g 100 g^{-1} citric acid + 0.50 g 100 g^{-1} ascorbic acid solution for 2 min at 10°C on biofilms and attachment of *E. coli* and *L. monocytogenes* on green leaf lettuce. The treatment was unable to efficiently reduce or detach bacterial cells inside the biofilms or those cells attached to inaccessible sites. Moreover, a better understanding of the mechanism involved in bacterial attachment and biofilm formation is needed.

Cell wall structure and firmness of plant commodities is related to calcium since it maintains them by combining with pectin to form calcium pectate. Moreover, calcium lactate avoids the off-flavours associated with chloride salt, being a good alternative to it (Luna-Guzmán and Barret, 2000). Studies with FC melon showed that treatment with calcium generally kept firmness better than control during storage (Aguayo et al., 2007; Luna-Guzmán et al., 1999). FC lettuce seems also to be favoured when treated with 120 ppm NaClO and 15 g L^{-1} calcium lactate at 20 and 50°C. Samples washed with calcium lactate showed higher crispness than those immersed in NaClO even when lower than those which received a heat shock treatment (50°C). Authors (Martín-Diana et al., 2006a) hypothesised that calcium lactate avoided the loss of turgor during the first days of storage while the combination with 50°C extended this benefit to 12 days. It is possible that it may occur the activation of texture-related enzymes, as the pectin methylesterase, affecting the washing solution absorption into the lettuce. This might in turn increase the amount of calcium retained by the vegetable (Martín-Diana et al., 2006a). Experiments with FC carrots showed that washing with 15 g L^{-1} calcium lactate at 50°C was effective in reducing lignification at the cutting-edge area while keeping turgor of cortex tissue cells. Data about microbial growth should be given in order to a better knowledge of the effects (Rico et al., 2007b). In that way, Aguayo et al. (2007) found that $CaCl_2$ and calcium lactate reduced the bacterial growth by 2 log CFU g^{-1} and the yeast growth by 1 log CFU g^{-1} when diced melon was stored up to 8 days at 5°C. Moreover, FC melon dipped into hot calcium solution (60°C, 1 min) as $CaCl_2$ (0.5%), or weak organic acid salts like calcium propionate (0.9%) or lactate (1.4%) had a lower microbial growth and a better firmness. This could be explained by an increment in bound calcium concentration (50%) which in turns produces firmer melon pieces.

2.5. Electrolyzed Water

EW is generated by electrolysis of a dilute salt (NaCl, usually about 0.1%) solution of pure water in an electrolysis chamber where anode and cathode electrodes are separated by a membrane. On the anode side, acidic EW is generated and has strong bactericidal effect on most known pathogenic bacteria, due to its low pH, high oxidation–reduction potential (ORP) (about 1100 mV) and the presence of hypochlorous acid. The cathode area produces alkaline reducing water (Ongeng et al., 2006).

The acidic type was the first form developed and accepted by the food industry in Japan (Izumi, 1999). It was found to be useful at killing bacteria and parasites on raw fish without altering its sensory characteristics. EW has a strong bactericidal effect against pathogens and spoilage microorganisms, more effective than NaClO due to its high redox potential (Izumi, 1999; Koseki and Itoh, 2001; Bari et al., 2003). Hypochlorous acid is present in EW at pH 6.8 and it is generated by electrolysis of NaCl solution, since HCl formed at the anode site neutralizes the NaOH at the cathode site. Advantages of using EW at neutral pH are that it does not affect pH, surface colour or general appearance of FC vegetables (Izumi, 1999).

In the recent years, several studies have reported that the use of EW in FC fruit and vegetables is greatly influenced by type of vegetable and minimal processing. (Izumi, 1999; Koseki and Itoh, 2001; Wang et al., 2004; Ongeng et al., 2006, Rico et al., 2007a). EW containing 15 to 50 ppm available NaClO was effective as a disinfectant for FC carrots, spinach, bell pepper, potato and cucumber, without discoloration and lowering microbial counts from 0.6 to 2.6 log units (Izumi, 1999).

Acetic EW disinfected the surfaces of lettuce by reducing the aerobic and coliform bacteria, moulds, and yeasts, with a similar effect of NaClO, resulting in a 1 log unit reduction for total aerobic counts and 0.5 log for coliforms in shredded lettuce pieces. After 7 days at 1-2°C, physical and sensory parameters did not change (Hua et al., 2004). In other study, it was reported that EW has no impact on taste, smell, and texture of shredded lettuce (Koseki et al., 2001). In the same way lettuce samples washed with EW containing 60 mg L^{-1} free chlorine (pH 6.5) resulted as effective as NaClO, showing a good quality retention (Rico et al., 2007a). Growth of *Salmonella* was reduced in alfalfa seeds and sprouts when treated with EW (Kim et al., 2003) while acetic EW has also been shown to effectively inactivate *E. coli* O157:H7, *S. enteritidis* and *L. monocytogenes* on lettuce (Park et al., 2001).

The use of acidic EW (AEW) and neutral EW (NEW) as disinfectant for food processing equipment has been studied (Venkitanarayanan et al., 1999a; Kim et al., 2001; Park et al., 2002; Guentzel et al., 2007). Ayebah and Hung (2005) indicated that AEW did not have any adverse effect on stainless steel. However, issues such us gas emission, strong acidity of AEW, metal corrosion, free chlorine content and by-products formation need to be supported for further research.

The efficacy of NEW and AEW to reduce natural microflora as well as the main foodborne pathogens associated to FC produce has been demonstrated. Using AEW alone resulted in moderate control of bacterial growth of FC cilantro during storage (Wang et al., 2004). Reductions of viable aerobes by 2 log CFU g^{-1} were obtained in lettuce washed with AEW (pH 2.6; oxidation reduction potential, 1140 mV; 30 ppm of available chlorine) for 10 min (Koseki et al., 2001). Also, AEW was effective to reduce *E. coli* O157:H7 (Venkitanarayanan et al., 1999b; Sharma and Demirci, 2003; Liao et al., 2007; Stopforth et al., 2007), *Salmonella* and *L. monocytogenes* (Venkitanarayanan et al., 1999b; Fabrizio and Cutter, 2003, Stopforth et al., 2007). Due to its neutral pH, NEW could be less aggressive to the corrosion of processing equipments compared to AEW. NEW (pH 6.8) containing 15 to 50 ppm available chlorine was effective as a disinfectant for FC vegetables without causing discoloration. NEW did not affect tissue pH, surface colour or general appearance (Izumi, 1999). Abadías et al. (2008) demonstrated that the bactericidal activity of diluted NEW (50 ppm of free chlorine, pH 8.60) against *E. coli* O157.H7, *Salmonella*, *L. innocua* and *E. carotovora* on FC lettuce was similar to that of 120 ppm free NaClO.

A theory for the inactivation mechanism of bacteria by EW proposed that ORP could firstly affect and damage the redox state of glutathione disulfide-glutathione couple, and then penetrate the outer and inner membranes of *E. coli* O157:H7, resulting the necrosis (Liao et al., 2007).

The application of EW in the food industry has been recently reviewed by Huang et al. (2008) as a disinfectant for food processing equipment, vegetables, fruits, poultry, meat and seafood. However, there is scarce information about the effects of NEW and AEW on the main quality attributes in FC vegetables. Martínez-Hernández et al., (2009) showed that the use of AEW (40 ppm free chlorine, pH 3.0) retained the total content of polyphenols and chlorophylls better than a conventional 100 ppm NaClO treatment during the shelf-life of FC Mizuna baby leaves stored at 5ºC up to 11 days.

EW has many potential uses for the food industry being advantageous because it involves on-site production of the disinfectant, which means there are no chemicals to store or handling costs for dealing with them. The main problem is that, in some cases, it has a short shelf life (about 2 weeks); therefore mostly only companies with access to an electrolysis machine could get profit from its benefits (Artés et al., 2009).

2.5. Steamer Jet Injection

Vapour or hot water could be seen as an alternative to replace NaClO. However, exposure to high temperatures may produce detrimental changes in the product lowering its nutritional value. By that reason, exposure time to steamer jet injection should be low and much research is needed to optimise exposure conditions for FC plant commodities. In that way, Martín-Diana et al. (2007) reported a lower RR and a partial inactivation of browning-related enzymes when FC lettuce was exposed to water steam. Moreover, mesophilic counts were maintained as low as when NaClO was applied. Independently of these positive effects, ascorbic acid and carotenoids content showed a reduction, even when samples look like it did not, showing similar values of acceptability and fresh appearance than control.

2.6. Biological Compounds, Natural Microbiota and/or their Antimicrobial Products

Nowadays efforts to reduce or eliminate chemically synthesized additives in foods are conducted to find natural alternatives to control bacterial and fungal growth in fresh-cut plant commodities (Soliva-Fortuny and Martín-Belloso, 2003). Commercial products based on natural biocide action such us lactate esters from lactic acid and alcohols (Purac), organic acids and flavonoids (Citrox), and lactoperoxidase, H_2O_2 and thiocyanate (Catallix) have been released to the market. Citrox and Purac applied at the commercial recommended doses did not prevent transfer of *E. coli* cells between inoculated and non inoculated FC lettuce and therefore indicating cross-contamination during washing (López-Gálvez et al., 2009). Other natural derivate compounds such us grapefruit seed extracts (Xu et al., 2009) and green tea extracts (Martín-Diana et al., 2008) has been studied due to their antimicrobial effect in order to extend the shelf-life of FC commodities.

Eugenol (from clove, *Syzygium* sp.) and thymol (from thyme and oregano) exhibited antioxidant and antimicrobial effects on MAP stored table grape (Burt, 2004; Valero et al., 2006). Rots in apple are better controlled by carvacrol than by eugenol (Neri et al, 2009). It is worth to mention that there is always an important risk of conferring off-odours when applying essential oils. Sensory evaluation, particularly for FC produces is a point to take seriously into account.

Essential oils from plant also have antioxidant and anti-inflammatory properties for humans. They could be used for extending shelf-life of FC products and, at the same time, to confer other benefits for consumers' health (Gutiérrez et al., 2009).

The use of natural aroma compounds to improve shelf-life and safety of FC fruits has been reviewed (Lanciotti et al., 2004). Plant volatiles have been widely used as food flavouring agents and most of them are generally recognized as safe (GRAS) (Newberne at al., 2000). The addition of hexanal at levels not exceeding 100 ppm in the storage atmosphere of FC apples reduced the growth rate of natural microflora during storage at 4°C (Lanciotti et al., 1999). Some research has been conducted on the effect of natural competitive microbiota, in most of the cases lactic acid bacteria and/or their derived antibacterial products for FC plant commodities. It has been suggested that they enhance the safety of FC vegetables (Bari et al., 2005) and apples (Leverentz et al., 2006). However this approach has not been yet fully explored.

An interesting option is the use of edible coatings as carriers of antimicrobial agents. In that way, the effect of lemongrass, oregano oil and vanillin incorporated in apple puree-alginate edible coatings, on the shelf-life of FC Fuji apples, has been investigated. Vanillin containing coatings (0.3% w/w) were the most effective in terms of sensory quality after 2 weeks at 4°C. All antimicrobial coatings significantly inhibited the growth of psychrophilic aerobes, yeasts and molds. At the same time, lemongrass (1.0 and 1.5% w/w) and oregano oil containing coatings (0.5% w/w) exhibited the strongest antimicrobial activity against *L. innocua* (4 log reduction) (Rojas-Graü et al., 2007). Raybaudi-Massilia et al. (2008) studied the effect of malic acid and essential oils of cinnamon, palmarosa and lemongrass and their main active compounds as natural antimicrobial substances incorporated into an alginate-based edible coating on the shelf-life and safety of FC Piel de Sapo melon. Palmarosa oil incorporated at 0.3% into the coating appear to be a promising alternative for FC melon, since it had a good acceptation by panellists, maintained the quality parameters, inhibited the native flora growth and reduced *S. enteritidis* population.

3. UV-C Radiation and Intense Light Pulses

UV light at sufficiently short wavelength could be used to break down microorganisms in food, air and water purification systems. The use of non-ionizing, germicidal and artificial UV at a wavelength of 190–280 nm (UV-C) could be effective for surface decontamination of FC products. This treatment offers some advantages to the food industry since it does not leave any residue, does not have legal restrictions; it is easy to use and lethal to most types of microorganisms (Bintsis et al., 2000). In addition, UV-C does not require extensive safety equipment to be implemented (Yaun et al., 2004).

The most important effect of UV-C is at the DNA level by inducing the formation of pyrimidine dimmers which alter the DNA helix and blocks microbial cell replication (Kuo et al., 1997; Lucht et al., 1998). Consequently, cells which are unable to repair radiation-damaged DNA die and sub-lethally injured cells are often subject to mutations (Lado and Yousef, 2002).

Strains sensitivity to UV-C changes depending on the microorganism (Gardner and Shama, 2000). For example *Pseudomonas aeruginosa* was more sensitive than *Micrococcus radiodurans* and *Candida albicans* (Abshire and Dunton, 1981). *In vitro* studies performed by Sumner et al. (1995) indicated that UV-C was hihgly effective in destroying *S. typhimurium* while *L. monocytogenes* and *S. Enterica,* also under *in vitro* conditions, were inhibited during 13 days at 5°C at doses of 2.4 kJ m^{-2} (Escalona et al., 2010).

Related to temperature, UV-C seems to be effective in a range from 5 to 37°C. However, the structure and surface of the vegetable have an influence on the incident irradiation (Bintsis et al., 2000; Lado and Yousef, 2002; Gardner and Shama, 2000). An interesting benefit of UV-C application is that it stimulates plant defence mechanisms increasing resistance to pathogens. But, from the other side, UV-C may modify cell permeability on vegetables increasing electrolytes, amino acids and carbohydrates leakage, which consequently may stimulate bacterial growth (Nigro et al., 1998). Based on Ben-Yehoshua and Mercier (2005), it is very important to find a safe dose able of greatly weaken microorganisms development without damaging the product.

UV-C light has been evaluated as an alternative for surface disinfection of FC fruit and vegetables (Erkan et al., 2001; Marquenie et al., 2002; Allende and Artés, 2003ab; Yaun et al., 2004). *In vivo* studies on zucchini squash slices exposed to UV-C during 10 and 20 min showed that decay and senescence were delayed by affecting microbial growth (Erkan et al., 2001). However, several products (tomato, Lu et al., 1987; Liu et al., 1993; strawberry; Marquenie et al., 2002; carrot, Mercier and Arul, 1993; table grape, Nigro et al., 1998 and sweet-potato, Stevens et al., 1999) showed an increase in the RR of the slices, while the C_2H_4 production and the sensitivity to chilling injury during storage at 5°C were unaffected. Similar results were found for. Related to the stimulation of defence mechanisms beneficial for health, it has been shown that 4 to 14 kJm^{-2} UV-C applied to broccoli heads delayed yellowing and chlorophyll degradation at 20ºC, with an increased total phenols and flavonoids content, along with higher antioxidant capacity and lower RR (Civello et al., 2006). Some negative effects of the use of UV-C has been observed in FC apple stored for 7 days at 5ºC indicated by an increase in surface browning when 5.6, 8.4 and 14.1 kJ m^{-2} doses were used. This process could be related to an increased enzymatic activity caused by membrane breakage with consequent loss of compartmentalization (Gómez et al., 2010).

The natural microflora of FC Red Oak Leaf lettuce up to 10 days at 5ºC was reduced by the use of two sided UV-C radiation (1.18, 2.37 or 7.11 kJm^{-2}) even when 7.11 kJm^{-2} induced tissue softening and browning after 7 days at 5ºC (Allende et al., 2006). The same trend was also observed for one sided UV-C radiation of shredded Red Oak Leaf and Lollo rosso lettuces throughout 10 d at 5ºC where 0.4 to 8.14 kJ m^{-2} decreased psychrotrophic and coliform bacteria, and yeast growth, but only significant reductions were found when the highest level was applied (Allende et al., 2003ab). The same authors reported an increase in the lactic acid bacteria (LAB) which seems to be stimulated by UV-C, probably due to reduced growth of competitive flora.

Sometimes results are not really consistent, like observed by López-Rubira et al. (2005). They analysed the effect of UV-C doses from 0.56 to 13.62 kJm^{-2} on microbial growth in FC pomegranate arils stored up to 15 d at 5ºC. UV-C doses did not affect RR, but some of them reduced mesophilic, psychrotrophic, LAB and enterobacteriaceae counts. Nevertheless, microbial counts were not systematically reduced throughout shelf life and yeast and moulds were unaffected.

Combination of MAP and UV-C could have a positive effect by reducing psychrotrophic bacteria, coliform, and yeast growth in FC lettuce without adversely affect sensory quality (Allende and Artés, 2003b). Pre-processing application of UV-C may have also an influence on FC produce's behaviour. Robles et al., (2007) indicated that tomatoes pre-treated with 4 KJm^{-2} UV-C followed by storage under 5 kPa O_2 + 1 kPa CO_2 at 12ºC for 21 days showed retarded ripening and kept better firmness and sensory attributes than air storage.

Decay caused by *Botrytis cinerea* of whole bell peppers harvested from integrated pest management production and treated with 2.27 kJm^{-2} UV-C was reduced after 21 days at 5ºC. A residual effect was found since inhibition remained after a further 5 days period at 15ºC. Bell peppers sticks, also from integrated cultivation, showed a high reduction in mesophilic counts when 2.27 kJm^{-2} UV-C were applied. On the other hand, sticks from organically cultivated bell peppers, which showed the highest *Enterobacteria* growth, only 2.76 kJ m^{-2} reduced these bacteria counts after shelf life (Artés et al., 2006a).

The effects UV-C doses to reduce pathogen and spoilage bacterial growth in FC baby spinach has also been studied (Escalona et al., 2010). UV-C doses relatively high (12 or 24 kJ m^{-2}) showed an inhibitory effect on initial microbial counts. However, this effect was lost during storage, reaching higher counts on radiated leaves compared to control. Results were confirmed by Artés-Hernández et al. (2009) where UV-C light (4.64, 7.94 and 11.35 kJ m^{-2}) reduced the initial mesophilic and psychrophilic counts on the processing day, but no residual inhibitory effect was found for the higher doses after 6-13 days at 5 and 8°C. It is possible that high doses of UV-C could stimulate microbial growth by a superficial damage caused on spinach leaves.

Recently, the effects of radiation processing on phytochemicals and antioxidants in plant produce have been reviewed (Alothman et al., 2009). It has been reported that abiotic stresses like UV light may enhance the nutraceutical content of fresh fruit and vegetables. It would affect the secondary metabolism of fresh produce and increase the synthesis of phytochemicals with nutraceutical activity or reduce the biosynthesis of undesirable compounds (Cisneros-Zevallos, 2003). Also, the impact of lighting on shredded cauliflower packaged in different film types and stored at 4°C in darkness and partial or continuous lighting has been evaluated. The light accelerated browning in the cut zones and the development of abnormal colouring in these areas marked the end of shelf life. Among the sensory attributes studied, colour was the most affected by exposure to light (Sanz-Cervera et al., 2007). UV-C doses at 0.25 and 1.0 kJ m^{-2} increased anthocyanins concentrations in fresh strawberries (Baka et al., 1999). Bell peppers exposed to 7 kJ m-2 UV-C light showed lower total phenolic content and higher antioxidant activity (DPPH scavenging activity) with insignificant effect on the carotenoids content (Vicente et al., 2005). Broccoli florets treated with 4-14 kJ m^{-2} UV-C light displayed lower total phenolic and total flavonoid content along with higher antioxidant capacity compared to control (Costa et al., 2006). Application of 2 or 4 kJ m^{-2} on blueberries did not change the total phenolic content while it increased the total anthocyanins content and FRAP values (Perkins-Veazie et al., 2008). Low UV-C doses (1.6 and 2.8 kJ m^{-2}) for keeping overall quality of FC watermelon have been reported. Low UV-C illumination preserved sensory attributes up to 11 days at 5°C, with maintenance of lycopene and ascorbic acid, and increased total antioxidant capacity (Artés-Hernández et al., 2010).

Intense pulses light (IPL) is an emerging nonthermal technology for the rapid inactivation of pathogenic and spoilage microorganisms in foods. It involves the use of intense pulses of short duration (from 85 ns to 0.3 ms) and a broad spectrum (from 0.45 to 15 Hz) to ensure microbial inactivation on the surface of either foods or packaging materials. The limited energy cost, the lack of residual compounds, and its great flexibility are some of the major benefits of this technique (Elmnasser et al., 2007; Gómez-López et al. 2005). It seems to induce structural changes of microbial DNA, comparable to the effect caused by continuous UV sources, but others mechanisms could to be involved (Takeshita et al., 2003). ILP effect is dependent on the light absorption by microorganisms. Moreover, certain food components could also absorb the effective wavelengths and decrease their efficiency. ILP has been used to successfully inactivate *E. coli* O157:H7 on alfalfa seeds (Sharma and Demirci, 2003), and *A. niger* spores on corn meal (Jun et al., 2003). Regarding shelf-life of FC vegetables, Hoornstra et al. (2002) showed that a 2 log CFU g^{-1} reduction in aerobic counts on selected vegetables can increase their shelf-life about 4 days. Gómez-López et al. (2005) reported that rich in carbohydrate foods such as fruit and vegetables seem to be more suitable for decontamination by ILP, although it was found an increase on the RR of treated produces. In

FC vegetables such as spinach, celery, green paprika, soybean sprouts, radicchio, carrot, iceberg lettuce, and white cabbage, a PL treatment of 7 J intensity for 45 and 180 s per side provided log reductions in mesophilic aerobic counts between 0.21 and 1.67, and between 0.56 and 2.04, respectively (Gómez-López et al., 2005). ILP did not prolong the sensory shelf-life of FC white cabbage or Iceberg lettuce, while from the microbial point of view, 1 extra storage day at 7ºC was achieved for lettuce. Additionally ILP seems to be useful for enzymes inactivation. IPL treatments (4.8, 12, 28 J cm^{-2}) extended the microbiological shelf-life of FC mushrooms by 2-3 days compared with untreated samples. However, high pulse light fluencies (25 and 28 J cm^{-2}) dramatically affected the texture of sliced mushrooms and promoted the enzymatic browning by an increase on polyphenol oxidase (PPO) activity at the highest dose (Oms-Oliu et al., 2010).

4. OZONE

Ozone is a gas at room temperature formed by the highly unstable tri-atomic oxygen molecule (O_3) obtained from the addition of an oxygen atom ($O^{\cdot}$) to a molecular diatomic oxygen (O_2) (Horvath et al., 1985). The gas is colorless with a pungent odor readily detectable at concentrations as low as 0.02 to 0.05 ppm (by volume), which is below levels of health concern. It is a powerful oxidant, second only to the hydroxyl free radical, among chemicals typically used in water treatment. Therefore, it is capable of oxidizing many organic and inorganic compounds in water.

The O_3 is commercially generated by passing O_2 through an electrical charge. Thus, molecular O_2 is split into two $O^{\cdot}$ which are highly reactive moieties. When a free $O^{\cdot}$ encounters O_2, it combines to form the O_3 molecule which rapidly degrades back to O_2 with the released $O^{\cdot}$ combining with another $O^{\cdot}$ to form O_2, or combining with other chemical moieties to cause oxidation (Rice et al., 1981). Upon release of $O^{\cdot}$, O_3 acts as a strong oxidizing agent being very effective in destroying microorganisms (Guzel-Seydim et al., 2004a). O_3 was approved for use as a disinfectant or sanitizer in foods and food processing in the United States (FDA, 1997).

Ozone is scarcely soluble in water and, even when O_3 is more soluble than O_2, NaClO is 12 times more soluble than O_3. Concentrations of O_3 used in water treatment are typically below 14 %, which limits the mass transfer driving force of gaseous O_3 into the water. Consequently, typical concentrations of O_3 found during water treatment range from <0.1 to 1mg L-1, although higher concentrations can be attained under optimum conditions (Hoigné and Bader, 1983). Basic chemistry research has shown that O_3 decomposes spontaneously during water treatment by a complex mechanism that involves the generation of hydroxyl free radicals (Glaze, 1987) which are among the most reactive oxidizing agents in water. The decomposition of O_3 is so rapid in the water phase of food that its antimicrobial action may take place mainly at the surface, leaving no residues (Achen, 2000). The bactericidal effects of O_3 have been shown on a wide variety of Gram+ and Gram- bacteria as well as spores and vegetative cells (Foegeding, 1985). O_3 destroys microorganisms by the progressive oxidation of vital cell components, preventing the microbial growth and extending the shelf-life of many fruit and vegetables, and their industrial use is increasing (Parish et al., 2003). No significant effect of temperature on the efficacy of O_3 has been noted in the range of 10-26ºC (Ölmez and Akbas, 2009).

The main systems for O_3 application include the gaseous phase storage (for cyclic or continuous exposure) or as ozonated dipping (Aguayo et al., 2006). An increase in the shelf-life of apples and oranges treated with 0.4 ppm O_3 was attributed to the oxidation of C_2H_4 in the cold room (Skog and Chu, 2000). Sarig et al. (1996) found that O_3 controlled *Rhizopus stolonifer* and induced resveratrol and pterostilbene phytoalexins synthesis in table grapes, the berries being more resistant to subsequent infection. In fact, 8 ppm O_3 increased up to three fold the resveratrol level in Napoleon table grape (Artés-Hernández et al., 2003) and also the continuous applications of low O_3 doses seem to enhance the total polyphenol content during long-term storage of seedless table grapes (Artés-Hernández et al., 2007). The effect of a flow of 4 ± 0.5 ppm O_3 enriched air applied cyclically (30 min every 3 h) on whole and sliced tomatoes prevented fructose, glucose, ascorbic and fumaric acid losses. In addition O_3 kept a good overall quality of tomato slices and reduced 1.1-1.2 log CFU g^{-1} for bacterial and 0.5 log CFU g^{-1} for fungal counts (Aguayo et al., 2006). The efficacy of gaseous O_3, applied under partial vacuum in a controlled reaction chamber, for the elimination of *Salmonella* inoculated on melon rind has been investigated (Selma et al., 2008). Gaseous O_3 (10,000 ppm for 30 min under vacuum) reduced viable, recoverable *Salmonella* from inoculated melons with a reduction between 4.2-2.8 log CFU/rind-disk (12 cm^2). A novel ozone-generation system capable of generating O_3 inside a sealed package at various geometries has been developed. Whole prepackaged spinach leaves inoculated with *E. coli* O157:H7 under this ozone-generation system showed that this treatment reduced *E. coli* O157:H7; however, minimizing quality changes after treatment requires further research (Klockow and Keener, 2009).

Conventional sanitizers could be replaced by the use of O_3 water due to its efficacy at low concentrations and short contact time requirements as well as the breakdown to non-toxic products (Graham, 1997). O_3 solubility has a strong influence on the efficacy of this treatment. Solubility depends on temperature, increasing with decreased water temperature (Bablon et al., 1991). It is important to provide certain turbulence to the water as a way of increasing bubble contact and subsequent O_3 solubilisation. The flow rate and contact time as well as purity and pH of water are also important and greatly affect the rate of O_3 solubilisation. Kim et al. (1999a) indicated that high pH destabilizes O_3 and decreases the apparent rate of solubilisation. Moreover, tap water may contain organic matter that consumes O_3.

Inoculating targeted microorganisms on pure cell suspensions or on the food surface and treating these surfaces with O_3 under conditions that simulate normal processing has been useful for studying the efficacy of O_3 water. In that way some authors (Restaino et al., 1995; Singh et al., 2002) have found a decrease in pathogens including *S. aureus, S. typhimurium, Y. enterocolitica, L. monocytogenes* and *E. coli* O157:H7. The use of 1.5 ppm O_3 water (pH = 6, 25°C) for 15 s reduced between 1.5 and 5 log CFU g^{-1} *E. coli* O157:H7, *P. fluorescens, L. mesenteroides* and *L. monocytogenes* counts (Kim et al.,1999b). Khadre and Yousef (2001) indicated that effectiveness of O_3 in disinfecting food-contact surfaces may be tested using spores of *B. stearothermophilus* as indicators, due to their high resistance to O_3.

Food composition may have a significant effect on the bactericidal power of O_3 against spore former, a Gram$^-$ rod, and a Gram$^-$ cocci. High levels of protein (caseinate) and fat (whipping cream) resulted in the greatest amount of protection to spores of *B. stearothermophilus* and vegetative cells of *E. coli* and *S. aureus* (Guzel-Seydim et al., 2004b). O_3 may have a positive influence on some unwanted changes like Zhang et al. (2005) reported for celery sticks. In this case, dipping 0.18 ppm O_3 water for 5 min reduced RR, inhibited browning and improved sensory quality. After 9 d at 4°C, bacterial population was reduced 1.69 log CFU g^{-1} compared to control water. Rico et al. (2006) found that O_3 water (1 mg L^{-1} at 18–20°C) reduced both enzyme activity and enzymatic browning of shredded lettuce. However, this enzyme inactivation showed a negative effect, as the reduction in activity of the texture-related pectin methylesterase was correlated with a lower crispiness. Selma et al. (2007) treated shredded lettuce with 5 ppm O_3 water for 5 min reaching a reduction of 1.8 log CFU g^{-1} in *Shigella sonnei* counts. Silveira et al. (2007; 2010) found a similar sanitizer effect between 150 ppm NaClO and O_3 dips (0.4 ppm, 3 min) on diced Galia melon. Both treatments reduced 1 log CFU g^{-1} compared to water alone and extended shelf-life to 10 days at 5°C.

In some cases O_3 appears to be lowly successful. Aguayo et al. (2003) washed 3 min FC Amarillo melon on 6.5 ppm O_3 water. After 10 days at 5°C, the mesophilic and psychrotrophic bacterial counts were reduced only on 1.2 and 0.2 log CFU g^{-1} respectively and on 0.4 and 1.5 log CFU g^{-1} respectively those of yeasts and moulds. In sliced Thomas tomato stored for 10 days at 5°C, the O_3 water (3.8 ppm, 3 min) reduced on 1.9, 1.6 and 0.7 the mesophilic, psychrotrophic and yeast counts respectively. In shredded Iceberg lettuce, Baur et al. (2005) maintained better sensory and microbiological quality, prolonging shelf-life, in samples treated with 200 mg L^{-1} free chlorine than in O_3 water (1 ppm, 4°C, 120 s). In contrast, the application of 2 ppm O_3 water for 2 min was found to be optimum processing conditions of shredded green leaf lettuce, in terms of reducing microbial load and maintaining the sensory quality compared with 100 ppm NaClO water during chilling storage (Ölmez and Akbas, 2009).

Beltrán et al. (2005) found no evidence of browning in FC potatoes dipped in O_3 water (20 mg L^{-1} 1 min) or in O_3 plus peroxyacetic acid (300 mg L^{-1}) and stored up to 14 d under partial vacuum at 4°C. These treatments kept the initial texture and aroma. However, the ozonated water alone was not effective in reducing total microbial populations. The combination of O_3–Tsunami 100 (15% PA and 11% H_2O_2) was the most effective treatment to control microbial growth achieving 3.3, 3.0 and 1.2 log-reductions for LAB, coliforms and anaerobic bacteria, respectively.

Regarding O_3 handling, even when it has been approved as food additive, workers who might have contact with O_3 must be careful since gas concentrations over 0.2 ppm may affect respiration and may produce dizziness and eyes and throat irritation (Hoof, 1982). Ambient O_3 levels at the industry facilities should be continuously monitored in view of the fact that high levels may be fatal to humans (Shirk, 2000)

5. Superatmospheric Oxygen

For most of the cases, a conventional MAP with 1 to 8 kPa O_2 combined with 10 to 20 kPa CO_2 at 0 to 5°C is frequently recommended for keeping quality of FC fruit and vegetables (Gorny 1997; Artés et al., 2006b). However, the combined effect of high O_2 concentrations (over 60 kPa) and CO_2 around 10 to 20 kPa may provide adequate suppression of microbial growth increasing produce shelf-life (Allende et al., 2004; Geysen et al., 2006; Conesa et al., 2007ab; Escalona et al., 2007). This superatmospheric O_2 MAP could be effective for inhibiting enzymatic browning, preventing anaerobic fermentation and moisture and odour losses while reducing aerobic and anaerobic microbial growth (Day, 2001).

A number of factors may explain the toxicity of hyperbaric O_2 like the adverse effects on the oxidation-reduction potential of the system, the oxidation of enzymes having sulfhydryl groups or disulfide bridges, and the accumulation of injurious reactive O_2 species (Kader and Ben-Yehoshua, 2000). CO_2 causes a decrease in intra and extracellular pH and interferes with the cellular metabolism with a stronger inhibitory effect at low temperature due to enhanced CO_2 solubility (Dixon and Kell, 1989).

Recently, some studies have been carried out for analysing the effect of high O_2 and CO_2 levels on quality and safety of FC plant commodities (Allende et al., 2004; Amanatidou et al. 1999, 2000; Geysen et al., 2006; Conesa et al., 2007a,b; Escalona et al., 2007; Jacxsens et al. 2001; Sánchez-Ballesta et al., 2007). In practice, the use of high O_2 in MAP for keeping quality of FC produces is still emergent and needs to be supported by more research. Kader and Ben-Yehoshua (2000) have found some contradictory results since superatmospheric O_2 may stimulate, have no effect, or reduce RR and C_2H_4 production, depending on the commodity, ripening stage, O_2, CO_2 and C_2H_4 concentrations in the atmosphere, storage time and temperature. Related to microbial growth, it seems to be that exposure to high O_2 did not strongly inhibited it, while high CO_2 reduced the growth to some extent in most cases. Allende et al. (2004) found a reduction of aerobic microbial growth in superatmospheric O_2 due to the accumulation of high CO_2 levels. The antimicrobial activity of CO_2 at high concentrations is already well known. Probably the same effect was observed by Van Der Steen et al. (2002) for raspberries and strawberries when applying a high O_2 level in a high-barrier film, showing a reduction in microbial growth while keeping sensory quality attributes. The addition of CO_2 is unnecessary when high O_2 atmospheres are injected in barrier film packages, because the respiratory activity of produces generate antimicrobial CO_2 levels reaching 20 kPa. By using this high O_2 MAP the shelf life for FC vegetables can be twice compared to conventional low O_2 MAP just by suppressing microbial growth (Jacxsens et al., 2001). In the case of *L. innocua* levels higher than 75 kPa O_2 were needed to reduce growth on shredded lettuce (Escalona et al., 2007), in agreement with other studies where *L. innocua* growth was almost not influenced by O_2 concentrations up to 100 kPa (Geysen et al., 2006). *In vitro* studies on yeast growth under superatmospheric O_2 showed contradictory results. Jacxsens et al. (2001), found a reduced growth of *Candida lambica* while a stimulated growth of *C. guilliermondii* and *C. sake* under around 80 kPa O_2 at 8°C was reported by Amanatidou et al. (1999). Yeast growth can be reduced or stimulated under high O_2 levels depending on the species or strain and their sensitivity to high O_2 (Van der Steen et al., 2003). Allende et al. (2002), reported yeast counts higher than 5 log CFU g^{-1} after 3 days at 4°C on mixed salad stored under superatmospheric O_2 MAP. For FC melon, high O_2 levels kept the initial colour and firmness retarding anaerobic fermentation better than low O_2 concentrations. In addition, yeast growth was delayed (Oms-Oliu et al., 2008b). Related to residual effect Zheng et al. (2008), found that berries kept in 60 to 100 kPa O_2 showed inhibited decay and, after additional 2 days in air at 20ºC, treated fruit also exhibited less decay rate, suggesting that high O_2 levels had residual effect on decay control.

Linked to the response of FC vegetables to high O_2, Jacxsens et al. (2001) reported a low sensitivity in grated celeriac and shredded chicory endive. In 3, 80 and 95 kPa O_2, grated celeriac reached 120, 97 and 124 nmol O_2 kg^{-1} s^{-1} respectively and for shredded endive the RR was 90, 107 and 132 nmol O_2 kg^{-1} s^{-1} also respectively. In addition, exposure of shredded carrot to 50 kPa O_2 + 30 kPa CO_2 showed an accelerated C_2H_4 production, possibly as an injury response at 8°C after 2 to 3 days (Amanatidou et al., 2000). Escalona et al. (2006)

reported a higher effect of CO_2 than that of O_2 in the RR of FC butter lettuce. A level of 10 kPa CO_2 reduced the RR and when raised over 20 kPa CO_2 the RR increased probably due to a shift to anaerobic metabolism. At 20 to 100 kPa O_2 and 1, 5 and 9ºC, the respiratory quotient was about 0.7 to 1. In FC carrot stored in initial 70 kPa O_2, the RR expressed in O_2 consumed was 3-fold higher than under initial 2.5 kPa O_2 + 7 kPa CO_2 as a consequence of oxidative processes. By using low permeability film a rapid depletion of O_2 and accumulation of CO_2 was obtained under initial low O_2 concentration (Oms-Oliu et al., 2008a).

Few reports indicate the effects of elevated O_2 on enzymatic browning. From in vitro studies of PPO kinetics, the main enzyme responsible of browning, it seems to be that the substrate concentration as well as the O_2 level had a clear inhibitory effect on the reaction rate. Moreover, the inhibitory effect of O_2 was more evident at low final product concentration (Gómez et al., 2006). In vivo studies showed that an atmosphere of 80 kPa O_2 + 20 kPa CO_2 delayed browning of shredded lettuce under MAP for 10 days at 5°C compared to air (Heimdal et al., 1995), although it is well know that browning susceptibility is highly cultivar dependent. Jacxsens et al. (2001) found that low O_2 caused a reduction in the enzymatic browning in FC vegetables since O_2 is a necessary substrate for the reaction. However, high O_2 atmospheres were more effective in reducing or inhibiting the enzymatic browning of Iceberg lettuce, radicchio and Lollo Rosso lettuce. In the case of FC bell pepper, 80 kPa O_2 combined with 15 kPa CO_2 maintained the main sensory quality attributes and inhibited growth of the spoilage microorganisms and *Enterobacteriaceae* (Conesa et al., 2007ab).

In the case of shredded Iceberg lettuce Heimdal et al. (1995) found that under 80 kPa O_2 + 20 kPa CO_2 browning was higher than under moderate vacuum MAP. For Spartan apple, Lu and Toivonen (2000) reported that a pretreatment with 100 kPa O_2 before cutting reduced surface browning, flesh softening, and off-flavour in slices. This inhibition of browning was associated with retention of cellular integrity.

Regarding sensory quality, acceptable scores were found in FC spinach leaves treated with 80 to 100 kPa O_2 + 20 kPa CO_2 compared to low O_2 and high CO_2, where the spinach was affected by fermentation (Allende et al., 2004). Beneficial effects of superatmospheric O_2 have been reported for shredded chicory endive and mixed vegetable salads (Jacxsens et al. 2001; Allende et al., 2002; Escalona et al., 2007).

In addition, FC carrots stored in 50 kPa O_2 + 30 kPa CO_2 had similar or better quality than those in 1 kPa O_2 + 10 kPa CO_2 after 12 days at 8°C. Therefore, FC carrots could tolerate high CO_2 levels in combination with high O_2 levels, and this innovative MAP can be used for keeping their fresh characteristics and lowering microbial growth during shelf life (Amanatidou et al., 2000). FC butter lettuce kept a good visual appearance with 75 kPa O_2 + 15 kPa CO_2 after 10 days at 7°C, reducing cut surface browning and loss of freshness (Escalona et al., 2007). Spinach leaves kept a better appearance when stored at higher than 50 kPa O_2, delaying anaerobic fermentation and increasing shelf life when compared to atmospheres with 0.5 and 25 kPa O_2 (Magalhães et al., 2007).

6. Innovative Gas Treatments

Packaging under non-conventional gases like Ar, He, Xe or N_2O has been proposed for improving quality of some FC plant commodities. Those gases may be chemically inert, but they had some antimicrobial and/or physiological effects, although it did not appear to be through modification of enzymes activity (Gorny and Agar, 1998). In fact, it has been shown that an Ar-enriched MAP reduced microbial growth and delayed quality loss of fresh broccoli and lettuce (Day, 1996; Jamie and Saltveit, 2002). In that way, a patent registered the use of Ar in the preservation of cut and segmented fresh fruits (Powrie et al., 1990). He or Ar-enriched atmospheres might modify the O_2, CO_2, and C_2H_4 diffusion in some commodities (Burg and Burg, 1965). Certainly, low O_2 atmospheres combined with high Ar, He or N_2 had different diffusive characteristics because both Ar and He are monoatomic gases and are smaller in size than the diatomic N_2 (Jamie and Saltveit, 2002). It has been found that 90 kPa Ar + 2 kPa O_2 did not delay the accumulation of phenolics in FC lettuce, or the loss of chlorophyll from broccoli florets beyond that of low O_2 atmospheres made with He or N_2 (Lougheed and Lee, 1991). According to Sidorkin et al. (1989) there was no difference in the viability of cells, the uptake of sucrose, or the density of tobacco suspension cultures grown for 8 days in 21 kPa O_2 and 79 kPa of Ar or N_2. On the other hand Xe showed good results on forming clathrate hydrates (crystalline water-based solids) under 1.5 MPa pressure in preservation of asparagus and cucumber, but the high cost may limit its application. Storing asparagus spears at 4 °C under 1.1 MPa mixture of Ar and Xe (2:9, v:v) for 24 h extended their shelf-life compared with a conventional MAP of 5 kPa O_2 + 5 kPa CO_2 (Zang et al., 2007).

N_2O has effects as an inhibitor of C_2H_4 production and RR on fruits (Benkeblia & Varaquaux, 2003; Palomer et al., 2005) as well as a fungistatic agent (Thom and Marquis, 1984; Qadir & Hashinaga, 2001a,b). N_2O has 77% solubility in fruit cell, although its absorption in tissues is completely reversible (Gouble et al., 1995). Therefore, N_2O have a direct effect by extending the shelf life of these products. Innovative MAP of 90 kPa Ar + 5 kPa O_2 and 90 kPa N_2O + 5 kPa CO_2 were used with kiwifruit slices at 4 °C for 12 days in comparison to air and N_2-enriched storage. MAP with 90 kPa N_2O maintained the best quality of slices, delaying firmness loss and browning in pericarp and core surfaces. Slight modifications in the most important discriminated quality factors for the slices under N_2O-enriched, a rapid quality loss for air and N_2-enriched atmosphere and an acceptable quality in Ar-enriched atmospheres after 8 days at 4°C were found (Rocculi et al., 2005). FC spinach leaves fertilized with 8 or 16 mmol N L^{-1} under a floating trays system and stored under N_2O-enriched MAP showed after 8 days at 5°C a reduced microbial growth, with good sensory quality preserving the total antioxidant capacity, phenolics and chlorophylls content (Rodríguez-Hidalgo et al., 2010)

Conclusion

The food industry is now seeking alternatives to chlorine which may assure the safety of FC produce and maintain the quality and shelf-life, while also reducing the rate of water consumption during processing. Sustainable strategies that would allow replacing it are promissory. Chlorine dioxide, ozone, organic acids, peroxyacetic acid, electrolyzed oxidizing

Table 1. Alternative sanitizers to chlorine for fresh cut fruit and vegetables

Sanitizer	Commodity	References
Peroxiacetic acid	Melon	Rodgers et al., 2004; Rocha-Bastos et al., 2005; Silveira et al., 2010
	Carrot, lettuce, leek, white cabbage	Ruiz-Cruz et al., 2007; Vandekinderen et al., 2009
	Tomato, sweet pepper, cucumber	Alvaro et al., 2009
	Rocket	Martínez-Sánchez et al., 2006
	Fresh-cut potato	Beltrán et al., 2005
	Escarole, lettuce	Rodgers et al., 2004; Allende et al., 2008; López-Gálvez et al., 2009
	Apples, strawberries	Rodgers et al., 2004
Chlorine dioxide	Cabbage	Sy et al., 2005; Gómez-López et al., 2008
	Lettuce	Rodgers et al., 2004; Sy et al., 2005, Gómez-López et al., 2008; Mahmoud and Linton, 2008a,b; López-Gálvez et al., 2010
	Carrot	Sy et al., 2005; Gómez-López et al., 2007
	Strawberry	Rodgers et al., 2004; Mahmoud et al., 2007
	Melon	Sy et al., 2005; Rodgers et al., 2004
	Apple	Rodgers et al., 2004
	Faba bean	Artés et al., 2007a,b
	Chard	Tomás-Callejas et al., 2010
	Green pepper	Han et al., 2001
Hydrogen peroxide	Melon	Sapers and Simons, 1998; Ukuku et al., 2004, 2005
	Lettuce	Gopal et al., 2010
	Green bell pepper, cucumber, zuchini	Sapers, 2003
Citric acid – ascorbic acid – calcium	Chinese cabagge	Kim and Klieber, 1997
	Melon	Luna-Guzmán et al., 1999; Aguayo et al., 2003, 2007
	Lettuce	Martín-Diana et al., 2006a; Ölmez and Temur, 2010
	Green celery	Gómez and Artés, 2004
	Cilantro	Allende et al., 2009
Electrolyzed water	Carrot	Izumi, 1999
	Spinach	Izumi, 1999
	Bell pepper	Izumi, 1999
	Potato	Izumi, 1999
	Cucumber	Izumi, 1999
	Lettuce	Hua et al., 2004; Koseki et al., 2001; Abadías et al., 2008; Rico et al., 2007a
	Alfalfa sprouts	Kim et al., 2003
	Cilantro	Wang et al., 2004
	Mizuna leaves	Martínez-Hernández et al., 2009

Table 1. (Continued)

Sanitizer	Commodity	References
Steamer jet injection	Lettuce	Martín-Diana et al., 2007
Biological compounds, natural microbiota and/or their antimicrobial compounds	Lettuce	López-Gálvez et al., 2009; Xu et al., 2007; Martín-Diana et al., 2008; Gutierrez et al., 2009
	Cucumber	Xu et al., 2007
	Apple	Lanciotti et al., 1999; Leverentz et al., 2006
	Carrot	Gutierrez et al., 2009
	Cabbage	Bari et al., 2005
	Brocoli	Bari et al., 2005
	Mung bean sprouts	Bari et al., 2005
UV-C radiation and intense light pulses	Spinach	Artés-Hernández et al., 2008; Escalona et al., 2010
	Zucchini squash	Erkan et al., 2001
	Tomato	Lu et al., 1987; Liu et al., 1993; Robles et al., 2007
	Strawberry	Marquenie et al., 2002; Allende et al., 2007
	Lettuce	Allende et al., 2003a,b, 2006; Gómez-López et al., 2005
	Pomegranate arils	López-Rubira et al., 2005
	Bell pepper	Artés et al., 2006
	Mushroom	Oms-Oliu et al., 2010
	Watermelon	Fonseca and Rushing, 2006; Artés-Hernández et al., 2010
Ozone	Lettuce	Baur et al., 2005; Rico et al., 2006; Selma et al., 2007; Ólmez and Akbas, 2009
	Apple	Skog and Chu, 2000
	Orange	Skog and Chu, 2000
	Table grape	Sarig et al., 1996
	Tomato	Aguayo et al., 2006
	Melon	Aguayo et al., 2003; Silveira et al, 2007, 2010; Selma et al., 2008
	Spinach	Klockow and Keener, 2009
	Celery	Zhang et al., 2005
	Potato	Beltrán et al., 2005
Superatmospheric oxygen	Mixed salads	Allende et al., 2002, 2003; Amanatidou et al., 1999, 2000
	Carrots	Oms-Oliu et al., 2008a
	Bell pepper	Conesa et al., 2006, 2007
	Table grapes	Sanchez-Ballesta et al., 2001
	Strawberry	Van Der Steen et al., 2002; Zheng et al., 2008
	Melon	Oms-Oliu et al., 2008b
Innovative gas treatments	Asparagus	Zang et al., 2007
	Apple	Qadir and Hashinaga, 2001
	Guava	Qadir and Hashinaga, 2001
	Mandarin	Qadir and Hashinaga, 2001
	Persimmon	Qadir and Hashinaga, 2001
	Strawberry	Qadir and Hashinaga, 2001
	Tomato	Qadir and Hashinaga, 2001

water, UV-C and hydrogen peroxide are the main alternative sanitizing agents that gained interest in recent years. The effects of these disinfecting agents on the microbiological, nutritional and sensory quality of FC produce, and also the possible environmental impact and the potential on minimizing water consumption rates in the food industry are aspects to be elucidated during the next years. The potential and limits of these innovative sustainable techniques must also be well defined and included in the regulations.

It is worth to mention that genetic *cv* selection must be oriented to slow ripening and senescence rates, with low C_2H_4 production and/or sensitivity, improved firmness, good adaptation to minimal processing and enhanced antioxidant systems.

Application of a sanitizer or a combination of disinfection treatments does not mean a sanitation program. It implies much more effort, from a well designed integrated production, handling and processing to proper distribution chains, respecting low storage temperatures and optimal MAP. In that way, food safety hazards associated with FC produce must be clearly identified. It should be taken into account that the produce moves from the field through the processing plant and then to the consumer's table. Consequently three highly important sources of contamination -microbial, chemical and physical - may represent a health hazard. Procedures like HACCP together with a specific action plan should be developed on each case in order to identify and control any of these hazards from entering the handling process.

How to interpret microbial results and which standardized procedures are best for consistent microbial analysis in testing FC produce must be a top priority. Consequently, new techniques for microbial detection like PCR and qPCR have to be promoted and fitted for different crops and handling conditions.

References

Abadias M, Usall J, Oliviera M, Alegre I, Viñas I., 2008. Efficacy of neutral electrolyzed water (NEW) for reducing microbial contamination on minimally-processed vegetables. *Inter. J. Food Mi.* 123, 151-158.

Abshire, R., Dunton, H., 1981. Resistance of selected strains of Pseudomonas aeruginosa to low-intensity ultraviolet radiation. *Appl. Environ. Microbiol.* 41, 1419–1423.

Achen, M., 2000. Efficacy of ozone in inactivating Escherichia coli O157:H7 in pure cell suspensions and on apples. *M.S. Thesis.* The Ohio State University, Columbus.

Adams, M., Hartley, A., Cox L., 1989. Factors affecting the efficacy of washing procedures used in the production of prepared salads. *Food Microbiol.* 6, 69-77.

Aguayo, E., Allende, A., Artés, F., 2003. Keeping quality and safety of minimally fresh processed melon. *Eur Food Res. Technol.* 216, 494–499.

Aguayo, E., Escalona, V., Artés, F., 2006. Effect of the cyclic exposure to ozone gas on phytochemical, sensorial and microbial quality in whole and sliced tomatoes. *Postharvest Biol. Technol.* 39, 166-177.

Aguayo, E., Escalona, V., Artés, F., 2007. Effect of hot water treatment and various calcium salts on quality of fresh-cut 'Amarillo' melon. *Postharvest Biol. Technol.* 47(3), 397-406.

Ahvenainen, R., 1996. New approaches in improving the shelf life of minimally processed fruit and vegetables. *Trends Food Sci. Technol.* 7, 179–187.

Allende, A., Jacxsens, L., Devlieghere, F., Debevere, J., Artés, F., 2002. Effect of super-atmospheric oxygen packaging on sensorial quality, spoilage, and Listeria monocytogenes and Aeromonas caviae growth in fresh processed mixed salads. *J. Food Protect.* 65, 1565-1573.

Allende, A., Artés, F., 2003a. UV-C radiation as a novel technique for keeping quality of fresh processed 'Lollo Rosso' lettuce. *Food Res. Int.* 36, 739–746.

Allende, A., Artés, F., 2003b. Combined ultraviolet-C and modified atmosphere packaging treatments for reducing microbial growth of fresh processed lettuce. Lebensm. *Wiss. Technol.* 36, 779–786.

Allende, A., Luo, Y., McEvoy, J., Artés, F., Wang, C., 2004. Microbial and quality changes in minimally processed baby spinach leaves stored under super atmospheric oxygen and modified atmosphere conditions. *Postharvest Biol. Technol.* 33, 51-59.

Allende, A., McEvoy, J., Luo, Y., Artés, F., Wang, C., 2006. Effectiveness of two-sided UV-C treatments in inhibiting natural microflora and extending the shelf-life of minimally processed 'Red Oak Leaf' lettuce. *Food Microbiol.* 23, 241–249.

Allende, A., Marína, A., Buendía, B., Tomás-Barberán, F., Gil, M.I., 2007. Impact of combined postharvest treatments (UV-C light, gaseous O_3, superatmospheric O_2 and high CO_2) on health promoting compounds and shelf-life of strawberries. *Postharvest Biol. Technol.* 46, 201-211.

Allende, A., Selma, M.V., López-Gálvez, F., Villaescusa, R., Gil, M.I., 2008. Role of comercial sanitizers washing systems on epiphytic microorganisms and sensory quality of fresh-cut escarole and lettuce. *Postharvest Biol. Technol.* 49, 155-163.

Allende, A., McEvoy, J.,Tao, Y., Luo, Y., 2009. Antimicrobial effect of acidified sodium chlorite, sodium chlorite, sodiumhypochlorite, and citric acid on Escherichia coli O157:H7 and natural microflora of fresh-cut cilantro. *Food Control,* 20, 230-234

Alothman, M., Bhat, R., Karim, A.A., 2009. Effects of radiation processing on phytochemicals and antioxidant in plant produce. *Trends in Food Sci. Technol.* 20, 201-212.

Alvaro, J.E., Moreno, S., Dianeza, F., Santosa, M., Carrasco, G., Urrestarazua, M., 2005. Effects of peracetic acid disinfectant on the postharvest of some fresh vegetables. *J. Food Engineering.* 95(1), 11-15.

Amanatidou, A., Smid, E., Gorris, G., 1999. Effect of elevated oxygen and carbon dioxide on the surface growth of vegetable-associated micro-organisms. *J. Appl. Microbiol.* 86, 429-438.

Amanatidou, A., Slump, R., Gorris, G. and Smid E., 2000. High oxygen and high carbon dioxide modified atmospheres for shelf-life extension of minimally processed carrots. *J. Food Sci.* 65, 61-66.

Artés, F., 2004. Refrigeration for preserving the quality and enhancing the safety of plant foods. Bulletin International Institute of Refrigeration. *LXXXIV,* 1, 5-25.

Artés, F., Allende, A., 2005. Minimal fresh processing of vegetables, fruits and juices. In: Emerging Technologies in Food Processing. Editor. D.W. *Sun. Edit. Elsevier.* 26, 675-715.

Artés F., Conesa, A., López-Rubira, V., Artés-Hernández, F., 2006a. UV-C treatments for improving microbial quality in whole and minimally processed bell peppers. In: Ben-Yehoshua S, D'Hallewin G., Erkan M, Rodov V, Lagunas M. (edits), *The use of UV as a postharvest treatment: Status and prospects.* Edit. Antalya Kros Ofset - Yakup Kublay. 12-17.

Artés, F., Gómez, P., Artés-Hernández, F., 2006b. Modified atmosphere packaging of fruits and vegetables. *Stewart Postharvest Rev.* 5(2), 1-13.

Artés, F., Gómez, P., Artés-Hernández, F., 2007. Physical, physiological and microbial deterioration of minimally fresh processed fruits and vegetables. *Food Sc. Technol. Intern.* 13 (3),177–188.

Artés, F., Gómez, P.A., Aguayo, E., Escalona, V.H., Artés-Hernández F., 2009. Sustainable sanitation techniques for keeping quality and safety of fresh-cut plant commodities. *Postharvest Biol. Technol.* 51, 287–296.

Artés-Hernández, F., Artés, F., Tomás-Barberán, F., 2003. Quality and enhancement of bioactive phenolics in cv. Napoleon table grapes exposed to different postharvest gaseous treatments. *J. Agric. Food Chem.* 51, 5290-5295.

Artés-Hernández F., Aguayo E., Artés F., Tomás-Barberán F., 2007. Enriched ozone atmosphere enhances bioactive phenolics in seedless table grapes after prolonged shelf life. *J. Sci. Food Agric.* 87, 824–831.

Artés-Hernández, F., Escalona, V.H., Robles, P.A., Martínez-Hernández, G.B., Artés, F. 2009. Effect of UV-C radiation on quality of minimally processed spinach leaves. *J. Sci. Food Agric.* 89, 414-421.

Artés-Hernández, F., Robles, P.A., Gómez, P.A., Tomás-Callejas, A., Artés, F., 2010. Low UV-C ilumination for keeping overall quality of fresh-cut watermelon. *Postharvest Biol. Technol.* 55, 114-120.

Ayebah, B., Hung, Y.C., 2005. Electrolyzed water and its corrosiveness on various surface materials commonly found in food processing facilities. *J. Food Process Engineering.* 28, 247–264.

Bablon, G., Belamy, W., Billen, G., Bourbigot, M., Daniel, F., 1991. Practical applications of ozone: principles and case studies. In: Langlais, G., Reckhow, D.A., Brink, D.R., (eds.). *Ozone in water treatment: application and engineering.* Chelsea, Mich., USA, Lewis Publisher, Inc. 133-316.

Baka, M., Mercier, J., Corcuff, R., Castaigne, F., Arul, J., 1999. Photochemical treatment to improve storability of fresh strawberries. *J. Food Sci.* 64, 1068-1072.

Bari, M., Sabina, Y., Isobe, S., Uemura, T., Isshiki, K., 2003. Effectiveness of electrolyzed acidic water in killing Escherichia coli O157:H7, Salmonella enteriitidis, and Listeria monocytogenes on the surfaces of tomatoes. *J. Food Protect.* 66, 542−548.

Bari, M., Ukuku, D., Kawasaki, T., Inatsu, Y., Isshiki, K., Kawamoto, S., 2005. Combined efficacy of nisin and pediocin with sodium lactate, citric acid, phytic acid, and potassium sorbate and EDTA in reducing the Listeria monocytogenes population of inoculated fresh-cut produce. *J. Food Protect.* 68, 1381-1387.

Baur, S., Klaibera, R., Hammesb, W.P., Carlea, R., 2005. Sensory and microbiological quality of shredded, packaged iceberg lettuce as affected by pre-washing procedures with chlorinated and ozonated water. *Innovat. Food Sci. Emerging Technol.* 6, 171-182.

Beltrán, D., Selma, M., Tudela, J., Gil, M., 2005. Effect of different sanitizers on microbial and sensory quality of fresh-cut potato strips stored under modified atmosphere or vacuum packaging. *Postharvest Biol. Technol.* 37, 37–46.

Benkeblia, N., Varoquaux, P., 2003. Effect of nitrous oxide (N_2O) on respiration rate, soluble sugars and quality attributes of onion bulbs Allium cepa cv. Rouge Amposta during storage. *Postharvest Biol. Technol.* 30, 161-168.

Ben-Yehoshua, S. and Mercier, J., 2005. UV irradiation, biological agents, and natural compounds for controlling postharvest decay in fresh fruits and vegetables. In: Ben-Yheoshua S. (ed.), *Environmentally friendly technologies for agricultural produce quality*. CRC Taylor & Francis. Boca Raton. Florida. 265-299.

Betts, G., and Everis, L., 2005. Alternatives to hypochlorite washing systems for the decontamination of fresh fruit and vegetables. In: W, Jongen (Ed.), *Improving the safety of fresh fruit and vegetables.* Wageningen, Netherlands.

Beuchat, L., Brackett R., 1990. Survival and growth of Listeria monocytogenes on lettuce as influenced by shredding, chlorine treatment, modified atmosphere packaging, and temperature. *J. Food Sci.* 55, 755–75.

Beuchat, L., 2000. Use of sanitizers in raw fruit and vegetable processing. In: Alzamcra S.M., Tapia M.S. and Lopez-Malo A. (eds), *Minimally processed fruits and vegetables, Aspen,* Maryland. 63–78.

Beuchat, L., 2002. Ecological factors influencing survival and growth of human pathogens on raw fruits and vegetables. *Microbiol. Infect.* 4, 413-423.

Bintsis, T., Litopoulou-Tzanetaki, E., Robinson, R., 2000. Existing and potential applications of ultraviolet light in the food industry - a critical review. *J. Sci. Food Agric.* 80, 637-645.

Boyette, M., Ritchie, D., Carballo, S., Blankenship, S. and Sanders, D., 1993. Chlorination and postharvest disease control. *HortTechnolgy*. 3, 395-400.

Bruhn, C., 2002. Consumer issues in quality and safety. In: A.A. Kader (ed.). *Postharvest technology of horticultural crops.* 3rd ed. University of California. 31-37.

Burg, S., Burg, E., 1965. Gas exchange in fruits. *Physiol. Plant.* 18, 870–884.

Burt, S. 2004. Essential oils: their antibacterial properties and potential applications in foods-a review. *Int. J. Food Microbiol.* 94, 223-253

Capece, P., 2001. Trihalomethanes and our water supply. *Southern Data Systems*, 20(3), 35 pp. http://www.southerndatastream.com/thm.index.htm.

Carlin, F., Nguyen-the, C., 1999. Minimally processed produce. Microbiological issues. In *Proceedings International Conference on Fresh-cut Produce,* Chipping Camden, UK: Campden and Chorleywood Food Research Association. 9-10 September, 1999.

Cisneros-Zevallos, L., 2003. The use of controlled postharvest abiotic stresses as a tool for enhancing the nutraceutical content and adding-value of fresh fruits and vegetables. *J. Food Sci.* 68, 1560-1565.

Civello, P., Chaves, A., Martínez, G., 2006. UV-C treatment delays postharvest senescence in broccoli florets. *Postharvest Biol. Technol.* 39, 204–210.

Code of Federal Regulations (CFR), 2007a. Title 21, Part 173.300. *Secondary Direct Food Additives Permitted in Food for Human consumption: Chlorine dioxide.* Available at http://frwebgate3.accedd.gpo.gov/cgi-bin.

Code of Federal Regulations (CFR), 2007b. Title 21, Part 173.315. *Secondary Direct Food Additives Permitted in Food for Human consumption: Chemicals used in washing or to assist in the peeling of fruits and vegetables.* Chlorine dioxide. Available at http://frwebgate3.accedd.gpo.gov/cgi-bin.

Conesa A., Artés-Hernández F., Geysen S., Nicolaï B., Artés F., 2007a. High oxygen combined with high carbon dioxide improves microbial and sensory quality of fresh-cut peppers. *Postharvest Biol. Technol.* 43, 230–237.

Conesa A., Verlinden B.E., Artés-Hernández F., Nicolaï B., Artés F., 2007b. Respiration rates of fresh-cut bell peppers under supertamospheric and low oxygen combined or not with high carbon dioxide. *Postharvest Biol. Technol.* 45, 81–88.

Costa, L., Vicente, A.R., Civello, P.M., Chaves, A.R., Martinez, G.A., 2006. UV-C treatment delays postharvest senescence in broccoli florets. *Postharvest Biology and Technology,* 39, 204-210.

Day, B., 1996. High oxygen modified atmosphere packaging for fresh prepared produce. *Postharvest News Inf.* 7, 31N–34N.

Day, B., 2001. Fresh prepared produce: GMP for high oxygen MAP and non-sulphite dipping. Guideline N° 31. *Chipping Campden: Campden and chorleywood food research association group.* 1-76.

Dixon, N., Kell, D., 1989. The inhibition by CO_2 of the growth and metabolism of micro-organisms: a review. *J. Appl. Bacteriol.* 67, 109-136.

Elmnasser, N., Guillou, S., Leroi, F., Orange, N., Bakhrouf, A. and Federighi, M., 2007. Pulsed-light system as a novel food decontamination technology: a review*, Can. J. Microbiol.* 53, 813–821.

Environmental Protection Agency (EPA), 1999. *Guidance Manual: Alternative disinfectants and oxidants: chlorine dioxide.* USA. 4-36.

Erkan, M., Wang, C., Krizek, D., 2001. UV-C radiation reduces microbial populations and deterioration in Cucurbita pepo fruit tissue. *Environ. Exp. Bot.* 45, 1–9.

Escalona, V., Verlinden, B., Geysen, S., Nicolaï, B., 2006. Changes in respiration of fresh-cut butterhead lettuce under controlled atmospheres using low and superatmospheric oxygen conditions with different carbon dioxide levels. *Postharvest Biol. Technol.* 39, 48-55.

Escalona, V., Geysen, S., Verlinden, B., Nicolaï, B., 2007. Microbial quality and browning of fresh-cut butter lettuce under superatmospheric oxygen condition. *Eur. J. Hort. Sci.* 72, 130-137.

Escalona, V.H., Aguayo, E., Mertínez-Hernández, G.B., Artés, F. 2010. UV-C doses to reduce pathogen and spoilage bacterial growth in vitro and in baby spinach. Postharvest Biol. Technol. In press. doi:10.1016/j.postharvbio.2010.01.008.

Fabrizio, K. A., Cutter, C. N., 2003. Stability of electrolyzed oxidizing water and its efficacy against cell suspensions of Salmonella Typhimurium and Listeria monocytogenes. *J. Food Protec.* 66, 1379–1384.

Foegeding, P., 1985. Ozone inactivation of Bacillus and Clostridium spore populations and the importance of the spore coat to resistance. *Food Microbiol.* 2, 123-134.

Fonseca, J., Rushing, J.W., 2006. Effect of ultraviolet-C light on quality and microbial population of fresh-cut watermelon. *Postharvest Biol. Technol.* 40, 256-261.

Food and Drug Administration (FDA), 1997. Substances generally recognized as safe, proposed rule, *Federal Register* 62 (74), 18937–18964.

Food and Drug Administration (FDA), 1998. *Department of Health and Human Services. Secondary Direct Food Additive for Human Consumption.* 21 CFR. Part 173.300 chlorine dioxide.

Gardner, D.W., Shama, G., 2000. Modeling UV-induced inactivation of microorganisms on surfaces. *J. Food Protect.* 63, 63–70.

Geysen, S., Escalona, V., Verlinden, B., Aertsen, A., Geeraerd, A., Michiels, C., Van Impe, J., Nicolaï, B., 2006. Validation of predictive growth models describing superatmospheric oxygen effects on Pseudomonas fluorescens and Listeria innocua on fresh-cut lettuce. *Int. J. Food Microbiol.* 111, 48–58.

Glaze, W.1987. The chemistry of water treatment processes involving ozone, hydrogen peroxide, and ultraviolet radiation. *Ozone Sci. Eng.* 9(4):335.

Gómez, P., Artés, F., 2004. Ascorbic and citric acids to preserve quality of minimally processed green celery. *Proceedings of IV Postharvest Iberian Symposium.* Oeiras, Portugal. 369-373.

Gómez, P., Geysen, S., Verlinden, B., Artés, F., Nicolaï, B., 2006. Modelling the effect of superatmospheric oxygen concentrations on in vitro mushroom PPO activity. *J. Sci. Food Agric.* 86, 2387–2394.

Gómez, P., Pagnon, M., Egea-Cortines, M., Artés, F. and Weiss, J. 2010. A fast molecular non-destructive protocol for evaluating aerobic bacterial load on fresh-cut lettuce. *Food Sci. Technol. Internat.* In press.

Gómez, P.L., Alzamora, S.M., Castro, M.A., Salvatori, D.M., 2010. Effect of ultraviolet-C light dose on quality of cut-apple: Microorganism, color and compression behavior. *J. Food Eng.* 98, 60-70.

Gómez-López, V., Devlieghere, F., Bonduellea, V., Debevere, J., 2005. Intense light pulses decontamination of minimally processed vegetables and their shelf-life. *Int. J. Food Microbiol.* 103, 79– 89.

Gómez-López, V.M., Rajkovic, A., Ragaert, P., Smigic, N., Devlieghere, F., 2009. Chlorine dioxide for minimally processed produce preservation: a review. *Trends Food Sci. Technol.* 20, 17-26.

Gopal, A., Coventry, J., wan, J., Roginski, H., Ajlouni, S., 2010. Alternative disinfection techniques to extend the shelf-life of minimally processed iceberg lettuce. *Food Microbiol.* 27, 210-219.

Gorny, J., 1997. A summary of CA and MA requirements and recommendations for fresh-cut (minimally processed) fruits and vegetables. In: Gorny, J. (ed.): Fresh-cut and vegetables and MAP. *Proceedings Seventh International Controlled Atmosphere Research Conference.* Davis, California. 30-33, 42.

Gorny J., Agar I., 1998. Are argon-enriched atmospheres beneficial? Perishables Handling Newsl. 94, 7–8.

Gouble, B., Fath, D., Soudain, P., 1995. Nitrous oxide inhibition of ethylene production in ripening and senescing climacteric fruits. *Postharvest Biol. Technol.* 5, 311–321.

Graham, D.M., 1997. Use of ozone for food processing. *Food Technol.* 51, 72–75.

Guentzel, J.L., Lam K.L., Callan, M.A., Emmons, S.A., Dunham, V.L., 2008. Reduction of bacteria on spinach, lettuce, and surfaces in food service areas using neutral electrolyzed water. *Food Microbiol.* 25, 36-41

Gutierrez, J., Bourke, P., lonchamp, J., Barry-Ryan, C., 2009. Impact of plant essential oils on microbiological, organoleptic and quality markers of minimally processed vegetables. *Innovative Food Sci. Emerging Technol.* 10, 195-202.

Guzel-Seydim, Z., Bever, P.I., Greene, A., 2004 a. Efficacy of ozone to reduce bacterial populations in the presence of food components. *Food Microbiol.* 21, 475–479.

Guzel-Seydim, Z., Greene, A., Seydim, A., 2004 b. Use of ozone in the food industry. *Lebensm.-Wiss. u.-Technol.* 37, 453-460.

Han, Y., linton, R.H., Nielsen, S.S., Nelson, P.E., 2001. Reduction of Listeria monocytogenes on green peppers (Capsicum annuum L.) by gaseous and aqueous chlorine dioxide and water washing and its growth at 7ºC. *J. Food Protec.* 64, 1730-1738.

Heimdal, H., Kuhn, B., Poll, L., Larsen, L., 1995. Biochemical changes and sensory quality of shredded and MA-packaged iceberg lettuce. *J. Food Sci.* 60, 1265-1268.

Hoigné, J. and Bader, H. 1983. Rate constants of reaction of ozone with organic and inorganic compounds in water – II. Dissociating organic compounds. *Water Res.* 17:185-194.

Hoof, F.V., 1982. Professional risks associated with ozone. In: W. J. Masschelein (ed.). *Ozonation manual for water and waste water treatment*. New York: Wiley-Interscience. 200-201

Hoornstra, E., de Jong, G., Notermans, S., 2002. Preservation of vegetables by light. In: Society for Applied Microbiology (Ed.), *Conference Frontiers in Microbial Fermentation and Preservation*; 9–11 January, Wageningen, The Netherlands. 75– 77.

Horvath, M.L., Bilitzky, L., Huttner, J., 1985. Fields of utilization of ozone. In: R. J. H. Clark (ed.). Ozone. Elsevier Science Publishing Co. New York, USA. 257-316.

Hua, W., Hao, F., Luo, Y., 2004. *Surface treatment of fresh cut lettuce with acidic electrolyzed water to extend shelf life.* Annual Meeting Institute of Food Technologists. Abstract.

Huang Y, Hung Y, Hsu S, Huang Y, Hwang D. 2008. Application of electrolyzed water in the food industry. *J. Food Control.* 19, 329-345.

Izumi, H., 1999. Electrolyzed water as a disinfectant for fresh-cut vegetables. *J.Food Sci.* 64, 536–539.

Jacxsens, L., Devlieghere, F., Van Der Steen, C., Debevere, J., 2001. Effect of high oxygen modified atmosphere packaging on microbial growth and sensorial qualities of fresh-cut produce. *Int. J. Food Microbiol.* 71, 197-210.

Jamie, P., Saltveit, M., 2002. Postharvest changes in broccoli and lettuce during storage in argon, helium, and nitrogen atmospheres containing 2% oxygen. *Postharvest Biol. Technol.* 26, 113–116.

Jin, R.-y., Hu, S.-q., Zhang, Y-g., Bo, T. 2009. Concentration-dependence of the explosion characteristics of chlorine dioxide gas. *J. Hazardous Materials.* 166(2-3), 842-847.

Jun, S., Irudayaraj, J., Demirci, A., Geiser, D., 2003. Pulsed UV light treatment of corn meal for inactivation of Aspergillus niger spores. *Int. J. Food Sci. Technol.* 38, 883–888.

Kader, A.A., Ben-Yehoshua, S., 2000. Effects of superatmospheric oxygen levels on postharvest physiology and quality of fresh fruits and vegetables. *Postharvest Biol. Technol.* 20, 1-13.

Khadre, M.A., Yousef A.E., 2001. Sporicidal action of ozone and hydrogen peroxide: a comparative study. *Int. J. Food Microbiol.* 71, 131–138.

Kim, B., Klieber, A., 1997. Quality maintenance of minimally processed Chinese cabbage with low temperature and citric acid dip. *J. Sci. Food Agric.* 75, 31-36.

Kim, C., Hung, Y. C., Brachett, R. E., Frank, J. F., 2001. Inactivation of Listeria monocytogenes biofilms by electrolyzed oxidizing water. *J. Food Proces. Preserv.* 25, 91–100.

Kim, C., Hung, Y., Brackett, R., Lin, C., 2003. Efficacy of electrolyzed oxidizing water in inactivating Salmonella on alfalfa seeds and sprouts. *J. Food Protect.* 66, 208–214.

Kim, H., Ryu J., Beuchat, L., 2006. Survival of Enterobacter sakazakii on fresh produce as affected by temperature, and effectiveness of sanitizers for its elimination. *Inter. J. Food Microbiol.* 111, 134–143.

Kim, J.G., Yousef, A., Chism, G., 1999a. Use of ozone to inactivate microorganism on lettuce. *J. Food Safety.* 19, 17-34.

Kim, J.G., Yousef, A., Dave, S., 1999b. Application of ozone for enhancing the microbiological safety and quality of food: a review. *J. Food Protect.* 6, 1071-1087.

Klockow, P.A., Keener, K.M., 2009. Safety and quality of packaged spinach treated with a novel ozone-generation system. LWT – *Food Sci. Technol.* 42, 1047-1053.

Koseki, S., Itoh, K., 2001. Prediction of microbial growth in fresh-cut vegetables treated with acidic electrolyzed water during storage under various temperature conditions. *J. Food Protec.* 64, 1935−1942.

Koseki, S., Yoshida, K., Isobe, S., Itoh, K., 2001. Decontamination of lettuce using acidic electrolyzed water. *J. Food Protect.* 64, 652–658.

Kuo, F., Ricke, S., Carey, J., 1997. Shell egg sanitation: UV radiation and egg rotation to effectively reduce populations of aerobes, yeasts, and molds. *J. Food Protect.,* 60, 694–697.

Lado, B., Yousef, A., 2002. Alternative food-preservation technologies: Efficacy and mechanisms. *Microbes and Infection*, 4, 433–440.

Lanciotti, R., Gardini, F., Sinigaglia, M., Guerzoni, M., 1999. Effect of hexanal on the shelf-life of fresh apple slices. *J. Agric. Food Chem.* 47, 4769-4776.

Lanciotti, R., Gianotti, A., Patrignani, F., Belleti, N., Guerzoni, M.E., Gardini, F., 2004. Use of natural aroma compounds to improve shelf-life and safety of minimally processed fruits. *Trends Food Sci. Technol.* 15, 201-208.

Leaper, S. 1984. Synergistic killing of spores of Bacillus subtilis by peracetic acid and alcohol. *J. Food Technol.* 19, 355-360.

Leistner, L., Gould, G., 2002. *Hurdle technologies: combination treatments for food stability, safety and quality.* Kluwer Academic/Plenum Publishers, New York.

Leverentz, B., Conway, W., Janisiewicz, W., Abadias, M., Kurtzman, C., Camp, M., 2006. Biocontrol of the food-borne pathogens Listeria monocytogenes and Salmonella enterica serovar poona on fresh-cut apples with naturally occurring bacterial and yeast antagonists. *Appl. Environ. Microbiol.* 72, 1135-1140.

Liao, L.B., Chen, W.M., Xiao, X.M., 2007. The generation and inactivation mechanism of oxidation-reduction potencial of electrolyzed oxidizing water. *J. Food Engineer.* 78, 1326-1332.

Liu, J., Stevens, C., Khan, V., Lu, J., Wilson, C., Adeyeye, O., Kabwe, M., Pusey, P., Chalutz, E., Sultana, T., Droby, S., 1993. Application of ultraviolet-C light on storage rots and ripening of tomatoes. *J. Food Prot.* 56, 868–872.

López-Gálvez, F., Allende, A., Selma, M.V., Gil, M.I., 2009. Prevention of Escherichia coli cross-contamination by different sanitizers during washing of fresh-cut lettuce. *Int. J. Food Microbiol.* 133, 167-171.

López-Gálvez, F., Allende, A., Truchado, P., Martínez-Sánchez, A., Tudela, J.A., Selma, M.V., Gil, M.I., 2010. Suitability of aqueous chlorine dioxide versus sodium hypochlorite as an effective sanitizer for preserving quality of fresh-cut lettuce while avoiding by-product formation. *Postharvest Biol. Technol.* 55, 53-60.

López-Rubira, V., Conesa, A., Allende, A., Artés, F., 2005. Shelf life and overall quality of minimally processed pomegranate arils modified atmosphere packaged and treated with UV-C. *Postharvest Biol. Technol.* 37, 174–185.

Lougheed E., Lee R., 1991. Ripening, CO_2 and C_2H_4 production, and quality of tomato fruits held in atmospheres containing nitrogen and argon. *Proc. Fifth Internat. Controlled Atmosphere Research Conference*. 1989, Wenatchee. USA.

Lu, J., Stevens, C., Yakabu, P., Loretan, P., Eakin, D., 1987. Gamma, electron beam and ultraviolet radiation on control of storage rots and quality of Walla onions. *J. Food Process. Pres.* 12, 53–62.

Lu, C., Toivonen P., 2000. Effect of 1 and 100 kPa O_2 atmospheric pretreatments of whole 'Spartan' apples on subsequent quality and shelf life of slices stored in modified atmosphere packages *Postharvest Biol. Technol.* 18, 99–107.

Lucht, L., Blank, G., Borsa, J., 1998. Recovery of food-borne microorganisms form potentially lethal radiation damage. *J. Food Protect.* 61, 586–590.

Luna-Guzmán, I., Cantwell, M., Barret, D., 1999. Fresh-cut cantaloupe: effects of $CaCl_2$ dips and heat treatments on firmness and metabolic activity. *Postharvest Biol. Technol.* 17, 201-213.

Luna-Guzmán, I., Barret, D., 2000. Comparison of calcium chloride and calcium lactate effectiveness in maintaining shelf stability and quality of fresh-cut cantaloupe. *Postharvest Biol. Technol.* 19, 61-72.

Magalhães, A.M., Escalona, V., Artés, F., 2007. Respiration rate of minimally processed spinach leaves under innovative storage atmospheres. CIGR Section VI. *International Symposium Food and Agricultural Products: Processing and Innovations*. Naples. Italy. September. CD rom.

Mahmoud, B.S.M., Bhagat, A.R., Linton, R.H., 2007. Inactivation kinetics of inoculated Escherichia coli O157:H7, Listeria monocytogenes and Salmonella enteric on strawberries by chlorine dioxide gas. *Food Microbiol.* 24, 736-744.

Mahmoud, B.S.M., Linton, R.H., 2008. Inactivation kinetics of inoculated Escherichia coli O157:H7 and Salmonella enteric on lettuce by chlorine dioxide gas. *Food Microbiol.* 25, 244-252.

Marquenie, D., Michiels, C., Geeraerd, A., Schenk, A., Soontjens, C., Van Impe, J., Nicolaï, B., 2002. Using survival analysis to investigate the effect of UV-C and heat treatment on storage rot of strawberry and sweet cherry. *Int. J. Food Microbiol.* 73, 187–196.

Martín-Diana, A., Rico, D., Frías, J., Henehan, G., Mulcahy, J., Barat, J., Barry-Ryan, C., 2006a. Effect of calcium lactate and heat-shock on texture in fresh-cut lettuce during storage. *Postharvest Biol. Technol.* 77, 1069-1077.

Martín-Diana, A., Rico, D., Frias, J., Mulcahy, J., Henehan, G., Barry-Ryan, C., 2006b. Whey permeate as a bio-preservative for shelf life maintenance of fresh-cut vegetables. *Innovative Food Sci. Emerging Tech.* 7, 112-123.

Martín-Diana, A., Rico, D., Barry-Ryan, C., Frías, J., Henehan, G., Barat, J., 2007. Efficacy of steamer jet-injection as alternative to chlorine in fresh-cut lettuce. *Postharvest Biol. Technol.* 45, 97-107.

Martín-Diana, A.B., Rico, D., Barry-Ryan, C. 2008. Green tea extract as a natural antioxidant to extend the shelf-life of fresh-cut lettuce. *Innovative Food Sci. Emerging Technol.* 9, 593-603.

Martínez-Hernández, G.B., Tomás-Callejas, A., Artés, F., Artés-Hernández, F., 2009. Retención de polifenoles y capacidad antioxidante total en mizuna mínimamente procesada tras su desinfección alternativa con agua ácida electrolizada. *V Congreso Ibérico y III Congreso Iberoamericano de Ciencias y Técnicas del Frío*. September 23-25th, Castellón, Spain. Oral communication.

Martínez-Sánchez, A., Allende, A., Bennett, R.N., Ferreres, F., Gil, M.I., 2006. Microbial, nutritional and sensory quality of rocket leaves as affected by different sanitizers. *Postharvest Biol. Technol.* 42, 86-97.

Mercier, J., Arul, J., 1993. Induction of systemic disease resistance in carrot roots by pre-inoculation with storage pathogens. *Can. J. Plant Pathol.* 15, 281–283.

Neri, F., Mari, M., Brigati, S., Bertolini, P. 2009. Control of Neofabraea alba by plant volatile compounds and hot water. *Postharvest Biol. Technol.,* 51(3), 425-430.

Newberne, P., Smith, R.L., Doull, J., Feron, V.J., Goodman, J.I., Murno, I.C., Portoghese, P.S., Waddle, W.J., Wagner, B.M., Weil, C.S., Adams, T.B., Hallagan, J.B., 2000. GRAS flavouring substances. *Food Technol.* 54, 66-83.

Nieuwenhuijsen, M.J., Toledano, M.B., Elliot, P., 2000. Uptake of chlorination disinfection by-products; a review and a discussion of its implications for exposure assessment in epidemiological studies. *J. Expo. Anal. Environ. Epidemiol.* 10, 586–599.

Nigro, F., Ippolito, A., Lima, G., 1998. Use of UV-C light to reduce Botrytis storage rot of table grapes. *Postharvest Biol. Technol.* 13, 171–181.

Ölmez, H., Akbas, M.Y., 2009. Optimization of ozone treatment of fresh-cut green leaf lettuce. *J. Food Eng.* 90, 487-494.

Ölmez, H., Kretzschmar, U. 2009. Potencial alternative disinfection methods for organic fresh-cut industry for minimizing water consumption and environmental impact. *Lebensm. Wiss. Technol.,* 42, 686-693.

Ölmez, H., Temur, S.D., 2010. *Effects of diferent sanitizing treatments on biofilms and attachment of Escherichia coli and Listeria monocytogenes on green leaf lettuce.* LWT – Food Sci. Technol. doi:10.1016/j.lwt.2010.02.005.

Oms-Oliu G., Raybaudi-Massilia Martínez R.M., Soliva-Fortuny R., Martín-Belloso O., 2008a. Effect of superatmospheric and low oxygen modified atmospheres on shelf-life extension of fresh-cut melon. *Food Control.* 19, 191–199.

Oms-Oliu G., Soliva-Fortuny R., Martín-Belloso O., 2008b. Modelling changes of headspace gas concentrations to describe the respiration of fresh-cut melon under low or superatmospheric oxygen atmospheres *J. Food Eng.* 85, 401–409.

Oms-Oliu G., Aguiló-Aguayo, I., Martín-Belloso O., Soliva-Fortuny R., 2010. Effects of pulsed light treatments on quality and antioxidant properties of fresh-cut mushrooms (*Agaricus bisporus*). *Postharvest Biol. Technol.* doi:10.1016/j.postharvbio2009.12.011.

Ongeng, D., Devlieghere, F., Devevere, J., Coosemans, J., Ryckeboer, J., 2006. The efficacy of electrolysed oxidising water for inactivating spoilage microorganisms in process water on minimally processed vegetables. *Int. J. Food Microbiol.* 109, 187–197.

Palomer X, Roig-Villanova I, Grima-Calvo D, Vendrell M., 2005. Effects of nitrous oxide (N_2O) treatment on the postharvest ripening of banana fruit. *Postharvest Biol. Technol.* 36, 167–175.

Parish, M., Beuchat, L., Suslow, T., Harris, L., Garret, E., Farber, J., Busta, F., 2003. Methods to reduce or eliminate pathogens from fresh and fresh-cut produce. *Comprehensive Reviews in Food Sci. and Food Safety.* 161-173.

Park, C.M., Beuchat, L.R., 1999. Evaluation of sanitizers for killing Escherichia coli O157:H7, Salmonella and naturally occurring microorganism on cantaloupes, honeydew melons, and asparagus. *Dairy Food Environ. Sanitation.* 19, 842-847.

Park, C., Hung, Y., Doyle, M., Ezeike, G., Kim, C., 2001. Pathogen reduction and quality of lettuce treated with electrolyzed oxidizing and acidified chlorinated water. *J. Food Sci.* 66, 1368–1372.

Park, H., Hung, Y. C., Kim, C., 2002. Effectiveness of electrolyzed water as a sanitizer for treating different surfaces. *J. Food Protect.* 65, 1276–1280.

Pedahzur, R., Shuval, H., Ulitzur, S., 1997. Silver and hydrogen peroxide as potential drinking water disinfectants: their potential bactericidal effects and possible modes of action. *Water Sci. Technol.* 35, 87-93.

Perkins-Veazie, P., Collins, J.K., Howard, L., 2008. Blueberry fruit response to postharvest application of ultraviolet radiation. *Postharvest Biol. Technol.* 47, 280-285.

Powrie W., Chiu R., Wu H., 1990. Preservation of cut and segmented fresh fruit pieces. *U.S. Patent Number 4,* 895-729.

Qadir, A., Hashinaga, F., 2001a. Inhibition of postharvest decay of fruits by nitrous oxide. *Postharvest Biol. Technol.* 22, 279–283.

Qadir, A., Hashinaga, F., 2001b. Nitrous oxide inhibits in vitro growth of multiple postharvest fungi. *HortScience* 36(7), 1302–1304. 2001

Raybaudi-Massilia, R., Mosqueda-Melgar, J. and Martín-Belloso, O. 2008. Edible alginate-based coating as carrier of antimicrobials to improve shelf-life and safety of fresh-cut melon. *Internat. J. Food Microbiol.* 121, 313–327.

Restaino, L., Frampton, E.W., Hemphill, J.B., Palnikar, P., 1995. Efficacy of ozonated water against various food-related microorganisms. *Appl. Environ. Microbiol.* 61, 3471–3475.

Rice, R.G., Robson, C.M., Miller, G.W., Hill, A.G., 1981. Uses of ozone in drinking water treatment. *J. Am. Water Works Assoc.* 73(1), 44–57.

Rico, D., Martín-Diana, A., Frías, J., Henehan, G., Barry-Ryan, C., 2006. Effect of ozone and calcium lactate treatments on browning and texture properties of fresh-cut lettuce. *Sci. Food Agric.* 86, 2179–2188.

Rico D., Martín-Diana A., Barry-Ryan C, Frías J., Henehan G., Barat J., 2007a. Use of neutral electrolysed water (EW) for quality maintenance and shelf-life extension of minimally processed lettuce. *Innovat. Food Sci. Emerging Technol.* doi:10.1016/j.ifset.2007.05.002

Rico, D., Martín-Diana, A., Frías, J., Barat, J., Henehan, G., Barry-Ryan, C., 2007b. Improvement in texture using calcium lactate and heat-shock treatments for stored ready to eat carrots. *J. Food Eng.* 79, 1196-1206.

Ritenour M., Crisosto C., 1996. Hydrocooler water sanitation in the San Joaquin Valley stone fruit industry. *Central Valley Postharvest Newsletter.* 5, 15-17.

Robles, P., de Campos, A., Artés-Hernández, F., Gómez, P., Calderón, A., Ferrer, M., Artés, F., 2007. Combined effect of UV-C radiation and controlled atmosphere storage to preserve tomato quality. *V Congreso Iberoamericano de Tecnología Postcosecha y Agroexportaciones.* Cartagena, Spain. 29 May-1 June. CDrom.

Rocculi P., Romani S., Dalla Rosa, M., 2005. Effect of MAP with argon and nitrous oxide on quality maintenance of minimally processed kiwifruit. *Postharvest Biol. Technol.* 35, 319–328.

Rocha Bastos, M.S., Ferreira Soaresa, N.F., De Andradea, N.J., Arrudaa, A.C., Elesbão Alves, R., 2005. The effect of the association of sanitizers and surfactant in the microbiota of the Cantaloupe (Cucumis melo L.) melon surface. *Food Control*. 16(4), 369-373

Rodgers, S., Cash, J., Siddiq, M., Ryser, E., 2004. A comparison of different chemical sanitizers for inactivating Escherichia coli O157:H7 and Listeria monocytogenes in solution and on apples, lettuce, strawberries, and cantaloupe. *J. Food Protect.* 67, 721–731.

Rodríguez Hidalgo, S., Artés-Hernández F., Gómez P.A., Fernández, J., Artés F. 2010. Quality of fresh-cut baby spinach grown under floating trays system as affected by N fertilization and innovative package treatments. *J. Sci. Food Agric.,* 90: DOI 10.1002 /jsfa.3926.

Rojas-Grau, M., Raybaudi-Massilia, R., Soliva-Fortuny, R., Avena-Bustillos, R., McHugh, T., Martin-Belloso, O. 2007. Apple puree-alginate edible coating as carrier of antimicrobial agents to prolong shelf-life of fresh-cut apples, *Postharvest Biol. Technol.* 45(2), 254-264.

Rosen, J., Kader, A., 1989. Postharvest physiology and quality maintenance of sliced pear and strawberry fruits. *J. Food Sci.* 54, 656-659.

Ruiz-Cruz, S., Acedo-Félix, E., Diaz-Cinco, M., Islas-Osuna, M.A., González-Aguilar, G.A., 2007. Efficacy of sanitizers in reducing Escherichia coli O157:H7, Salmonella ssp. and Listeria monocytogenes populations on fresh-cut carrots. *Food Control.* 18, 1383-1390.

Sánchez-Ballesta, M., Romero, I., Bernardo, J., Orea, J., González-Ureña, A., Escribano, M., Merodio, C., 2007. Involvement of the phenylpropanoid pathway in the response of table grapes to low temperature and high CO_2 levels. *Postharvest Biol. Technol.* 46, 29-35.

Sanz-Cervera, S., Olarte, C. Echavarri, F., Ayala, F., 2007. Influence of exposure to light on the sensorial quality of minimally processed cauliflower. *J. Food Sci.* 72, S12-S13.

Sapers, G., 1993. Scientific status summary browning of foods: control by sulphites, antioxidants and other means. *Food Technol.* 47, 75-84.

Sapers, G., 2003. *Hydrogen peroxide as an alternative to chlorine for sanitizing fruits and vegetables.* IFIS Pub. Food Sci. Central. www.foodsciencecentral.com/library.html.

Sapers, G., Simmons, G., 1998. Hydrogen peroxide disinfection of minimally processed fruits and vegetables. *Food Technol.* 52, 48-52.

Sarig, P., Zahavi, T., Zutkhi, Y., Yannai, S., Lisker, N., Ben-Arie, R., 1996. Ozone for control of post-harvest decay of table grapes caused by Rhizopus stolonifer. *Physiol. Mol. Pl. Pathol.* 48, 403–415.

Selma, M., Beltrán, D., Allende, A., Gil, M., 2007. Elimination by ozone of Shigella sonnei in shredded lettuce and water. *Food Microbiol.* 24, 492–499.

Selma, M.V., Ibañez, A.M., Cantwell, M., Suslow, T., 2008. Reduction by gaseous ozone of Salmonella and microbial flora associated with fresh-cut cantaloupe. *Food Microbiol.* 25, 558-565.

Sharma, R., Demirci, A., 2003. Inactivation of *Escherichia coli* O157:H7 on inoculated alfalfa seeds with pulsed ultraviolet light and response surface modelling. *J. Food Sci.* 68, 1448–1453.

Shewfelt RL, Prussia, SE., (eds.), 1993. *Postharvest handling: a systems approach.* Academic Press, Orlando, FL. 358 pp.

Shirk, O., 2000. *Las mediciones del ozono.* Dräger Sicherheitstechnik GmbH. Dräger Hispania Mapfre Seguridad 77, 17 – 21.

Sidorkin, V., Volskii, E., Utenkova, T., 1989. Chamber for the suspension culture of plant cells. *Fiziol. Biokhim. Kult. Rast.* 20, 602–607.

Silveira, A., Aguayo, E., Leglise, A., Artés, F., 2007. Emerging sanitizers and clean room improved the microbial quality of fresh-cut 'Galia' melon. CIGR 3rd International Symposium. *Food and Agricultural Products: Processing and Innovations.* Naples, Italy, 24-26 September. CDrom.

Silveira, A., Aguayo, E., Artés, F. 2010. Emerging sanitizers and Clean Room packaging for improving the microbial quality of fresh-cut 'Galia' melón. *Food Control.* 21, 863-871.

Simons, L.K., Sanguansri, P., 1997. Advances in the washing of minimally processed vegetables. *Food Aust.* 49, 75-80.

Singh, N., Singh, R., Bhunia, A., Stroshine, R., 2002. Efficacy of chlorine dioxide, ozone, and thyme essential oil or a sequential washing in killing Escherichia coli O157:H7 on lettuce and baby carrots. Lebensm. *Wiss Technol.* 35, 720-729.

Skog, L. and Chu, C., 2000. Ozone technology for shelf life extension of fruits and vegetables. *Acta Horticulturae.* 553, 285-291.

Soliva-Fortuny, R.C., Martín-Belloso, O., 2003. New advances in extending the shelf-life of fresh-cut fruits: a review. *Trends Food Sci. Technol.* 14, 341-353.

Stampi, S., De Luca, G., Zanetti, F., 2001. Evaluation of the efficiency of peracetic acid in the disinfection of sewage effluents. *J. Appl. Microbiol.* 91, 833– 838.

Stevens, C., Khan, V., Lu, J., Wilson, C., Chalutz, E., Droby, S., Kabwe, M.K., Haung, Z., Adeyeye, O., Pusey, L., Tang, A., 1999. Induced resistance of sweetpotato to Fusarium root rot by UV-C hormesis. *Crop Prot.* 18, 463–470.

Stopforth J. D., Mai, T., Kottapalli, B., Samadpour, M., 2008. Effect of acidified sodium chlorite, chlorine, and acidic electrolyzed water on Escherichia coli O157:H7, Salmonella, and Listeria monocytogenes inoculated onto leafy greens. *J. Food Protec.* 71(3), 625-628

Sumner, S., Wallner-Pendleton, E., Froning, G., Stetson, L., 1995. Inhibition of Salmonella typhimurium on agar medium and poultry skin by ultraviolet energy. *J. Food Protec.* 59, 319–321.

Suslow, T., 1997. Postharvest chlorination. Basic properties and key points for effective disinfection. University of California. *Division of Agriculture and Natural Resources.* Publication 8003. 8 pp.

Sy, K.V., Murray, M.B., Harrison, M.D., Beuchat, L.R., 2005. Evaluation of gaseous chlorine dioxide as a sanitizer for killing Salmonella, Escherichia coli O157:H7, Listeria monocytogenes, and yeasts and molds on fresh-cut produce. *J. Food Protect.* 68, 1176-1187.

Takeshita, K., Shibato, J., Sameshima, T., Fukunaga, S., Isobe, S., Arihara, K., Itoh, M., 2003. Damage of yeast cells induced by pulsed light irradiation. *Int. J. Food Microbiol.* 85, 151– 158.

Taormina, P., Beuchat, L., 1999. Comparison of chemical treatments to eliminate enterohemorrhagic Escherichia coli O157:H7 on alfalfa seeds. *J. Food Protect.* 62, 318–324.

Thom, S.R., Marquis, R.E., 1984. Microbial growth modification by compressed gases and hydrostatic pressure. *Applied Environ. Microbiol.* 47, 780-783.

Tomás-Callejas, A., López-Gálvez, F., Sbodio, A., Artés, F., Artés-Hernández, F., Suslow, T.V. 2010. Comparative effectiveness of common commercial levels of chlorine and chlorine dioxide to prevent Salmonella spp cross contamination on inoculated fresh cut Red Chard. *International Association for Food Protection Annual Meeting,* Anaheim, CA, August 1-4st.

Ukuku, D.O., 2004. Effect of hydrogen peroxide treatment on microbial quality and appereance of whole and fresh-cut melons contaminated with Salmonell spp. Internat. *J. Food Microbiol.* 95, 137-146.

Ukuku, D.O., Bari, M.L., Kawamoto, S., Isshiki, K., 2005. Use of hydrogen peroxide in combination with nisin, sodium lactate and citric acid for reducing transfer of bacterial pathogens from whole melons surfaces to fresh-cut pieces. *Internat. J. Food Microbiol.* 104, 225-233.

Valero, D., Valverde, J., Martíınez-Romero, D., Guillén, F., Castillo, S., Serrano, M., 2006. The combination of modified atmosphere packaging with eugenol or thymol to maintain quality, safety and functional properties of table grapes. *Postharvest Biol. Technol.* 41, 317-327.

Van Der Steen, C., Jacxsens, L., Devlieghere, F., Debevere, J., 2002 Combining high oxygen atmospheres with low oxygen modified atmosphere packaging to improve the keeping quality of strawberries and raspberries. *Postharvest Biol. Technol.* 26, 49-58.

Van Der Steen, C., Devlieghere, F., Debevere, J., 2003. High oxygen concentration in combination with elevated carbon dioxide to affect growth of fresh-cut produce microorganisms. *Acta Horticulturae.* 599, 141–147.

Vandekinderen, I., Devlieghere, F., Van Camp, J., Denon, Q., Sanchez Alarcon, S., Ragaert, P., De Meulenaer, B., 2009. Impact of a decontamination step with peroxyacetic acid on the shelf-life, sensory quality and nutrient content of grated carrots packed under equilibrium modified atmpsphere and stored at 7°C. *Postharvest Biol. Technol.* 54, 141-152.

Venkitanarayanan, K. S., Ezeike, G. O., Hung, Y. C., Doyle, M. P.,1999a. Inactivation of Escherichia coli O157:H7 and Listera moncytogenes on plastic kitchen cutting boards by electrolyzed oxidizing water. *J. Food Protec.* 62, 857–860.

Venkitanarayanan, K. S., Ezeike, G. O., Hung, Y. C., Doyle, M. P., 1999b. Efficacy of electrolyzed oxidizing water for inactivating Escherichia coli O157:H7, Salmonella Enteritidis and Listeria monocytogenes. *Applied Environ. Microbiol.* 65, 4276–4279.

Vicente, A.R., Pineda, C., Lemoine, L., Civello, P.M., Martinez, G.A., Chaves, A.R., 2005. UV-C treatments reduce decay, retain quality and alleviate chilling injury in pepper. *Postharvest Biol. Technol.* 35, 69-78.

Wang, H., Feng, H., Luo, Y., 2004. Microbial reduction and storage quality of fresh-cut cilantro washed with acidic electrolyzed water and aqueous ozone. *Food Research Internat.* l, 949–956.

Wiley, R C., 1994. Preservation methods for minimally processed refrigerated fruits and vegetables. In: Wiley R C. (ed.). *Minimally processed refrigerated fruits and vegetables.* Chapman & Hall, New York, USA, 66-134.

Xie, F., 2003. *Disinfection byproducts in drinking water: form, analysis and control.* CRC Press Technology & Industrial Arts. 176 pp.

Xu, W., Qu, W., Huang, K., Guo, F., Yang, J., Zhao, H., Luo, Y., 2007. Antibacterial effect of Grapefruit Seed Extract on foodborne pathogens and its application in the preservation of minimally processed vegetables. *Postharvest Biol. Technol.* 45, 126-133.

Yaun, B., Sumner, S., Eifert, J., Marcy, J., 2004. Inhibition of pathogens on fresh produce by ultraviolet energy. *Int. J. Food Microbiol.* 90, 1-8.

Yildiz, F., 1994. Initial preparation, handling and distribution of minimally processed refrigerated fruits and vegetables. In: Wiley R C. (ed.). *Minimally processed refrigerated fruits and vegetables.* Chapman & Hall, New York, USA. 15-49.

Zhang, L., Lu, Z., Yu, Z., Gao, X., 2005. Preservation of fresh-cut celery by treatment of ozonated water. *Food Control.* 16, 279-289.

Zhang M., Zhan Z.G., Wang S.J., Tang J., 2007. Extending the shelf-life of asparagus spears with a compressed mix of argon and xenon gases. *LWT – Food Sci. Technol.* doi:10.1016/j.lwt.2007.04.011.

Zheng Y, Yang Z, Chen X., 2008. Effect of high oxygen atmospheres on fruit decay and quality in Chinese bayberries, strawberries and blueberries. *Food Control.* 19, 470–474.

In: Potable Water and Sanitation
Editor: Joel M. McMann ISBN: 978-1-61122-319-4

Chapter 2

HEALTH IMPACT OF SANITATION IN INDIA

Brijesh C. Purohit
Department of Economics, Central University of Rajasthan, c/o Shri R.K. Patni PG College, City Road, Madanganj, Kishangarh-30582, District Ajmer Rajasthan, India

ABSTRACT

Despite significant investments over the last 20 years, India still faces the most daunting sanitation challenge in the world. In comparison to other countries, its sanitation is rated as the second worst in the world after China. The current situation depicts that only 28 per cent of people in rural areas, who comprise two-thirds of India's billion plus people, had access to toilets. Though in view of Millennium Development Goals, it aims at ending the appalling state of sanitation to half by 2015 which might have put severe burden on community, health services and cause losses to productivity to the nation.

This paper aims at assessing the status of sanitation and its direct and indirect impact on health in India. It integrates impact of water and sanitation and the efficacy of health systems and emphasizes that even in the presence of a public health system, these factors play an important role in influencing health outcomes. The study is divided into three sections. An overview of evolution of water and sanitation sector programmes in the first section of the paper is followed by an assessment of its direct impact on incidence of selected diseases in major Indian states. Using sub-state level (or district level) information, section third provides case studies of three Indian states belonging to rich, middle income and poor category respectively. The results of stochastic frontier analysis for these states indicate that the differential impact of health systems in terms of health outcomes could be explained by regional disparities and sanitation factors both at the state and district level. Based on our estimates, the final section provides policy measures which emphasize that more efficient health outcomes could be achieved by overcoming investment needs through different innovative measures. These may, depending upon regions and locations across the country, include better coordination between social sector and economic policies, higher subsidy to sanitation, improved management of existing programmes, community participation and better opportunities for women in health, education and other rights.

Keywords: Regional Disparities, Sanitation, Efficiency; health care system, sub-state level, India, stochastic frontier model, life expectancy, actual, estimated, poorer state, districts

I. INTRODUCTION

Safe water supply and sanitation (WSS) are now accepted as basic necessities for healthy living. The discharge of untreated sewage into water resources provides a vector for pathogens capable of sickening humans and animals. Pathogenic bacteria are able to survive in bodies of water for days or weeks, and eating contaminated seafood can cause typhoid fever, infectious hepatitis A and B, polio, and cholera (GESAMP, 2001). In general, a classification of water related diseases could be provided as follows[1] ((Table 1).

However, the measurement of health benefits from the availability of WSS remains controversial. Generally either the incidence of diseases or infant mortality rate has been considered as dependent variable. The magnitude of impact on health status has been measured in different ways. One of the earliest review of 144 studies by Esrey et al(1985) analysed the impact of improved water supply and sanitation facilities on ascariasis, diarrohea, drancunculiasis, hookworm infection, scistosomiasis and trachoma. These diseases were chosen because of their widespread nature and to point out the variety of mechanisms through which an improved water and sanitation could help the people. Through the estimates obtained from the above review it was found that median reduction in morbidity for diarrohea, trachoma and ascariasis by WSS was around 26, 27 and 29 percent respectively. By contrast the similar reduction in scistosomiasis and drancunculiasis was around 77 and 78 percent. The reduction in hookworm infection was put around 4 percent only (Esrey et al., 1985). By contrast some studies indicate that improved WSS facilities are not efficacious in improving health status and not particularly cost-effective (Walsh & Warren, 1979). Whereas, another review of 67 studies from 28 countries found that WSS investments can reduce diarrhoea morbidity and mortality rates by a median of 22% and 21%, respectively (Esrey et al, 1985). Most of the studies reviewed, however, were also influenced by the methodologies adopted, inadequate health indicators and lack of control for confounding variables (Blum & Feachem, 1983). Moreover the role of selective primary health care (SPHC) and other health facilities also have played an increasing role in controlling the incidence of water and sanitation related diseases.

In view of the fact that the benefits of WSS extend beyond its role in improving health status (Okun, 1988), attempts have been made to measure the health benefits resulting from investment in WSS through case-control studies. Such studies have been carried out in Malawi (Young. & Briscoe, 1988), the Philippines (Baltazar, J. et al., 1988) and Lesotho (Daniels, D. et al. 1990). It was reported in these studies that WSS investments can produce a 20%, 20%, and 24% reduction, respectively, in the incidence of diarrhoea. Several other studies have also examined the link between the water quality and child health in different parts of the developing world. For instance, Guilkey and Riphahn (1998) used a longitudinal data from metropolitan Cebu-Philippines from 1983 to 1986 to analyse the determinants of child mortality up to 2 years using structural discrete time hazard model. The study indicated

---

[1] For a full list of water-related infections which is large and varied, see, Annexure 1. Also see, for instance, Annette Prüss et al. 2002.

that child mortality varies significantly between birth weight and nutritional status. For Bangladesh and Philippines using semi-parametric model, another study analysed the impact of water quality, sanitation and socioeconomic factors on child health (Lee, Rosenzweig and Pitt, 1997) and it was found that there is no significant effect between water supply or source of drinking water, sanitation and child health. Using stratified partial likelihood estimation, in the Malaysian context, it was observed that there is no significant link between water quality and sanitation and child's health (Ridder and Tunali, 1999). WHO (2000) based on a primary survey using a logit regression model found a negative relationship between arsenicosis and household income. Increasing levels of household income are associated with a lower prevalence of arsenicosis. Galiani et al (2005) in a survey during the period of 1990-1999 in Argentina evaluated the mortality rates per 1000 for privatized and non-privatized water companies. And they found that the privatization of water services is associated with 33% reduction in the mortality rate, which amounts to a 5.3% reduction of the baseline rate. The privatization of water systems does not affect mortality in those municipalities with low levels of poverty. (UBN lower than 25%). The effect on the remaining treated municipalities is increasing in the level of poverty and highly significant. They found that child mortality fell by approximately 8% in the areas where water systems were privatized.

Table 1. Classification of Water Related Diseases

Transmission Route	**Description**	**Disease Group**	**Examples**
Water-borne	The pathogen is in water, which is ingested	Feco-oral	Diarrhea, Dysenteries, Typhoid fever, Scabies, Trachoma
Water-washed (or water scarce) Person-to-person transmission because of a lack of water for hygiene.		Skin and eye infections	
Water-based	Transmission via an aquatic intermediate host (e.g., snail)	Water-based	Schistosomiasis, Guinea worm
Water-related insect vector	Transmission by insects, which breed in water or bite near water.	Water-related insect vector	Dengue, Malaria, Trypanosomiasis

In the Indian context, studies indicate that around 105 million children under 5 years die each year due to water borne diseases and that India experienced a loss of 200 million man-hours a day every year because of these diseases (Shanmuganandan, 1999). It has been pointed out that India loses between Rs.36, 000 to Rs 366 billion crores per year (or approximately US$8.3 billion) due to water related diseases (Shanmuganandan, 1999). Deaths due to diarrhea could be reduced by 65 per cent, while overall child mortality could be reduced by 55 per cent. A survey of three villages Gudimallur, Devathanam, Vannivedu of Tamilnadu in October-November 1997 (Sankar, 2001) indicated that the treated wastewater led to water contamination and which in turn affected water related diseases. Another study indicated that up to 15 million people in rural Andhra Pradesh are using water obtained from unsafe water sources and of these, nearly 7 million rely upon water sources that are dangerously contaminated to cause to have identifiable health effects (Hughes et al., 2001). Jalan and Ravallion (2003) using primary survey data for the period 1993-94 found that the overall prevalence of diarrhoea is 10.1% in the sample, with an average of .33 days of illness and mean expenditure of 0.74 rupees per episode of diarrhea. Disease prevalence and length of illness fall with higher income and education. It is also observed that access to piped water significantly reduces diarrhoea prevalence and duration. Diseases prevalent amongst those with piped water would be 21% higher without it. Another study using factoral analysis also found significant impact of water and sanitation facilities in rural, urban and combined sectors on IMR, CDR and different diseases (Purohit and Siddiqui, 2004) A study in rural Uttarakhand using data from a primary survey of 1530 households in 2004-05 explored sanitation and health link by focusing on factors affecting diarrhoea episodes and latrine availability (Murugesan, Dayal and Chugh, 2008). It was found that latrine availability affects episodes of diarrhea negatively, while the availability of water has a positive effect on latrine availability and use. Education, poverty and the Swajal programme had significant effects on latrine availability. Another study in Gujarat investigated the impact of 100 newly built toilets in villages across Dahod District by a non-government agency during 2005-2007 (Agoramoorthy and Hsu, 2009). It was found that the toilets significantly reduced not only the cost of medical treatments but also the loss of wages induced by sanitation-related diseases. Money saved from sanitation illnesses for one person for a period of two years could cover the cost of a toilet. The study, therefore, recommended increasing the subsidy for toilets so that future health care to treat sanitation illnesses can be significantly reduced. A study in Tamil Nadu examined the factors affecting child health due to drinking water quality and sanitation in Chromepet and Pallavaram Township. Using Probit Model and Cox-Proportional Model with Primary data, the results indicated that drinking water quality, sanitation, kind of fuel used and precautionary measures taken by the household significantly affect the health (Srinivasulu and Haripriya, 2004). In Cox proportional Model, the drinking water quality and sanitation were not significant but precautionary measures were found highly significant indicating child mortality. The study suggested a need for stringent regulatory mechanism to supply clear drinking water as it is to poor who are usually affected by water related diseases. A study by Sulabh NGO in West Bengal in the districts of Murshidabad and Bankura, to assess the health impact of sanitation improvements in a few villages, declared as "Nirmal Grams". (100% Sanitation) indicated that by providing only toilets in the individual houses, the disease burden may not reduce substantially (Sulabh Int. Serv. Org, 2007). Improvements on drainage condition, general sanitation, personal hygiene, food sanitation are also to be taken, simultaneously to minimize the disease burden among the villagers. The issues of sustainability and use of toilets, constructed in the so-called Nirmal Grams need also close examination. The pilot study findings on perception and practice of personal and domestic hygiene, among the woman in selected urban and rural household, further strengthened the hypothesis, that hygiene behavior of the community do have a strong influence on the health gains from water-sanitation programmes. In order to optimize the health benefits from community water supply and sanitation projects, it is necessary to improve the knowledge of the community on their perception on safe water, sanitation and hygiene and their linkage of health and change their practice of personal and home hygiene, concurrently with the WSS programmes.

Thus, keeping in view the vital link of water and sanitation with the health status of the people, this study aims at assessing the status of sanitation and its direct and indirect impact on health in India. It integrates impact of water and sanitation and the efficacy of health systems and emphasizes that even in the presence of a public health system, these factors play an important role in influencing health outcomes. The study is divided into four sections. An overview of evolution of water and sanitation sector programmes in the second section of the paper is followed by an assessment of its direct impact on incidence of five selected diseases in major Indian states. Using sub-state level (or district level) information, section fourth provides case studies of three Indian states belonging to rich, middle income and poor category respectively.

II. Evolution of Water and Sanitation Sector Programmes

Water supply and sanitation were added to the Indian national agenda during the first five-year planning period (1951-1956). In 1954, the first national water supply programme was launched as part of the government's health plan. This was funded by both Central and state administrations through equal funding. The initial years were marked by limited achievements due to lack of qualified work forces and short supply of materials. In the subsequent five year plans, more funding was allocated for the development and strengthening of state public health engineering departments. In 1968, states were granted limited financial authority to sanction rural water supply schemes. In this duration (1954-68), the programme aimed at local community development and the welfare of the backward classes. A major policy initiative, namely, National Water Policy was enunciated in 1987 to guide the planning and development of water resources throughout the country. This policy led the Indian states to focus on (i) water resource management with a top priority to domestic water supply (ii) design standards for groundwater structures to protect groundwater sources (iii) water quality monitoring and mapping, and (iv) data management and valuation., Another National Water Policy in 2002 was adopted emphasizing a primacy to drinking water. As a result some states such as Karnataka, Madhya Pradesh, Orissa, Rajasthan and Tamil Nadu also enunciated state level policies. Thus the national policy guiding the water and sanitation sector in India in the Eighth Five-Year Plan (1992-97), emphasized that "Safe drinking water and basic sanitation are vital human needs for health and efficiency" and [given that] death and disease, particularly of children,... and the drudgery of women are directly attributable to the lack of these essentials." (GoI, 2002) High priority was given to serving villages that did not have adequate sources of safe water and to improving the level of service for villages classified as only partially covered[2]. The Ninth and Tenth Plan also broadly followed the directions set by the Eighth Plan.

[2] The Eighth Five-Year Plan also identified several points of emphasis including management of water as a commodity, delivery of water services based on principles of effective demand, standards of service corresponding to the level that users are willing to maintain, etc.

It should be noted that the primary responsibility for providing drinking water and sanitation facilities in the country rests with the State Governments[3], and, more specifically, the local bodies in the urban areas. The Centre provides or allocates funds and also ensures that funds are provided in State budgets, and progressively larger allocations have been made for water supply and sanitation in the various Five Year Plans.

Overall, India's approach to water supply and sanitation in the Eighth, Ninth and the Tenth Plan broadly follow the New Delhi declaration, adopted by the United Nations General Assembly in December 1990 which emphasizes protection of the environment and safeguarding of health through the integrated management of water resources and liquid and solid waste as well as other pertinent features[4]. Over the years, the focus in the rural sector both in water and sanitation has been on decentralization, demand responsiveness and people centered approach, increasingly taking the government from a direct service provider to a facilitator. In the urban sector, over the plan periods, an emphasis on Urban Water Supply and Sanitation (UWSS) and reforms mark notable improvements in the policy[5]. At present, States generally plan, design and execute water supply schemes (and often continue to operate) through their State Public Health Engineering Departments (or Panchayati Raj Engineering Departments or Rural Development Engineering Departments) and Water Boards. In addition, a variety of different government institutions at the Centre have a role in the management of drinking water supply, pollution control and financing of infrastructure development. For instance, the Central Water Commission (CWC) in the MoWR has responsibilities for regulating the use of surface water for irrigation, industry and drinking water purposes. The Central Groundwater Board (CGWB) of the same Ministry has an overseeing responsibility for the monitoring of groundwater levels and rates of depletion, as well as production of water resource inventories and maps. The National Rivers Conservation Directorate (NRCD) under the Ministry of Environment and Forests (MoEF) oversees the implementation of Action Plans to improve the quality of the nation's rivers (previously their activities were confined to the Ganga Action Plan, but now extends to the polluted stretches of 27 major rivers with works spread over 149 towns in 16 states). Likewise, the Central Pollution Control Board (CPCB) promotes basin-wide pollution control strategies. The CPCB liaises with State Water

---

[3] Water supply and sanitation is a State responsibility under the Constitution of India and following the 73rd and 74th Constitutional Amendments, the States may give the responsibility and powers to the Panchayati Raj institutions(PRIs) and Urban Local Bodies(ULBs).

[4] These include (a) organisation of reforms, promoting an integrated approach including changes in procedures, attitudes, and behaviour, and the full participation of women at all levels; (b) community management of services, backed by measures to strengthen local institutions in implementing and sustaining water and sanitation programmes; and (c) sound financial practices, achieved by better management of existing assets and extensive use of appropriate technologies. The Tenth Plan envisages 100 per cent coverage of rural and urban population with safe drinking water as per the stipulated norms and standards (40 lpcd of safe drinking water within a walking distance of 1.6 Kms or elevation difference of 100 metres in hilly areas, to be relaxed as per field conditions; at least one hand pump/spot source for every 250 persons).

[5] For instance, in the tenth plan, the objectives of universal coverage, adequacy in terms of minimum per capita consumption norms, regularity of supply, avoidance of excessive withdrawal leading to depletion and inaccessibility of water sources, and the need to conserve and make conjunctive use of water resources are highlighted. The Plan recognises that unsatisfactory service standards has lead to low tariff structures, which in turn results in poor resource positions, maintenance and service – a vicious circle – and that while surveys show consumer willingness to pay higher tariffs, such increases in tariff structures would have to be accompanied by substantial improvements in service quality. The Plan also recognises that given the above vicious circle, the main problem in financing of UWSS is the (un)sustainability of the current model, which is heavily dependent on state governments' willingness to provide guarantees for institutional finance and meeting the state share of project costs.

Pollution Control Boards and lays down standards for treatment of sewage and effluents. The Board is also responsible for action in the case on non-compliance[6].

MoUD launched the Accelerated Urban Water Supply Programme (AUWSP) in 1993-94 to implement water supply schemes in towns with population below 20,000 (as per 1991 census). Under this programme, 50 per cent of the estimated cost of the schemes is provided by the Ministry as a grant based on detailed project report.and the remaining 50 per cent provided by the State government including a 5 per cent beneficiary/town contribution. In special cases, 100 per cent finance is available as the central share[7]. There has also been a centrally supported water quality monitoring programme again with substantial funding for water quality laboratories and monitoring staff in each state. The new National Health Policy 2002 adopted by the Ministry recognizes that water supply and sanitation is part of an inter-connected set of factors that need to be addressed holistically and in coordination between various institutions.

At the state level, most states have separate public health engineering departments charged with planning, investigation and design. Following the Third Five-Year Plan recommendation to form statutory water and sewerage boards, in some states (such as

[6] Other government agencies involved either directly or indirectly with water supply and sanitation in India include the Ministry of Agriculture (MoA), which is involved in planning, formulation, monitoring and reviewing of various watershed based developmental project activities. At national level, the RGNDWM (under the Department of Drinking Water Supply, MoRD) formulates guiding policy, sets standards, and provides funds and technical assistance to the states for rural water supply and sanitation (RWSS). It allocates funds under the Centrally Sponsored Accelerated Rural Water Supply Programme (ARWSP) and supervises the RCRSP. The RGNDWM has also a National Human Resource Development Programme (NHRDP launched in 1994), which aims at training at least one grass root level The HRD programme also aims at empowerment of PRIs/Local Bodies with the objective of enabling them to take up operation and maintenance activities related to rural water supply systems. It also aims at capacity building of local communities by giving requisite training to mechanics/ health motivators/ masons etc. especially women to operate and maintain hand-pumps and the components of other water supply systems as well as to generate demand for adequate sanitation facilities. Under the NHRDP, several states have set up State level HRD cells for planning, designing, implementing, monitoring and evaluating an appropriate and need based HRD programme in a village through district level trainers who in turn may be trained at selected institutions forming the Indian Training Network (ITN). The MoUD is the nodal Ministry for policy formulation and guidance for the UWSS sector. The Ministry's responsibilities include broad policy formulation, institutional and legal frameworks, setting standards and norms, monitoring, promotion of new strategies, coordination and support to State Programmes through institutional expertise and finance. The Ministry is also responsible for managing international sources of finance. The Central Public Health and Environmental Engineering Organisation (CPHEEO), created in 1953, is the technical wing of the MoUD, which advises the Ministry in all technical matters and collaborates with the State Agencies about water supply and sanitation activities. CPHEEO plays a critical role in agreeing (giving technical sanction to) externally funded and special programmes and those parts funded by the Life Insurance Corporation of India (LIC). CPHEEO also plays a central role in setting design standards and norm setting for urban water supply and sanitation. The Ministry also supports the States by sponsoring research relevant to the sector, largely through NEERI, the National Environmental Engineering Research Institute in Nagpur, Maharashtra. Areas of research include compost plants, urban water treatment plant evaluations, urban water supply tariffs, water distribution system performance evaluation, low cost methods of flocculation, optimisation of water treatment, drinking water in urban slums, leak detection equipment etc.

[7] The AUWSP is being administered through the CPHEEO at the Centre. Another major area of Central Government assistance to the States is in human resources development. Concern with human resources development led to the Public Health Engineering (PHE) Training Programme starting in 1956. The programme is training 108 in-service engineers every year through 11 postgraduate courses in engineering. By 1996, a total of 1,950 had been trained on postgraduate courses. In addition, the CPHEEO arranges Short Term Courses. The Ministry has also jointly sponsored a Management Programme for Senior Public Health Officials with DFID funding for the past 25 five years. In addition to the countrywide Human Resources Development programmes, the Ministry is also supporting the establishment of HRD & Training Cells for the water sector in each State with a one-third grant for capital costs where proposals are approved.

Gujarat, Kerala, Maharashtra and Tamil Nadu) the WSS Boards or corporation (for instance, in Uttar Pradesh) have been constituted to handle urban and rural water. With the recent move toward decentralization, a mix of agencies have evolved, including state public health engineering departments, state boards and district engineering agencies[8].

Besides, NGOs (both national and international) have played a significant role in the sector in India over the past three decades. They have exhibited comparative advantages that other organizations lack[9]. Further, private initiatives also continue to drive the construction and maintenance of traditional open wells and household toilets. Nationwide, the number of privately constructed toilets is almost twice the number constructed through government programmes. During the 11th Five Year Plan, Government has identified 100% sanitation as a goal. In October 2008, government of India also announced a National Urban Sanitation Policy. The ultimate objective is that all urban dwellers will have access to and be able to use safe and hygienic sanitation facilities and arrangements so that no one defecates in the open. The vision of the policy is that all Indian cities and towns become totally sanitized, healthy and livable and ensure and sustain good public health and environmental outcomes for all their citizens with a special focus on hygienic and affordable sanitation facilities for the urban poor and women. The focus of the Policy is on Awareness Generation and Behavioural Change by generating awareness about sanitation and its linkages with public and environmental health amongst communities and institutions and also promoting mechanisms to bring about and sustain behavioural changes aimed at adoption of healthy sanitation practices

Despite this concerted planning effort in India, a recent study by UNICEF (2008) indicates that even with the progress of water supply in relation to MDG targets being on track, nearly 4% of its urban population and 14% of its rural population uses unimproved source of water. In terms of sanitation, nearly 18% and 58% in urban and rural areas are devoid of any facility. In fact in 2006, only 28 per cent of people in rural areas, who make up

[8] In some states, such as Maharashtra, the organisational matrix is complicated. The Groundwater Survey and Development Agency is responsible for hand pump programmes, the Maharashtra Water Supply and Sanitation Board is responsible for piped water supply, the district is responsible for operating and maintaining all drinking water installations and the Irrigation Department is responsible for drinking water sources downstream of command areas. Despite the apparent organisational complexity, the national trend is to decentralise capital investment responsibilities to Zilla Parishad engineering departments at district and block levels, and operations and maintenance activities to district and, in many cases, gram panchayat levels. Andhra Pradesh is the only state in which water supply and sanitation is the exclusive responsibility of the Panchayat Raj Engineering Department. At the local levels, gram panchayats are the lowest tier in the local administrative framework and may be responsible for one or more villages or habitations. The block is the intermediate tier in the local administration framework, serving, in the vast majority of states, a population ranging between 100,000 and 150,000. The district is the top tier of local government. Hygiene education is largely delivered via the Education Departments and the school sanitation programme, and via RGNDWM's programmes on IEC, especially in the TSC areas.

[9] These include the capacity to: (i) reach the rural poor and remote areas (ii) promote local participation (iii) operate at low costs, and (iv) adapt and be innovative when needed. The initial involvement of NGOs in the sector came in the sixties with the famines in Bihar (1964) and Maharashtra (1969). In this early period, NGOs functioned as drilling contractors to state governments and donor-financed programmes. The efforts of these NGOs convinced the donor community and the government to invest in hand pump programmes for drinking water, especially in the chronically drought-prone areas of the country. More recently, NGOs have operated successfully within donor assisted projects where the institutional context is conducive, such as the World Bank funded project in Karnataka and the World Bank-assisted Uttar Pradesh RWSS project. In these projects, NGOs have been used mostly as an intermediary between the community and the water agency, facilitating project activities related to community mobilisation, cost sharing, health education, and other non-technical activities.

two-thirds of India's billion plus people, had access to toilets. In urban areas, 30.66 million urban households which form 35.49% of the urban households suffer inadequate access to sanitation facilities. 7.87% households defecate in the open, 5.48% use community latrines and 19.49% use shared latrines. More than 37% of the human excreta generated are unsafely disposed. The percentage of notified and non-notified slums without latrines is 17% and 51% respectively. Besides, there are districts in different states which also continue to be known as diarrohea endemic[10].

III. Direct Impact of WSS on Incidence of Selected Diseases in Major Indian States

From the review of studies highlighted in the introductory section of this paper, we hypothesize a link among WSS, health facilities and other socio-economic variables which is depicted in the following flow chart:

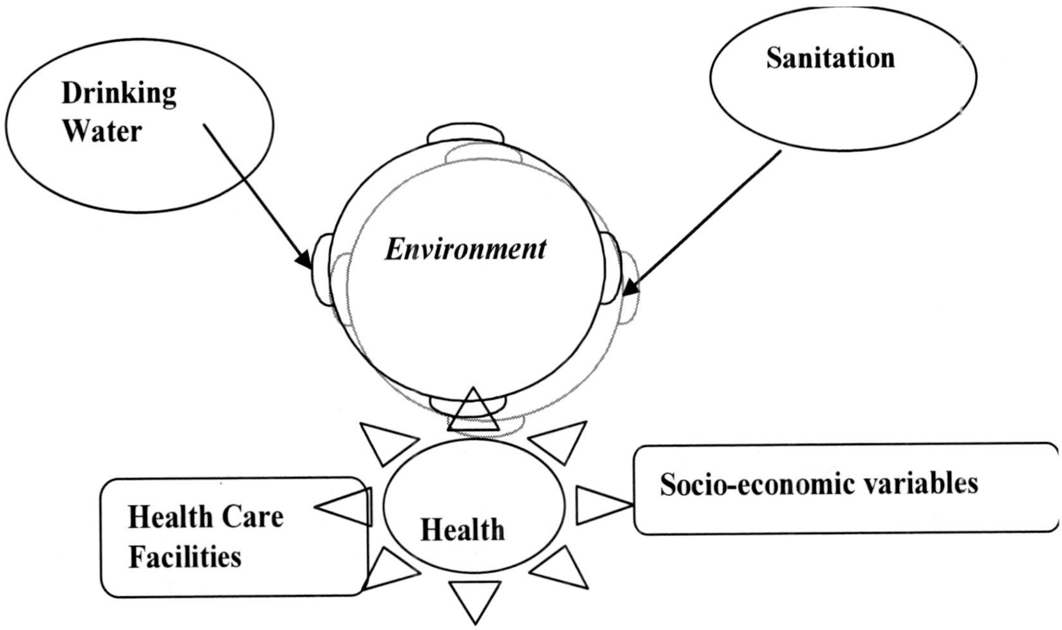

Chart 1. Link among WSS, Health Facilities and Other Socio-Economic Variables.

Data Base

To estimate direct and indirect impact of WSS, this study is based on secondary data. Information is collected from health information in India, Ministry of Rural Development and Water supply, Planning Commission document on water and sanitation policies (GoI, 2002), *Economic Survey* (GoI, 2007), indiastat.com, Human Development Reports of Maharashtra

[10] Sulabh International Social Service Organization (2007), pp.96.

(GoM, 2002), Karanataka (GoK, 2006), and Madhya Pradesh (GoMP, 2002); websites of the states and other published sources. Main variables used in the study are number of cases of different diseases, life expectancy (LEXP), infant mortality rates (IMR), per capita income (PCI) and parameters related to human development indicators (HDI), gender development indicator (GDI) and health facilities. The information relates to 27 Indian States and district level data for the three states.

In order to estimate the direct impact of WSS, we utilize regression analysis. The dependent variable considered by us is incidence of a particular disease denoted by no. of cases of the disease in 27 Indian states. The states included for this analysis are: Andhra Pradesh, Arunachal Pradesh, Assam, Bihar, Jharkhand, Goa, Gujarat, Haryana, Himachal Pradesh, Jammu & Kashmir, Karnataka, Kerala, Madhya Pradesh, Chhattisgarh, Maharashtra, Meghalaya, Mizoram, Nagaland, Orissa, Punjab, Rajasthan, Sikkim, Tamil Nadu, Tripura, Uttar Pradesh, Uttarakhand, West Bengal and Pondicherry. It should be mentioned that many of the WSS borne disease are now nearly non-existent or only prevalent in some states. These extinct varieties in the country include cholera, guinea worm, fileria, plaque and dengue. Thus we have focused in this section on five major diseases which are associated with WSS and continue to prevail in most of the states. These include acute diarrohea, hepatitis, typhoid, malaria and acute respiratory infections (ARI). The explanatory variables in all these regressions include four sets of variables representing drinking water facilities, sanitation amenities, health care facilities and socio-economic factors. Thus we have tried to use the variables, namely, total habitats covered by safe water supply (safdrwattotal), total habitats covered by safe water supply in rural and urban areas respectively (safdrwatrural and safdrwatrurban), total habitats covered partly or fully (partcover, fullcover), total habitats not as per all India accelerated rural water supply norms (Total hab not as per ARWSP %), total habitats affected by Physical & Biological contamination (Total affected NC/PC habitations including Physical & Biological contamination%), water being at a far off place (watfaraway), improvement in sanitary facilities as envisaged in the investment plan by the department of rural water supply and sanitation (Acheivement in sanitary complex %), no drainage facilities (nodrainage), no toilet facilities(nolatrines). The health facilities are covered by means of either of the explanatory variables representing government hospitals or hospital beds(govtH or govtB), hospitals or beds provided by local bodies (Lbhospi or Lbbeds), private hospitals or beds (pvthosp or pvtbeds), total hospitals or beds including government and private(tothosp or totbeds), total no. of hospital beds including allopathic and other systems of medicines (bedsalltype), population covered per bed, all hospitals or all types of beds (popperbed, popallhosp or popallbeds). Per capita income at constant prices is taken to represent the influence of socio-economic status (PCIC 2003-04). The results of our analysis are presented in the following Table 2-2(b).

The results of regressions indicate that water and sanitation facilities have indeed influenced the incidence of all the five diseases considered by us. The total no. of cases for acute diarrohea and typhoid has increased due to physical and biological contamination. The respective coefficient of this variable representing contamination for these diseases has been 2.639 and .347 (columns 2, Tables 2 and 2(a) respectively. However, in terms of male and female cases, this impact has been different than total no. of cases (columns 3 and 4, same tables). In regard to hepatitis, the results denote that cases of this disease have reduced due to more coverage of habitation under safe water supply and thus overall contamination has been significant with a negative sign (columns 5-7, Table 2). For Malaria cases (Table 2(a), last column) as expected the lack of drainage facilities had a positive influence in increasing the incidence of the disease. In case of ARI (Table 2b), the inadequate achievement of targets to cover areas through sanitation facilities (Physcsn= .249) has led to more no. of total cases. Also, partial coverage of habitations through safe water supply (Partcover=.500) had added influence to increase the incidence of ARI. Further the positive impact of non-availability of toilet facilities is also seen for typhoid case (nolatrines=.472, Table 2(a)). In both diseases, namely, typhoid and ARI, the impact has been higher on male no. of cases relative to females which may be due to differences in reporting for them. Pertinently, in all the regressions the impact of income variable (PCIC) has not been statistically significant suggesting that any increase in income has not been able to compensate for the WSS facilities. An important role of hospital facilities or health system (depicted either as total hospitals; totalhosp for typhoid and malaria and total beds of all types; bedsallty for hepatitis and ARI), as seen through these regressions is an increased reporting through health system thus a positive coefficient of this variable.

Table 2. Regression Results: Impact of Water and sanitation facilities on Acute Diarrhea and Hepatitis

Dependent Variable→	Acute Diarrhea			Hepatitis		
	Total Cases	Male Cases	Female Cases	Total Cases	Male Cases	Female Cases
Explanatory Variable\Statistic ↓						
Intercept	301542.219 (-1.030)	-108612.262 (-.650)	-98162.110 (-.605)	1055.585 (.519)	774.268 (.670)	281.317 (.313)
Total_affected_NC/PC_habitations_by Physical and Biological Contamination	2.639 (2.639*)	.404 (2.203*)	.406 (2.256*)	-.390 (-2.189*)	-.369 (-2.071**)	-.412 (-2.294)
tothosp	2.059 (2.059**)	-	-	-	-	-
nolatrines	1.574 (1.574)	.450 (2. 086*)	.483 (2.283*)	-	-	-
popperbed	-	-.262 (-1.212)	-.333 (-1.570)	-	-	-
PCIC_200304	-	-	-	-.078 (-.452)	-.080 (-.463)	-.074 (-.428)
bedsalltype	-	-	-	.648 (3.803†)	.657 (3.850†)	.628 (3.660†)
safdrwattotal	-	-	-	.142 (.780)	.117 (.641)	173 (.941)
R^{-2}	.337	.196	225	.339	.338	.329
F Statistic and DF	5.24†, 25	2.945**, 24	3.328*, 24	4.203*, 25	4.186*, 25	4.069*, 25

Source: Estimated.

Note: Figures in the parentheses denote "t" ratios. Level of significance: † =1%, *=5%, **=10%.

IV. Impact of Health Systems, Regional Disparities and Sanitation Factors; State and District Level Analysis

The above discussion of the direct impact of WSS on incidence of selected diseases in major Indian states indicates that health facilities have played a considerable role in reporting the incidence of WSS related diseases.

In this section we attempt an assessment of the indirect impact of sanitation on health in India. We integrate impact of water and sanitation and the efficacy of health systems and emphasize that even in the presence of a public health system, these factors play an important role in influencing health outcomes. Using sub-state level (or district level) information, this section provides case studies of three Indian states belonging to rich, middle income and poor category respectively. First we make use of stochastic frontier analysis for these states to assess the efficiency of these health systems (Annexure 2). From sub-state level data for each of the three states, namely rich, middle income and poor state, we derive the efficiency estimates and use these estimates to explain that the differential impact of health systems in terms of health outcomes could be explained by intra-state regional disparities and sanitation factors both at the state and district level. Thus in this estimation, we presume that differences in technical efficiency pertaining to health system could be discerned at district level health facility planning by non-health related parameters. Thus, we explain the dispersion in technical efficiency (or health system efficiency) by a set of variables which includes per capita income (PCI), male and female income respectively, literacy, rural and urban literacy respectively, urbanization, water supply and sanitation facilities, gender development index (GDI) and persons below poverty line (BPL). Having estimated the efficiency of health system in our first stage of estimation, our model in the second stage is thus:

Dispersion in Technical efficiency = f (PCI, male and female income respectively, literacy, rural and urban literacy separately, urbanization, water supply and sanitation facilities, infrastructure variable such as road per square km., BPL, GDI) + error term…..(1)

The three Indian states considered in the above income categories of states are Maharashtra, Karnataka and Madhya Pradesh belonging to rich, middle income and poor states. Among these, Maharashtra, as per 2001 census, is the second most populous state of India with a population of 96.7 million of which 42.4 percent reside in urban areas. In terms of per capita income, it is the second richest state in India with its per capita income 40 percent higher than all-India average. Its capital Mumbai is principal financial centre and a major commercial hub of the country. A comparison of key indicators reveals that Maharashtra's high income level has not been matched by its attainments in social development (Table 3). The middle income Indian state, namely, Karnataka considered here has its per capita income slightly above all India average (Table 3). Madhya Pradesh, the poor income state is marked by a per capita level much below all-India average (Table3)

As against all-India life expectancy of 61.8 years and 63.5 years for males and females (1999-2003), the respective figures for Maharashtra are 65.2 years and 67.6 years. IMR in the state (in 2005) is also lower (at 36 per 1000) relative to all-India IMR of 58 per thousand. However, it differs considerably between rural and urban areas at 58 per thousand and 31 per thousand. This differential in IMR across districts of the state is also glaring. IMR varies at 31/1000 in Mumbai to as high as 106/1000 in Gadchiroli. According to Human Development Report of Maharashtra (GoM, 2002), "Both qualitatively and quantitatively, there is a wide gap in healthcare infrastructure available in rural and urban areas".

Table 2(a). Regression Results: Impact of Water and sanitation facilities on Typhoid and Malaria

Dependent Variable→	Typhoid			Malaria
	Total Cases	Male Cases	Female Cases	Total Cases
Explanatory Variable\Statistic↓				
Intercept	-21250.614 (-.793)	-10410.91 (-.770)	-10839.70 (-.813)	2.575 (1.203)
Total_affected_NC/ PC_habitations_by Physical and Biological Contamination	.347 (2.125*)	.348 (2.167*)	.346 (2.077**)	-
tothosp	.306 (1.810**)	.305 (1.84**)	.306 (1.774**)	.417 (2.538$^{\dagger\,\dagger}$)
Nolatrines	.472 (2.854†)	.485 (2.989†)	.457 (2.709††)	-
popperbed		-	-	-
PCIC_200304	-.092 (-.544)	-.108 (-.651)	-.075 (-.434)	-
nodrain		-	-	.355 (2.161*)
R^{-2}	.402	.424	.378	.281
F Statistic and DF	4.866†, 23	5.226†, 23	4.49†, 23	6.276†, 27

Source: Estimated; Note: Figures in the parentheses denote "t" ratios. Level of significance: †=1%, † †=2%, *=5%, **=10%.

Table 2(b). Regression Results: Impact of Water and sanitation facilities on ARI

Dependent Variable→	ARI		
	Total Cases	Male Cases	Female Cases
Explanatory Variable\Statistic↓			
Intercept	-504482.21 (-1.39)	-187707.98 (-1.132)	-316774.22 (-1.616)
Bedallty	.391 (2.624$^{\dagger\,\dagger}$)	.405 (2.658$^{\dagger\,\dagger}$)	.378 (2.58$^{\dagger\,\dagger}$)
Physcsn	.249 ((1.688)	.258 (1.713)	.240 (1.658)
Partcove	.500 (3.516†)	.460 ((3.161†)	.531 (3.797†)
R^{-2}	.513	.490	.529
F Statistic and DF	9.774†, 25	9.007†, 25	10.337†, 25

Source: Estimated; Note: Figures in the parentheses denote "t" ratios. Level of significance: †=1%, † †=2%, *=5%, **=10%.

Table 3. Maharashtra, Karnataka and Madhya Pradesh in a Comparative Framework

State	Per capita Income in 2003-04 (Rs. at 1993-94 prices)	Life expectancy at birth (1999–2003)			Literacy Rate(%)(2001)	Percentage of population below poverty line (30 day recall period)
		Male	Female	Total		
Maharashtra	16479	65.2	67.6	66.4	76.88	25.02
Karnataka	13141	62.9	66.4	64.6	66.64	20.04
Punjab	15800	67.6	69.6	68.6	69.65	6.1
Gujarat	16779	62.5	64.6	63.5	69.14	14.07
Madhya Pradesh	8284	57.2	56.9	57.1	63.74	37.43
Tamil Nadu	12976	64.3	66.5	65.4	73.4	21.12
All India	11799	61.8	63.5	62.7	64.84	26.1

Sources: Economic Survey, 2007, Economic Times, 2005, GoM (2002).

Table 4. Number of Hospitals, Beds and Dispensaries per lakh population across States and by Rural and Urban Areas

States	Hospitals			Beds			Dispensaries		
	Rural	Urban	Total	Rural	Urban	Total	Rural	Urban	Total
Andhra Pradesh	1.7	10.6	4.2	19.8	189.8	65.5	0.4	0.4	0.4
Gujarat	0.6	15.1	5.6	24.3	422.5	161.7	8.2	31.2	16.1
Haryana	0.1	1.5	0.4	4	152.8	40.6	0.3	4.1	1.2
Karnataka	0.1	1.7	0.6	10.4	233.5	79.4	1.8	1.6	1.7
Kerala	6.5	7.5	6.7	198	414.2	255.1	6.4	6.4	6.4
Maharashtra	0.9	8.3	3.8	20.7	219.7	97.7	0.7	24.4	9.9
Punjab	1.8	2.2	1.9	62.9	156.2	90.4	7.7	3.6	6.5
Tamil Nadu	0.2	1.7	0.7	11.8	232.9	87.5	0.4	1.9	0.9
West Bengal	0.2	1.4	0.5	11.7	237	73.7	0.8	0.7	0.7
All India	0.7	4.3	1.6	19.4	212.7	69.1	1.7	6.7	3

Sources:GoM(2002).

There is a concentration of hospitals, nursing homes and qualified doctors both in public and private sector. In fact, in the year 2000, number of doctors per 1,00,000 population in urban areas is 139.8 as compared to only 23.7 in rural areas. Likewise, over 80% of the beds in the public hospitals are in urban areas which cover only 42.4% of the total population. This skewed distribution is also reflected in sub-state level across districts and regions in Maharashtra. In general, districts like Mumbai, Pune, Wardha and Nagpur have a better ratio of population to facilities. Rural infrastructure like PHCs and SCs are not adequately supported by necessary inputs. Due to inadequate and declining public investment and health expenditure in 1990's, Maharashtra's position in comparison to some major Indian states has worsened (Table 4).

Partly this could be due to population growth and in-migration which has increased the density of population between 1961-2001 by two and half times to current level of 314 persons per square km.

Results and Discussion for Maharashtra

Results of our frontier model for Maharashtra are presented in Table 5. It could be observed that all the independent variables to explain inverse of IMR have emerged with appropriate signs and are statistically significant. Notably immunization coverage (THREE DOSES) has the highest magnitude (.626), followed by percent of institutional delivery (coeff. of INSTDELPERCENT=.173) and urban availability of health facilities (URBANBEDS=.057).

Using the results of frontier model, actual and estimated IMR are presented in Table 6. These depict that Mumbai is the most efficient district (MED) with its lowest actual IMR (32 per thousand) coinciding with estimated IMR. Three other districts, namely, Ahmednagar, Sangli and Parbhani, with their actual IMR respectively at 47, 41 and 50 depict their actual coinciding with potential. However, owing to their overall higher level of IMR relative to Mumbai, these districts cannot be considered as MED. Thus in terms of actual IMR being higher than potential IMR, except for districts with zero deviations, Thane districts is fourth efficient district (with deviations at 8.75% from its potential) and Yavatmal is the lowest efficient district with actual as nearly 2.6 times higher than the potential (Table 6)

Performance of these districts relative to MED is presented in terms of their deviation from MED in Table 7. Pertinently, ranks computed based on these deviations depict that next to Mumbai, second and third efficient districts are Sangali and Thane. The lowest rank denoting the lowest performance is depicted for Yavatmal (Table 7).

Reasons for these inter-district disparities could be deciphered from Table 8 which depicts major inputs for health sector in the state. Notably, the distribution of dispensaries and ISM institutions in the state is highly inequitable. This is obvious from high values of Gini Coeficients for these parameters which is respectively .716 and .502. Moreover, there is a considerable difference between maximum and minimum values for each of the parameters (last two rows, Table 8).

Notable features that emerge for the highest efficiency of Mumbai are: the highest number of ANMs per lakh of population in Mumbai (352 per lakh), a very low percentage of hospitals in the public sector in the district (8%) and urban nature of the district depicted by 100% of the beds, hospitals and dispensaries located in urban areas. By contrast, the lowest efficient district of Yavatmal has less than half (that of Mumbai) per lakh availability of ANMs (108 per lakh), highest percentage of hospitals in public sector (nearly 65%) and nearly 53% beds in urban areas. This is suggestive of a better utilization of private sector facilities with urbanization and better efficiency in health outcomes and reduced IMR. It also indicates the importance of per capita availability of paramedical staff. Looking from the parameters of population served per bed, Mumbai's highest efficiency may be due to its very high utilization of beds which serves 3000 population and it is next only to Gadchiroli (3422 population covered per bed) and Latur (3005 population served per bed).

Table 5. Stochastic frontier normal/half-normal model for Maharashtra

Number of observations = 30; Wald chi2(3) = 4.722e+10		
Log likelihood = .45024468 Prob > chi2 = 0.0000		
Dependent Variable: Inverse IMR		
Explanatory Variable	Coefficient	Standard Error
URBANBEDS	0.057*	.000011
INSTDELPERCENT	0.173*	4.30e-06
THREE DOSES	0.626*	.0000201
Constant	-7.514*	.0000623
Other Statistics		
lnsig2v	-42.193	1654.818
lnsig2u	-1.481*	.2581989
sigma_v	6.89e-10	5.70e-07
sigma_u	.4767326	.0615459
sigma2	.227274	.0586819
lambda	6.92e+08	.0615459
Likelihood-ratio test of sigma u =0; chibar2(01) = 7.06 Prob>=chibar2 = 0.004		

Source: Estimated; Note: * indicates 1% level of significance.

Besides the factors within the health system, there are influences external to the system that may lead to differential in efficiency at the district level. In order to explore such external factors, as explained above (in eq.1) we used dispersion in efficiency as a dependent variable in the second stage of our regression exercise. Among the set of variables used by us as explanatory variables we included per capita income, number of families below poverty line (BPL), total literacy, male and female literacy separately, number of primary and secondary educational institutions separately, drop out rate, gender development index (GDI), rural habitat fully covered by water supply, percentage coverage through: safe drinking water, toilet facilities (TOILETS) and human development indicator (HDI). The results of best fit are presented in Table 9.

Two variables, namely, BPL and TOILETS have emerged statistically significant. The negative sign of BPL indicates that efforts to reduce poverty have helped to reduce regional disparity in efficiency of health system across districts. However, the lack of sanitation facilities as depicted by the positive sign of TOILETS, indicates that insufficient effort and public investment in the development of rural sanitation has hampered the outcome of direct intervention through health delivery system in the state. It may be useful if this inadequacy could be overcome through appropriate efforts.

Thus, the results of frontier model using district level information in Maharashtra indicate that despite urbanization and direct health sector intervention in the state, the efficiency of public health delivery system remains low. Considerable deficiency of public health manpower inputs and prevailing disparities across districts has adverse impact on improving the life expectancy in the state. Main factors within the health system for discrepancy in inter district performance are inequitable distribution of beds and dispensaries, availability of skilled attention at birth and inadequate staffing relative to patient load in less urbanized districts of the state. Overcoming these factors may lead to an improvement in the district level efficiency of public health system. This may also help the state to reduce IMR in deficient districts like Yavatmal and improve life expectancy speedily and more equitably. This may require a considerable increase in medical and public health expenditure which has not kept an increasing pace in the state relative to other better off states in the country. This calls for resource mobilization for the sector through innovative steps and more effective use of funds. Besides the direct inputs from health sector, a more concerted effort to improve rural sanitation in poorer performing districts may add to overall reduction in IMR and improved life expectancy in the state.

Table 6. Actual and Estimated IMR for Districts in Maharashtra

District	Actual IMR	Estimated IMR	Excess of Actual IMR over Estimated (%)
Mumbai	37	37	0
Thane	46.00	42.30	8.75
Raigad	63.00	39.81	58.25
Ratnagiri	75.00	41.18	82.14
Sindhudurg	70.00	37.63	86.02
Nashik	79.00	45.91	72.09
Dhule	73.00	55.01	32.69
Jalgaon	71.00	46.73	51.95
Ahmednagar	47.00	47.00	0
Pune	52.00	46.25	12.43
Satara	51.00	38.75	31.61
Sangli	41	41	0
Solapur	68.00	50.46	34.76
Kolhapur	55.00	40.35	36.32
Aurangabad	56.00	46.63	20.09
Jalna	76.00	46.79	62.42
Parbhani	50	50	0
Beed	52.00	44.05	18.05
Nanded	68.00	42.64	59.48
Osmanabad	70.00	46.33	51.08
Latur	57.00	50.59	12.68
Buldhana	82.00	46.15	77.68
Akola	101.00	39.19	157.73
Amaravati	94.00	42.63	120.50
Yavatmal	124.00	47.54	160.86
Wardha	88.00	40.25	118.64
Nagpur	75.00	50.13	49.61
Bhandara	81.00	47.76	69.60
Chandrapur	96.00	50.19	91.26
Gadchiroli	106.00	78.78	34.55

Source: GoM, 2002 and Our Estimates.

Table 7. Deviations from Maximum Efficient District in Maharashtra

	Deviation From Maximum Efficient District	Rank
Mumbai	0	1
Thane	9.00	3
Raigad	26.00	12
Ratnagiri	38.00	19
Sindhudurg	33.00	15
Nashik	42.00	22
Dhule	36.00	18
Jalgaon	34.00	17
Ahmednagar	10.00	4
Pune	15.00	7
Satara	14.00	6
Sangli	4.00	2
Solapur	31.00	13
Kolhapur	18.00	9
Aurangabad	19.00	10
Jalna	39.00	21
Parbhani	13.00	5
Beed	15.00	7
Nanded	31.00	13
Osmanabad	33.00	15
Latur	20.00	11
Buldhana	45.00	24
Akola	64.00	28
Amaravati	57.00	26
Yavatmal	87.00	30
Wardha	51.00	25
Nagpur	38.00	19
Bhandara	44.00	23
Chandrapur	59.00	27
Gadchiroli	69.00	29

Source: GoM, 2002 and Our Estimates.

Karnataka: Results and Discussion

Results of our panel data estimation using frontier model for Karnataka are presented in Table 10. It could be observed that all the independent variables to explain life expectancy (LEXP) have emerged with appropriate signs and are statistically significant. Notably the variables representing total number of medical institutions in the area (MEDINST) and rural population served per sub-centre (PERSC) have emerged with positive signs. This indicates the positive impact of governmental intervention in expansion of hospital facilities and the desirable impact of institutional delivery coverage in enhancing life expectancy. These variables are partly also indicative of adequacy of the various inputs provided through medical institutions and rural health sub-centers.

Table 8. Health Facilities in Maharashtra

	Population Served Per						Percentage in Public Sector		Percentage in Urban Areas		
	Bed	Hospital	Dispensary	ISM Institutions	All Medical Institutions	ANM Per Lakh	Hospitals	Dispensaries	Hospitals	Dispensaries	Beds
Mumbai	3000	13764	5251	65617	3593	351.67	8	100	100	100	100
Thane	1420	21468	3720	73562	3040	205.56	42.9	47.2	97.1	50.8	90.9
Raigad	2615	20192	12296	200000	7378	143.03	18.1	5.9	96.5	96.4	99.2
Ratnagiri	1017	34529	324577	300000	28472	127.52	15.7	10.7	88.4	95.5	66
Sindhudurg	2569	20337	7110	9016	3325	155.02	22	47.5	79.3	100	81.8
Nashik	1715	21192	8144	44975	5203	107.78	18.3	13.2	87.8	100	69.3
Dhule	2004	14809	12227	63466	6058	103.68	15.9	2.9	95	100	69
Jalgaon	2256	6742	17453	63224	4516	92.29	7.9	6.3	71.7	95.3	76.8
Ahmednagar	1987	25691	12892	31940	6766	94.61	7.8	7.8	87.7	99.6	60.7
Pune	539	11163	17257	300000	6631	218.51	19.7	100	90.4	100	67.9
Satara	2758	13703	184016	100000	11710	120.95	33.7	11.6	95.1	99.5	83.4
Sangli	1536	11438	12093	22542	4662	178.05	10	17.6	74.5	100	68.5
Solapur	1540	19631	26124	200000	10514	124.26	8.6	100	14	3.3	35.8
Kolhapur	2504	36544	14755	42470	8426	139.82	24.4	3.8	84.9	3.8	39.7
Aurangabad	1374	21350	2502	200000	2212	121.2	27.7	100	72.5	26.5	73.1
Jalna	2351	25152	477897	89606	18864	82.16	19.7	13.8	82.5	66.7	75.3
Parbhani	2710	23672	21192	42792	8865	86.25	25.7	10.7	80	100	55.3
Beed	2130	26973	36134	23355	9297	104.94	18.6	11.4	88.7	100	56.7
Nanded	1840	27212	17876	44528	8685	98.44	16	5.7	98.1	61.9	93.4
Osmanabad	1564	38320	47900	41912	14118	104.44	17.8	12.4	92.1	99.4	76.8
Latur	3005	28896	440670	62953	18954	122.15	12.8	2.5	93.4	100	73.5
Buldhana	2349	27533	12957	17089	5813	135.45	16.1	7.8	93.1	94.8	76.3
Akola	948	22815	8619	47492	5528	156.03	11.7	2.4	92.2	92.2	76.8
Amaravati	761	11064	7557	7812	2851	122.59	16.8	10.3	94.7	87.9	71.9
Yavatmal	1795	26303	19492	41191	8803	107.98	64.7	43.2	85.4	52.5	53.4
Wardha	483	27354	19009	86270	9925	124.98	19.4	7.3	96.9	99.5	91.8
Nagpur	715	9624	41130	29279	6159	181.95	22.8	100	44.7	84.3	69.9
Bhandara	2164	52755	79132	29154	15176	120.09	22.5	5.5	61.9	47.2	51.4
Chandrapur	972	20027	9701	16630	4692	134.26	7.6	4.2	81.7	67.2	64.7
Gadchiroli	3422	48659	22357	22357	9090	147.39	11.3	23.1	47.1	56.8	14.8
Gini Coefficient	0.238	0.242	0.716	0.502	0.329	0.174	0.281	0.638	0.108	0.200	0.143
Maximum Value	3422	52755	477897	300000	28472	351.67	64.7	100	100	100	100
Minimum value	483	6742	2502	7812	2212	82.16	7.6	2.4	14	3.3	14.8

Source: GoM, 2002 and Sharma, IEG.

Table 9. Regression Results for Dispersion in Maharashtra

Dependent Variable: Dispersion	Unstandardized Coefficients		standardized Coefficients	"t" values	Significance level	Adjusted R Square= 0.613 df2=29; F=23.93*
	B	Std. Error	Beta			
Constant	-2.1E+32	5.46E+32		-0.390	0.700	
BPL	-8.7E+27	3.43E+27	-0.298	-2.548**	0.017	
TOILETS	7.75E+1	1.29E+31	0.699	5.986*	2.19E-06	

Source: Estimated; *denotes significance at 1% level, **denotes significance at 2% level.

Table 10. Stochastic Frontier Panel Data Model for Karnataka (Time-invariant inefficiency model)

Number of observations = 54 Number of groups = 27		
Wald chi2(3) = 90.52, Log likelihood = 120.64605, Prob > chi2 = 0.0000		
Dependent Variable→ LEXP	Coefficient	Std. Err.
→Explanatory Variables		
MEDINST	0.094*	0.011
PERSC	0.057**	0.026
CONSTANT	3.342*	0.199
→Stochastic Parameters		
MU	0.095*	0.024
LNSIGMA2	-6.323*	0.342
ILGTGAMMA	1.406**	0.560
SIGMA2	0.001	0.000
GAMMA	0.922	0.040
SIGMA_U2	0.001	0.000
SIGMA_V2	0.000	0.000

Source: Estimated; Note: * denotes significance at 1% level ** denotes significance at 5% level.
MEDINST= total number of medical institutions in the area.
PERSC= rural population served per sub-centre.

In explaining these results it is worth mentioning that some of our variables used as explanatory variables might have been influenced by endogeneity. The latter is said to occur in models in which economically endogenous variables are determined by each other and some additional economically exogenous variables. The simultaneity gives rise to empirical models with variables that do not satisfy the zero conditional mean assumption (Baum, 2006, p.186-215). To derive consistent estimates we generally use an instrument variable method (IVM). This method consists in finding an instrument variable (IV) that satisfies two properties: the IV must be uncorrelated with error term and must be highly correlated with the variable which is influenced by endogeneity. Keeping in view this problem, we identified the endogenous variable by using instrument variable method[11]. Thus, in the specification for measuring efficiency of health system, we found that three variables were significant, which

[11] For details, see Purohit Brijesh C. (2010), *Health Care System in India,* Gayatri Publications, New Delhi.

included number of medical institutions in an area (MEDINST), rural population served per sub-centre (PERSC) and deliveries attended in medical institutions (TOTDELINST). The last of these variables, namely deliveries attended in medical institutions, could have been possibly endogenous within a given health system comprising of infrastructure facilities of hospitals, sub centres (SCs) and primary health centres (PHCs). Thus in the Instrument Variable Method (IVM), we used PERSC separately as instruments for TOTDELINST and total government beds per lakh population[12] (GOVTBEDSLAKH) in an area.

Table 11. Actual and Estimated (Potential) Life Expectancy for Districts in Karnataka

		Actual Life Expectancy (2004)	Potential Life Expectancy (2004)	Actual as % of Potential (2004)
1	Bagalkot	60.80	70.03	86.82
2	Bangalore Rural	66.50	72.83	91.31
3	Bangalore Urban	67.30	76.44	88.04
4	Belgaum	67.70	75.99	89.09
5	Bellary	66.10	71.63	92.28
6	Bidar	63.30	68.62	92.25
7	Bijapur	62.60	69.99	89.44
8	Chamarajnagar	63.50	67.68	93.82
9	Chikmaglur	63.20	69.32	91.17
10	Chitradurga	64.60	72.50	89.10
11	Dakshina Kannada	67.40	68.17	98.87
12	Davangere	65.80	72.37	90.92
13	Dharwad	61.90	65.91	93.92
14	Gadag	62.70	66.83	93.82
15	Gulbarga	62.90	74.58	84.34
16	Hassan	65.20	72.28	90.21
17	Haveri	62.20	69.43	89.58
18	Kodagu	63.30	65.20	97.09
19	Kolar	64.20	74.05	86.70
20	Koppal	63.50	68.55	92.63
21	Mandya	62.90	71.96	87.40
22	Mysore	64.80	73.64	87.99
23	Raichur	63.90	69.42	92.05
24	Shimoga	67.40	70.22	95.99
25	Tumkur	65.30	74.02	88.22
26	Udupi	67.80	68.82	98.52
27	Uttara Kannada	62.90	69.90	89.99

Source: Estimated.

[12] lakh =100, 000

Using Instrument variable method and applying 2SLS, our estimates indicated an endogeneity between PERSC and TOTDELINST as well as separately between PERSC and availability of government beds (GOVTBEDSLAKH)[13] Thus using this diagnostic check, the final specification for frontier efficiency measurement using panel data comprised of two variables, namely, number of medical institutions in an area (MEDINST) and rural population served per sub-centre (PERSC), in which the latter variable indeed captures the influence of deliveries attended in medical institutions (TOTDELINST) and availability of government beds. Further, the suitability of fixed effect model with panel data in frontier estimation for the preferred specification was verified by us by using Hausman specification test which confirmed that fixed effect model is more consistent relative to random effect model.

Using the results of frontier model, actual and estimated life expectancy (LEXP) are presented for the year 2004 in Table 11. These depict Dakshina Kannada as the most efficient district (MED) with its actual LEXP moving highest towards its estimated LEXP in the year 2004. However, in terms of highest percentage increase in actual life expectancy we find Hassan as the best performer in the duration 1991-2004 (last column, Table 12).

Reasons for these inter-district disparities could be deciphered from Table 13 which depicts major inputs for health sector in the state. Notably, the distribution of per capita hospitals, primary health centers (PHCs), beds and ANMs in the state is highly inequitable. In fact, there is a considerable difference between maximum and minimum values for each of the parameters (columns 2-4, Table 13). Pertinently number of hospital beds per lakh of population is the highest (234) in Kodagu, followed by Mysore (137), Bangalore Urban (123), Chikmaglur (113), Dharwar (112) and so on. In terms of staff per lakh, it is again Kodagu (83.67) followed by Chikmaglur (57.67) and Uttar Kannada (53.04). In terms of population served per PHC, maximum and minimum is for Bidar(28.76 thousand) and Udupi (14.55 thousand). In terms of population served per sub-centre, the corresponding maximum and minimum is for Raichur (6.25 thousand) and Dakshina Kannada (2.66 thousand). These values of inputs do not indicate that it is the low work load at sub-center in Dakshina Kannada which is helping it to achieve a better life expectancy. By contrast, Kodagu district with the lowest population served per medical institution (10.17 thousand) and the maximum number of hospital beds (234) and staff per lakh (83.67) has 3rd rank both in 2004 and 1991 in terms of its distance of potential and unlike Hassan where the highest actual increase in life expectancy took place (9.58%) between 1991-2004, Kodagu has depicted only 3.77% increase in life expectancy in the duration. Thus in most of the districts neither the adequate availability of healthcare sector inputs nor does merely efficient utilization of these inputs explain the differentials in achievements in life expectancy. However, to some extent, among other inputs, this differential pattern in efficiency is explained by percentage of population not having facilities of safe drinking water, toilets and electricity in rural and urban areas separately (Table 14). In fact, the best achiever district, namely Hassan, has a much lower non-availability in rural areas (17.06%) relative to either Kodagu (30.35%), Bidar (28.96%) Koppal (33.24%), Udupi((29.69%), Raichur(39.32%) or Dakshina Kannada (32.18%).
This clearly suggests that the availability of basic infrastructure facilities of safe drinking water, toilets and electricity for the rural population has played a strong supportive role in enhancing overall efficiency of health system, besides the adequate utilization of basic health inputs of hospitals, PHCs, SCs and medical and paramedical staff. Our analysis thus suggests

[13] For details, see Purohit Brijesh C. (2010), *Health Care System in India,* Gayatri Publications, New Delhi.

that improvement in the LEXP in the less efficient districts may be feasible expeditiously if these facilities are made available to reduce rural-urban disparities in the state.

Table 12. Ranks of Districts and Increase in LEXP in Karnataka

	District/ Year	Ranks of Districts according to realization of potential Life Expectancy		% increase in actual LEXP in 1991-2004
		1991	2004	
1	Bagalkot	19	25	3.05
2	Bangalore Rural	9	12	3.26
3	Bangalore Urban	24	22	3.86
4	Belgaum	21	20	5.12
5	Bellary	13	9	5.25
6	Bidar	8	10	3.77
7	Bijapur	16	18	5.74
8	Chamarajnagar	5	6	1.60
9	Chikmaglur	17	13	5.16
10	Chitradurga	14	19	2.87
11	Dakshina Kannada	2	1	2.12
12	Davangere	12	14	4.44
13	Dharwad	10	5	4.74
14	Gadag	7	7	4.50
15	Gulbarga	27	27	5.71
16	Hassan	26	15	9.58
17	Haveri	18	17	4.36
18	Kodagu	3	3	3.77
19	Kolar	25	26	3.55
20	Koppal	6	8	5.83
21	Mandya	23	24	3.28
22	Mysore	20	23	3.02
23	Raichur	11	11	5.79
24	Shimoga	4	4	2.43
25	Tumkur	22	21	3.65
26	Udupi	1	2	2.57
27	Uttara Kannada	15	15	3.28

Source: Estimated; GoK, 2006; LEXP= Life Expectancy in years.

Besides the factors within the health system, as noted by us in the beginning of this section, there are influences external to the system that may lead to differential in efficiency at the district level. In order to explore such external factors, we used dispersion in efficiency as a dependent variable in the second stage of our regression exercise using data for the district level. Among the set of variables used by us as explanatory variables we included per capita income (PCI), gross domestic product per worker, total literacy, male and female literacy separately, total enrolment, proportion of out of school children, population density and gender development index. The results of best fit are presented in Table 15.

Table 13. Health Facilities and Related Parameters in Karnataka

District	Population served Per Medical Institution (in'000)	Number Of Hospital Beds Per Lakh* Population	Staff Per Lakh*	Population served Per PHC(in'000)	Population served Per SC(in'000)	Children Of One Year Age Received Complete Immunization (%)
Bagalkot	26.32	47	26.58	26.81	7.66	64.4
Bangalore Rural	15.46	51	36.73	20.74	5.29	92.7
Bangalore Urban	37.19	123	14.23	27.46	6.08	72.1
Belgaum	23.14	50	24.04	24.54	5.58	74.3
Bellary	20.36	91	25.31	25.32	5.28	77.3
Bidar	23.39	67	34.35	28.76	5.23	89
Bijapur	22.15	67	28.11	22.63	4.99	77.6
Chamarajnagar	14.63	86	40.08	15.96	4.11	84.8
Chikmaglur	10.94	113	57.67	18.29	2.84	95.9
Chitradurga	14.45	88	51.19	22.41	6.23	80.8
Dakshina Kannada	20.97	96	37.62	18.68	2.66	93.3
Davangere	15.34	99	31.71	18.33	5.07	93.4
Dharwad	34.12	112	24.75	25.95	4.33	86.3
Gadag	19.69	57	29.94	22.45	5.17	86.9
Gulbarga	19.06	66	26.16	22.66	4.69	81
Hassan	11.02	110	45.13	17.58	3.11	84.4
Haveri	16.39	54	33.01	23.37	3.95	81.2
Kodagu	10.17	234	83.67	16.63	2.96	109.7
Kolar	17.41	99	35.57	23.68	5.24	79.3
Koppal	21.18	51	22.16	24.08	6.16	72.8
Mandya	13.58	91	40.09	20.46	3.97	81.5
Mysore	14.90	137	44.15	17.36	3.49	90.2
Raichur	25.38	58	19.76	26.82	6.25	83.8
Shimoga	13.88	110	44.63	19.38	2.91	109.5
Tumkur	17.40	61	34.01	21.69	5.09	78.4
Udupi	13.10	89	38.75	14.55	3.64	93.8
Uttara Kannada	12.51	105	53.04	15.72	3.14	80
Karnataka	18.56	88	26.58	21.42	4.46	81.94

Source: GoK, 2006; Lakh= 100,000.

In the regression results using Dispersion in health system efficiency (DISPERSION), endogeneity was doubted for gender development. To identify this possible endogeneity, we used female literacy (FEMLIT) (or other variables like, gender enrolment and proportion of out of school children) as instrument[14]. The results using IVM and 2SLS indicated that female literacy could be used as an instrument for gender development and this is retained in the final regression.

[14] See, for details, Purohit Brijesh C. (2010), *Health Care System in India,* Gayatri Publications, New Delhi.

Table 14. Population Having None of The Three Facilities of Safe Drinking Water, Toilets and Electricity in Karnataka(%)(2001)

District	Total	Rural	Urban
Bagalkot	29.03	33.42	18.24
Bangalore Rural	12.04	13.99	4.72
Bangalore Urban	3.07	9.18	2.28
Belgaum	21.52	25.66	8.5
Bellary	24.61	30.95	13.17
Bidar	24.73	28.96	8.6
Bijapur	33.36	38.84	13.61
Chamarajnagar	32.93	36.14	14.29
Chikmaglur	19.82	22.65	8.26
Chitradurga	20.28	22.67	9.65
Dakshina Kannada	21.67	32.18	5.51
Davangere	16.73	20.38	8.44
Dharwad	14.97	18.97	11.74
Gadag	19.97	20.84	18.32
Gulbarga	30.78	38.18	9.6
Hassan	14.94	17.06	5.1
Haveri	22.29	24.58	13.09
Kodagu	27.03	30.35	5.8
Kolar	12.32	14.88	4.49
Koppal	31.24	33.24	21.51
Mandya	19.2	20.77	10.87
Mysore	18.7	27.67	3.78
Raichur	34.06	39.32	18.68
Shimoga	18.63	24.79	7.42
Tumkur	18.32	21.15	6.61
Udupi	25.25	29.69	6.13
Uttara Kannada	18.34	22.53	8.14
Karnataka (State Average)	19.13	25.66	6.88

Source: GoK, 2006.

Only two of the variables, namely, female literacy (FEMLITR) and percentage of total Population Having None of the Three Facilities of Safe Drinking Water, Toilets and Electricity (%) (TOTA3) have emerged statistically significant. The negative sign of these variables indicates that over the decade, planned efforts to impart female literacy and to reduce lack of three basic facilities have partly helped to reduce regional disparity in efficiency of health system across districts (Table 15). However, as could be observed from the low coefficients of these variables, this has not been able to compensate for deficiency in pubic investment and health sector investment policies in the state which have not taken into account changing population densities in different regions and districts of the state (Table 16, column 3).

Table 15. Instrumental variables (2SLS) Regression results for Dispersion as Dependent Variable in Karnataka

Number of observations = 27;		
F(2, 24) = 11.84* R-squared = 0.4754, Adj. R-squared = 0.4317 Root MSE = .03242		
DISPERSION	Coefficient.	Standard Error
FEMLITR	-.003*	.000731
TOTA3	-.004*	.001
CONSTANT	1.390*	.059

Instrumented: FEMLITR; Instruments: TOTA3, GDI.

Source: Estimated; Note: * denotes significance at 1% level.

** denotes significance at 5% level.

Dispersion= dispersion in health system efficiency as estimated from the results presented in Table 3; FEMLITR = female literacy; TOTA3= Population Having None of the Three Facilities of Safe Drinking Water, Toilets and Electricity (%); GDI= gender development index.

In fact, for some of these districts increase in population density has constrained their achievements. Besides Bangalore Urban, for instance, examples of these districts also include Bellary, Koppal, Gulbarga and Raichur (Table 16, column 3).Pertinently, high level of rural poverty (ranging above 40%) in some of the districts like Gulbarga, Kolar, Bellary, Chitradurga and Raichur has also played constraining role in overcoming district level disparities in life expectancy. A number of these districts (like Bellary, Chitradurga, Kolar and Koppal) with high rural poverty have shown a high increase in NDDP in 1991-2002 (last column, Table 16). However, a further break-up of this income growth across sectors indicates that this growth in their NDDP has been achieved due to higher growth in their secondary and tertiary sectoral incomes (columns 6-7, Table 16). Consequently, rural poverty and disparities in income have influenced their health outcomes and life expectancy which the direct health sector inputs have not been able to compensate. Besides these constraining factors, a positive aspect of the development in the state is observed through a positive impact of improvement in Gender Development Index. In fact, better performing districts like, Hassan, Koppal, and Raichur in terms of achievements in their life expectancy have depicted higher improvement in GDI in 1991-2001 (Table 16, column 2).

Overall, thus it is evident from the results that besides differentials in availability and efficient utilization of health system inputs, inequitable distribution of income across rural and urban sectors, changing population densities in respective districts as well as lack of gender specific focus of public sector intervention in terms of education and other opportunities has lead to disparities in health outcomes (or life expectancy). This also suggests need for appropriate links and coordination between economic and social sector policies particularly at district level that could avoid sub-optimal health outcomes for the poorer districts in the middle income state in the country.

Table 16. Population Densities in Different Regions and District

District/Sectors	Improvement in GDI (1991-2001)	Increase in Population Density (1991-2001)	No. Of Rural Families Below Poverty Line (2001)	Growth rate of NDDP (1991-02) (at constant prices)			
				Primary sector	Secondary sector	Tertiary sector	All Sectors
Bagalkot	18.22	18.96	23.5	6.8	4.3	7.6	6.4
Bangalore Rural	22.14	12.15	35.75	8.9	7.6	11.9	9.7
Bangalore Urban	23.48	34.80	15.67	2	6.2	11.3	9.1
Belgaum	20.95	17.60	23.7	4.2	4.5	7.2	5.4
Bellary	21.44	22.45	44.57	4	6.2	7.5	5.8
Bidar	19.92	19.48	39.6	2.5	5.5	6.9	4.9
Bijapur	17.90	17.01	42	1.6	5.9	6.7	4.3
Chamarajnagar	18.01	9.25	36	4.4	2.6	6.7	4.8
Chikmaglur	15.64	12.06	27	2.6	4.2	5.5	3.8
Chitradurga	20.23	14.74	41.5	3.9	4.7	6.8	5.1
Dakshina Kannada	10.70	14.60	15.4	4.2	-1	9.7	5.1
Davangere	17.17	14.83	20	4	3.1	6.8	4.8
Dharwad	17.89	12.91	39	3.9	5.2	5.9	5.3
Gadag	24.70	13.59	46.4	5.1	3.6	9.8	6.7
Gulbarga	25.69	21.38	33.7	2.6	6.6	7.4	5.3
Hassan	24.26	10.00	27.13	3.6	6.2	6.9	5.2
Haveri	24.17	13.31	32	5.3	5.4	8.2	6.4
Kodagu	11.83	11.76	19	1.3	5.1	4.9	2.8
Kolar	21.39	13.70	40.27	5.4	3	6.3	5.3
Koppal	31.07	24.81	42.5	3.2	10	9.7	7.1
Mandya	20.77	7.25	29.86	3.5	4.8	6.6	4.9
Mysore	21.98	15.02	28.14	5.4	5.9	8.3	6.9
Raichur	25.59	21.72	43.2	1.5	6.7	4.9	3.5
Shimoga	15.56	12.87	36	3.8	3.9	7.6	5.4
Tumkur	17.05	11.47	31.4	3.9	4.1	7.1	5.1
Udupi	9.32	6.72	24.67	4.2	0.4	7	4.6
Uttara Kannada	16.61	10.92	30.45	1.5	7.7	6.3	5.1

Source: GoK, 2006; GDI= gender development index; NDDP= net district domestic product.

The results of our analysis of efficiency variation at sub-state level using a case study of a health care system in a middle income Indian state, namely, Karnataka (India), thus, indicate that the efficiency of public health delivery system remains low. Considerable disparities across districts in terms of per capita availability of hospitals, beds and manpower inputs has adverse impact on improving the life expectancy in the state. Overcoming these factoral disparities within the health system may lead to an improvement in the district level efficiency of public health system. This may also help the state to improve life expectancy speedily and more equitably in deficient districts. However, this has to combine with adequate infrastructure facilities like safe drinking water supply, toilets and electricity to improve the outcomes in deficient districts of Bagalkot, Kolar, Kodagu and Uttar Kannada. This may also require a considerable increase in medical and public health expenditure in rural areas in the state. This could be attempted partly through funds from National Rural Health Mission (NRHM) and also by improving rural sanitation in poorer districts. Besides it, rural poverty and disparities in income, overall skewed distribution of income across rural and urban sectors, changing population densities in respective districts as well as lack of gender specific focus of public sector intervention in terms of education and other opportunities have influenced the health outcomes and life expectancy which the direct health sector inputs have not been able to compensate. The results also suggests need for strengthening of links and coordination between economic and social sector policies particularly at district level, the absence of which has led to sub-optimal health outcomes for the poorer districts in the middle income state in the country. In this regard our study has thus contributed to a better understanding of the reasons for relative performance of health system of different districts within a middle income state of Karnataka in a low income country by using panel data and frontier efficiency model for the state.

Table 17. Stochastic Frontier Normal/Half-Normal Model for Madhya Pradesh

Log Likelihood = 70.548		
Number Of Observations = 45		
Wald Chi2(3) = 8.214e+09		
Prob > Chi2 = 0.0000		
LEXP	Coefficient	Std. Err.
POPERPHC	0.174*	2.60E-06
SHCPERLAKH	0.325*	9.65E-06
ADQSTAFF	0.016*	2.53E-06
Constant	1.600*	5.24E-05
/lnsig2v	-37.408	242.9984
/lnsig2u	-4.587*	0.210819
sigma_v	7.53E-09	9.15E-07
sigma_u	0.101	0.011
sigma2	0.0102	0.002
Lambda	1.34E+07	0.011
Likelihood-Ratio Test Of Sigma_u=0: chibar2 (01) = 10.19 Prob>=chibar2 = 0.001		

Source: Estimated; Note: * indicates 1% level of significance.

Madhya Pradesh: Results and Discussion

The results of our frontier model for Madhya Pradesh are presented in Table 17[15]. All the variables, namely, population per primary health centers (POPPERPHC), primary health centers (PHC) with adequate staff (ADQSTAFF) and sub-centers per lakh of population (SHCPERLAKH), have emerged significant with expected positive signs. The actual and estimated Life Expectancy at Birth (LEXP) for the districts (sub-state level units) of MP is presented in Table 18. It can be observed that Indore with its actual LEXP and Estimated LEXP coinciding at 69.7 years has emerged as the most efficient district (MED). Although actual and estimated LEXP with the given level of inputs also coincide for two other districts, namely, Dewas (LEXP 63.3 years) and Mandsaur (LEXP 58.5 years). Despite this matching of actual and estimated LEXP, these districts cannot be considered as MED due to their lower LEXP relative to Indore. The lowest performing district is Shivpuri with low actual LEXP (47.3 years), which remains only 79.36% of its potential LEXP (59.6 years) (Table 18). The ranks of these districts according to their deviation from the MED are presented in Table 19.

It should be noted that two of the districts, namely, Bhopal and Jabalpur have a potential for increasing LEXP even higher than the MED. However, unlike Indore, shortfall of actual realization of the potential LEXP for these districts does not indicate them as efficient ones. Therefore, rank for these two districts (namely, Bhopal and Jabalpur) using deviations from MED are not computed since it may not give appropriate picture of their efficiency. Thus, except for these districts, second most efficient district is Gwalior (with deviation at 2.11 years from MED) and lowest efficient district is Rewa with deviation at 11.66 years from MED.

Reasons for this difference in district level efficiency could partly be traced in Table 20 that presents input levels for each of the districts in MP. Pertinently, it could be observed from Table 20 that Gwalior has higher load of rural population at PHC (43103 patients) relative to Indore (28270 patients) but it is lower compared to Rewa (55062 patients). Further, it is the number of adequately staffed PHCs (76.7%) and PHCs with adequate supplies (76.7%) that makes this district more efficient in performance relative to Rewa. The latter has merely 1/12th of input levels in contrast to Gwalior for both of these parameters. In case of Indore, lower rural population load per PHC makes up for the less number of adequately staffed and less numbers of adequately supplied PHCs. Notably, the skilled attention at birth is much higher in Indore (72%) relative to Gwalior (61.8%) and Rewa (20.8%). Also it is noteworthy that in most of the inputs, difference between lowest and highest values is very high. The distance between these values is ten times for population per PHC, more than two times for rural population served per SHC, nearly nine times for skilled attention at birth, 15 times for number of adequately staffed PHC and 25 times for adequately supplied PHC (Table 20). Inequity as depicted by Gini coefficient is notably high (.409) for the inputs of supplies (Table 20).

Thus the main factors within the health system responsible for discrepancy in performance of individual districts include inequitable distribution of supplies, availability of skilled attention at birth and inadequate staffing relative to patient load of rural population at PHCs. This is also corroborated by other studies which indicate that: i) there is a shortage of doctors in the state which is 28% and 19% respectively for class I and class II doctors ii) attendance by staff in rural health centers is poor (Bajpai et al, 2005) iii) some of the districts like Bhopal, Gwalior, Datia, Morena, Damoh and Tikamgarh have no vacancies particularly on main doctors and health officers. By contrast, eight districts including Katni, Ratlam, Barwani, Jhabua, Shajapur, Shadol, Dindori and Sidhi have over 30% vacancies (GoMP, 2002).

---

[15] This description of MP results is based on Purohit (2010a)

Table 18. Actual and Estimated Life Expectancy of the Districts in MP

	Actual Life Expectancy	Estimated Life Expectancy	Actual Life Expectancy (as % to potential)
Balaghat	58.50	61.82	94.63
Barwani	59.60	63.71	93.55
Betul	54.70	63.22	86.52
Bhind	59.90	60.75	98.61
Bhopal	64.90	76.13	85.24
Chhatarpur	50.40	59.82	84.26
Chhindwara	60.40	63.21	95.56
Damoh	54.00	62.00	87.10
Datia	55.20	61.11	90.33
Dewas	63.30	63.30	100.00
Dhar	63.40	64.65	98.07
Dindori	62.60	62.60	100.00
EastNimar	57.70	62.47	92.36
Guna	53.50	60.52	88.40
Gwalior	65.30	67.59	96.62
Harda	55.60	60.26	92.27
Hoshangabad	55.60	62.39	89.11
Indore	69.70	69.70	100.00
Jabalpur	57.50	70.51	81.55
Jhabua	55.80	63.71	87.58
Katni	57.50	61.05	94.18
Mandla	62.60	63.32	98.86
Mandsaur	58.50	58.50	100.00
Morena	57.40	60.51	94.87
Narsimhapur	56.90	60.41	94.20
Neemuch	58.50	62.77	93.20
Panna	53.00	59.02	89.79
Raisen	55.30	60.36	91.61
Rajgarh	54.20	58.94	91.96
Ratlam	58.10	63.14	92.02
Rewa	53.60	58.04	92.35
Sagar	54.00	61.93	87.19
Satna	49.60	59.97	82.71
Sehore	54.50	60.80	89.63
Seoni	59.90	64.23	93.26
Shahdol	57.10	67.40	84.71
Shajapur	58.30	58.50	99.66
Sheopur	57.40	61.31	93.62
Shivpuri	47.30	59.60	79.36
Sidhi	58.00	59.74	97.09
Tikamgarh	52.80	59.24	89.13
Ujjain	59.80	62.55	95.60
Umaria	57.10	61.14	93.39
Vidisha	54.70	59.43	92.04
West Nimar	59.60	62.33	95.63

Source: GoMP, 2002 and Our Estimates.

Table 19. Deviations and Ranks of Districts in MP

District	Deviation from Maximum Efficient District (MED)	Rank
Balaghat	-7.88	21
Barwani	-5.99	7
Betul	-6.48	10
Bhind	-8.95	27
Bhopal	6.43	n.c
Chhatarpur	-9.88	34
Chhindwara	-6.49	11
Damoh	-7.70	19
Datia	-8.59	24
Dewas	-6.40	9
Dhar	-5.05	4
Dindori	-7.10	14
EastNimar	-7.23	16
Guna	-9.18	28
Gwalior	-2.11	2
Harda	-9.44	32
Hoshangabad	-7.31	17
Indore	0.00	1
Jabalpur	0.81	n.c
Jhabua	-5.99	6
Katni	-8.65	25
Mandla	-6.38	8
Mandsaur	-11.20	41
Morena	-9.19	29
Narsimhapur	-9.29	30
Neemuch	-6.93	13
Panna	-10.68	39
Raisen	-9.34	31
Rajgarh	-10.76	40
Ratlam	-6.56	12
Rewa	-11.66	43
Sagar	-7.77	20
Satna	-9.73	33
Sehore	-8.90	26
Seoni	-5.47	5
Shahdol	-2.30	3
Shajapur	-11.20	42
Sheopur	-8.39	22
Shivpuri	-10.10	36
Sidhi	-9.96	35
Tikamgarh	-10.46	38
Ujjain	-7.15	15
Umaria	-8.56	23
Vidisha	-10.27	37
West Nimar	-7.37	18

Source: Estimated; n.c = not computed.

Table 20. Health Sector Inputs at District Level in MP

District	Popula-tion per Health Centre	Rural Popula-tion per PHC	Rural Popu-lation Served per SHC	Deliveries Attended by Skilled Staff(%)	Percentage of PHCs with adequate		
					Staff	Infrastructure	Supply
Balaghat	4394	35767	4377	29.4	21.115	6.110	3.004
Barwani	3889	29776	3846	34.5	37.713	13.736	13.736
Betul	4469	33398	4190	34.2	29.371	5.871	26.576
Bhind	6795	57305	5950	18.9	31.500	15.800	5.312
Bhopal	23549	35767	5677	69.7	22.198	n.a	11.134
Chhatarpur	6275	28057	6087	24.4	45.150	5.003	17.462
Chhindwara	4645	20845	4406	43.3	52.457	2.509	47.465
Damoh	6044	58506	5417	29.1	59.740	33.448	6.686
Datia	6155	44570	5635	32.9	44.256	n.a	22.198
Dewas	5886	37970	4944	99.9	45.604	18.174	59.145
Dhar	3776	29047	3631	35.7	40.045	13.330	66.686
Dindori	2694	23019	3019	17.6	5.003	n.a	9.974
EastNimar	5240	27127	4605	42.8	16.777	9.974	6.686
Guna	6716	54615	6068	36.3	42.098	5.312	15.800
Gwalior	13251	43103	6401	61.8	76.708	7.691	76.708
Harda	6679	53298	6116	42.6	40.045	n.a	59.740
Hoshangabad	6497	46828	5240	42.6	20.697	16.777	29.079
Indore	17830	28270	6622	72	41.679	16.777	37.338
Jabalpur	9988	57560	4772	44.4	59.740	13.330	13.330
Jhabua	3581	39864	3687	22.4	39.252	21.328	46.525
Katni	5688	44144	5177	44.4	17.637	n.a	11.822
Mandla	3561	28639	3730	17.6	35.874	3.597	10.697
Mandsaur	5858	20929	6292	46.4	46.525	13.330	56.826
Morena	7182	65477	6347	33	49.899	n.a	37.338
Narsimhapur	5632	42331	5585	27.8	46.993	11.822	n.a
Neemuch	5712	29067	4983	46.4	43.816	18.728	56.261
Panna	5339	49739	5329	13.5	7.691	n.a	15.333
Raisen	5491	39710	5219	29.4	15.029	20.086	49.899
Rajgarh	6143	33422	6242	34.2	37.713	13.736	30.877
Ratlam	6103	32582	5103	60.4	49.899	13.599	31.817
Rewa	6425	55062	6164	20.8	6.890	3.387	6.890
Sagar	7094	49325	5838	43	44.701	3.387	27.660
Satna	6009	32957	5748	16.9	26.576	6.686	20.086
Sehore	6129	52020	5818	38.5	64.715	17.637	76.708
Seoni	3621	36041	3680	42.6	55.147	3.387	41.264
Shahdol	3339	24992	2844	51.7	14.585	n.a	20.697
Shajapur	6294	50042	6075	39.7	10.486	10.486	n.a
Sheopur	5597	58892	5294	33	42.948	14.296	56.826
Shivpuri	6578	92384	6035	47.3	18.174	9.116	27.385
Sidhi	5186	37361	5196	10.4	13.330	3.287	3.287
Tikamgarh	6684	55044	6351	29.5	29.371	n.a	41.264
Ujjain	8592	49884	6162	68.2	31.187	18.728	37.338
Umaria	5159	39358	4976	51.7	25.028	n.a	12.554
Vidisha	7104	47724	6628	27.8	44.256	16.777	55.701
West Nimar	4262	24388	4352	34.5	43.380	9.974	16.777
Gini Coefficient	0.223	0.182	0.105	0.235	0.264	0.396	0.409
Minimum Value	2694	20845	2844	10.4	5.003	2.51	3.004
Maximum Value	23549	92384	6628	99.9	76.708	33.448	76.708

Source: GoMP, 2002 and GoI, 2006.

Further there exist considerable vacancies in the entire state relative to various categories of doctors. Its proportion varies between 100% (for civil surgeons) to 56% (for MO&H) and to 28.2% (for specialist). This scarcity in rural areas is glaring even relative to all-India average. In contrast to all India average shortfall (as % of sanctioned strength) in Total Specialists (Surgeon, OB&GY, Physicians, Pediatricians) (62.92%), Doctors at Primary Health Centres (9.43%) and nurse midwife(46.86%), the state has a shortfall in these categories of health manpower as 70.76%, 16.53% and 76.35% respectively, which are considerably higher.

Besides the factors within the health system, in line with our methodology for other two states, we explored influences external to the system leading to differentials in efficiency at the district level. Thus we used dispersion in efficiency (or Deviations)as a dependent variable in the second stage of our regression exercise. Among the set of variables used by us as explanatory variables, for MP it included per capita income, number of families below poverty line (BPL), total literacy, male and female literacy separately, number of primary and secondary educational institutions separately, drop out rate, gender development index (GDI), rural habitat fully covered by water supply, percentage coverage through safe drinking water and human development indicator (HDI). The results of best fit are presented in Table 21. Three variables, namely, GDI, safe water supply coverage and HDI have emerged statistically significant (Table 21). Among these, except GDI, other variables have negative sign. It depicts that the efforts to improve HDI and proper water facilities have helped the state to reduce disparities in efficiency at district level.

Table 21. Regression results for Deviations as Dependent Variable in Madhya Pradesh

Dependent Variable: DEVIATIONS	Unstandardized Coefficients		Standardized Coefficients	"t" values	Significa-nce Level	Adjusted R Square=0.227 df2=41; F=5.298*
	B	Std. Error	Beta			
(Constant)	12.701	4.814		2.638**	0.012	
GDI	29.140	12.740	0.444	2.287**	0.027	
SAFWATER	-0.040	0.022	-0.253	-1.821+	0.076	
HDI	-32.344	10.222	-0.621	-3.164*	0.003	

Source: Estimated; * denotes significance at 1% level, ** denotes significance at 5% level, + denotes significance at 10% level.

However, the positive sign of GDI suggests that the level of gender development among districts has not helped to reduce district level disparities in health system performance. This phenomenon of diverse impact of GDI and HDI could be explained if we consider the movement of HDI and GDI[16]. It could be observed that despite high HDI and GDI, movements of GDI and HDI do not commensurate with the MED. For instance, Indore, the MED, is first rank in HDI but in GDI it is 13th rank. Dewas district, which has a lower efficiency ranking has the first rank in GDI. This kind of different movement in these two parameters may be justified in view of the fact that despite various measures such as

[16] see, for instance, Purohit, Brijesh C. (2010a)

reservations for women in Panchayats, advocacy for women's rights and promotion of institutional deliveries, the women in the State continue to lag behind in health, education and liberty to rights[17]. Possibly these factors (other than LEXP) have had an impact to result in the observed movements in GDI. Thus to overcome the impact of GDI on district level disparities in LEXP and increase the efficiency of the health system, more efforts to overcome disparities in GDI at district level may prove useful.

The results of frontier model using district level information in MP indeed indicate that despite decentralization and health sector reforms in the state, the efficiency of public health delivery system remains low. Considerable deficiency of public health manpower inputs and prevailing disparities across districts has adverse impact on improving the life expectancy in the state. Main factors within the health system for discrepancy in inter-district performance are inequitable distribution of supplies, availability of skilled attention at birth and inadequate staffing relative to patient load of rural population at primary health centers. Overcoming these factors may lead to an improvement in the district level efficiency of public health system. This may also help the state to improve life expectancy speedily and more equitably. This may require a considerable increase in medical and public health expenditure from the current level of Rs. 164 per capita (in 2004-05) which is much lower than all-India average of Rs. 304 per capita (GoI, 2005). This calls for resource mobilization for and within the sector. Partly this may be coming forth with the recommendations of the Twelfth Finance Commission from a grant in aid of Rs. 1.8 billion over the period 2006-10 (GoI, 2004). Remaining gap may require innovative steps and more effective use of funds from patient welfare committees (RKS). Besides the direct inputs from health sector, a more conducive environment for gender development by reducing inequality in opportunities for women in health, education and other rights may provide the necessary impetus towards reducing maternal morbidity and mortality and add to overall life expectancy in the state. The results also indicate that such kind of analysis if extended to cover other states may further help to improve health facility planning to bring about a better overall regional distribution of health inputs in the country. Pertinently these results suggest that the impact and importance of other socio-economic factors on overall health facility utilization is realized through their linkages and this fact may be incorporated in planning future expansion of health facilities particularly in backward areas of MP and other states.

CONCLUSIONS AND POLICY IMPERATIVES

Thus, the results of stochastic frontier analysis for these states indicate that the differential impact of health systems in terms of health outcomes could be explained by regional disparities and sanitation factors both at the state and district level. In view of these results, we suggest that policy measures should aim at more efficient health outcomes by overcoming investment needs through different innovative measures. These may vary depending upon regions and locations across the country.

For instance, the results of frontier model using district level information in Maharashtra indicate that despite urbanization and direct health sector intervention in the state, the low efficiency of public health delivery system could be overcome by improved public health

[17] http://in.news.yahoo.com/070308/43/6d09p.html

manpower inputs and better equitable distribution of beds and dispensaries, availability of skilled attention at birth and adequate staffing relative to patient load in less urbanized districts of the state. These steps may lead to an improvement in the district level efficiency of public health system and help the state to reduce IMR in deficient districts like Yavatmal and improve life expectancy speedily and more equitably. This may require a considerable increase in medical and public health expenditure and necessitate resource mobilization for the sector through innovative steps and more effective use of funds. Besides the direct inputs from health sector, an effort to improve rural sanitation in poorer performing districts may add to overall reduction in IMR and improved life expectancy in the state.

In case of the middle income Indian state, namely, Karnataka, a higher efficiency of health system could be achieved through overcoming disparities across districts in terms of per capita availability of hospitals, beds and manpower inputs which had an adverse impact on improving the life expectancy in the state. However, this has to be supplemented with other inputs like adequate infrastructure facilities of safe drinking water supply, toilets and electricity to improve the outcomes in deficient districts of Bagalkot, Kolar, Kodagu and Uttar Kannada. Towards this end, an increase in medical and public health expenditure in rural areas in the state could be attempted partly through funds from National Rural Health Mission (NRHM) and also by improving rural sanitation in poorer districts. Additionally, in view of: rural poverty and disparities in income, overall skewed distribution of income across rural and urban sectors, changing population densities in respective districts as well as lack of gender specific focus of public sector intervention in terms of education and other opportunities, strengthening of links and coordination between economic and social sector policies particularly at district level, may lead to more optimal health outcomes for the poorer districts in the middle income state in the country.

The results of poorer state, namely MP, indeed indicate that despite decentralization and health sector reforms in the state, the low efficiency of public health delivery system may require to focus on remedial measures to correct for discrepancy in inter-district performance through achieving a better equity in distribution of supplies, availability of skilled attention at birth and inadequate staffing relative to patient load of rural population at primary health centers. Besides such steps, the state may require a considerable increase in medical and public health expenditure from the current level of Rs. 164 per capita (in 2004-05) which is much lower than all-India average of Rs. 304 per capita (GoI, 2005). In this regard, partly resources may be mobilized from a grant in aid of Rs. 1.8 billion over the period 2006-10 with the recommendations of the Twelfth Finance Commission and rest of resource gap may be fulfilled through innovative steps and more effective use of funds from patient welfare committees (RKS). Besides the direct inputs from health sector, efforts for a more conducive environment for women in terms of better equality in opportunities for them in health, education and other rights could lead to further reduction in maternal morbidity and mortality and add to overall life expectancy in the state. The results thus indicate that such kind of analysis if extended to cover other states may further help in health facility planning and achieve better overall regional distribution of health inputs in the country. Pertinently these results indicate the impact and importance of other socio-economic factors on overall health facility utilization and their linkages which may be incorporated in planning future expansion of health facilities particularly in backward areas of MP and other states.

Annexure 1: Water Related Infections

1. Anaemia
2. Arsenicosis
3. Ascariasis
4.Campylobacteriosis
5. Cyanobacterial Toxins
6. Dengue
7. Diarrhoea
8. Dracunculosis
9. Drowning
10. Filariasis
11. Fluorosis
12. Guinea Worm
13. Hepatitis A
14. Hepatitis E and Fb
15. Hookworm
16. Impetigo
17.Japanese Encephalitis
18. Lead Poisoning
19. Legionellosis
20. Leptospirosis
21. Malaria
22. Malnutrition
23. Methamoglobinamia
24. Onchocerciasis(River Blindness)
25. Ringworm
26.Scabies
27. Schistosomiasis
28. Spinal Injury
29. Trachoma
30. Trichuriasis Scabies
31. Typhoid and Paratyphoid Fevers
32. Yellow Fever
33. Leishmaniasis

Annexure 2. Stochastic Frontier Model for Estimating the Health System Efficiency

Model Specification

We consider a general stochastic frontier model[18] that is presented as:

$$\ln q_j = f(\ln x) + v_j - u_j \tag{1}$$

Where $\ln q_j$ is the health output (life expectancy or inverse of IMR) produced by a health system "j"

X is a vector of factor inputs represented by per capita health facilities (including per capita availability of hospital beds, per capita primary health centers (or sub centers), per capita doctors, per capita paramedical staff, per capita skilled attention for birth.

v_j is the stochastic (white noise) error term

u_j is one sided error term representing the technical inefficiency of health system "j"

Both v_j and u_j are assumed to be independently and identically distributed (iid) with variance σ_v^2 and σ_u^2 respectively

From the estimated relationship $\ln \hat{q}_j = f(\ln x) - u_j$

The efficient level of health outcome (with zero technical inefficiency) is defined as:

[18] For more details on econometric aspect, see for instance, Coelli, T., D. S. P. Rao, and G. Battese(1998) and a review of health related applications on this aspect see, for instance, Worthington, A.C. (2004) and Purohit(2010).

$\ln q^* = f(\ln x)$
This implies $\ln TE_j = \ln \hat{q}_j - \ln q^* = -u_j$
Hence $TE_j = e^{-u}{}_j$, $0 <= e^{-u}{}_j <= 1$
If $u_j = 0$ it implies $e^{-u}{}_j = 1$

Health system is technically efficient.

This implies that technical efficiency of jth health system is a relative measure of its output as a proportion of the corresponding frontier output.

A health system is technically efficient if its output level is on the frontier which in turn means that q/q^* equals one in value. At the district level only cross sectional data are available and the distribution of the inefficiency term is assumed to be a standard truncated normal distribution.

REFERENCES

Agoramoorthy Govindasamy and Minna Hsu (2009), India needs sanitation policy reform to enhance public health, *Journal of Economic Policy Reform,* vol. 12, issue 4, pp. 333-342.

Annette Prüss, David Kay, Lorna Fewtrell and Jamie Bartram (2002), Estimating the Burden of Disease from Water, Sanitation, and Hygiene at a Global Level, *Environment Health Perspectives*, Volume 110, Number 5, May, 537-542.

Baltazar, J. et al. (1988), Can the case-control method be used to assess the impact of water supply and sanitation on diarrhoea? A study in the Philippines, *Bulletin of the World Health Organization*, 66: 627-635.

Bajpai N., Dholkia R. H. and Sach J. (2005), Scaling Up Primary Health Services in Rural India, *CGSD Working Paper No. 29*, the Earth Institute at Columbia University, November.

Baum Christopher F. (2006), *An Introduction to Modern Econometrics Using Stata*, Stata Press, Texas.

Blum, D. & Feachem, R. (1983), Measuring the impact of water supply and sanitation investments on diarrhoeal diseases: problems of methodology, *International journal of epidemiology*, 12: 357-365.

Coelli, T., D. S. P. Rao, and G. Battese (1998), *An Introduction to Efficiency and Productivity Analysis*, Kluwer, Boston.

Daniels, D. et al. (1990), A case-control study of the impact of improved sanitation on diarrhoea morbidity in Lesotho. *Bulletin of the World Health Organization*, 68: 455-463.

Esrey, S. et al. (1985), Interventions for the control of diarrhoeal diseases among young children: improving water supplies and excreta disposal facilities, *Bulletin of the World Health Organization*, 63: 757-772.

Galiani Sebastian & Paul Gertler & Ernesto Schargrodsky (2005), Water for Life: The Impact of the Privatization of Water Services on Child Mortality, *Journal of Political Economy*, University of Chicago Press, vol. 113(1), pages 83-120, February.

GESAMP (Joint Group of Experts on the Scientific Aspects of Marine Environmental Protection). (2001), Protecting the oceans from land-based activities: Land based sources and activities affecting the quality and uses of the marine, coastal and associated fresh water environment *(GESAMP No. 71)*, UN Environment Programme, New York.

Hughes Gordon, Kseniya Lvovsky and Meghan Dunleavy (2001), *Environmental Health In India: Priorities In Andhra Pradesh*, South Asia Environment and Social Development Unit, World Bank, Washington D.C.

Government of India (2007), *Economic Survey*, Ministry of Finance, New Delhi.

Government of India (2006), *Reproductive and Child Health Survey*, 2006, Ministry of Health and Family Welfare, New Delhi.

Government of India (2004), *Report of the Twelfth Finance Commission (2005-2010)*, Ministry of Finance, New Delhi.

Government of India (2002*), India Assessment 2002, Water Supply and Sanitation*, Planning Commission of India, New Delhi.

Government of Madhya Pradesh (2002), *The Madhya Pradesh Human Development Report 2002,* Bhopal.

Government of Maharashtra (2002), *Human Development Report, Maharashtra, 2002*, Mumbai.

Government of Karnataka (2006), *Karnataka Human Development Report, 2005*, Planning and Statistics Department, Bangalore.

Guilkey, D. K. and R. T. Riphahn (1998), The determinants of child mortality in the Philippines: estimation of a structural model, *Journal of Development Economics* 56: 281-305.

Jalan J. and M. Ravallion (2003), Does Piped Water Reduce Diarrhea for Children in Rural India?, *Journal of Econometrics 112*: 15-173.

Murugesan Anand, Vikram Dayal, Saurabh Chugh (2008), An Empirical Study of Sanitation and Health in Rural Uttarakhand, India, *International Journal of Ecological Economics & Statistics,* Winter 2008, Volume 10, Number W08, 91-99.

Lee Lung-fei, Mark R. Rosenzweig and Mark M. Pitt (1997), The Effects of Improved Nutrition, Sanitation, and Water Quality on Child Health in High-Mortality Populations, *Journal of Econometrics*, 77:209-235.

Okun, D. (1988), The value of water supply and sanitation in development: an assessment. *American journal of public health*, 78: 1463-1467.

Purohit Brijesh C. (2010), *Health Care System in India,* Gayatri Publications, New Delhi.

Purohit Brijesh C(2010a), Efficiency of Health Care System at Sub-State Level in Madhya Pradesh (India), *Social Work in Public Health,* 25:42–58, Taylor & Francis, Philadelphia, USA.

Purohit Brijesh C. and Siddiqui, Tasleem (2004), Environmental Determinants of Disease and Mortality Pattern in India: A Factor analytic Exploratory Approach, pp.721-731 in Mallikarjun, M and Pawan K. Chugan (Eds.), *Trade, Technology and Environment*, Excel Books, New Delhi.

Ridder, G. and I.Tunali (1999), Stratified Partial Likelihood Estimation, *Journal of Econometrics,* 92, pp. 193-232.

Sankar. U., (2001), *Economic Analysis of Environmental Problem In Tanneries And Textile Bleaching and Dying Units And Suggestions For Policy Action*, Allied Publishers Ltd., New Delhi.

Shanumganandan, S. (1999), Water Quality, Water Resources and Health Water Quality in Relation to Human Health: A Study With Reference To Water-Borne Diseases And Major Environmental Issues in Indian Subcontinent, Working paper, Madurai Kamaraj University, India.

Sharma Suresh, *Reproductive and Child Health Status In India: District Level Analysis*, Mimeograph, Population Research Center, Institute of Economic Growth, University Enclave, Delhi-8, URL: http://www.sasnet.lu.se/EASASpapers/11SureshSharma.pdf.

Srinivasulu R.and G. Haripriya (2004), The Impact of Drinking Water Quality and Sanitation on Child Health, SSRN-id1152234.

Sulabh International Social Service Organization (2007), *Study on Disease Burden to Inadequate Water & Sanitation Facilities in India*, New Delhi *The Economic Times* (2005), September, 5, P.9, Kolkata.

UNICEF (2008), The Drinking Water and Sanitation Situation in Asia and the Pacific, 2006, *Prepared for the UNICEF Asia and Pacific Representatives Meeting,* Cha'am, Thailand 1-3 November, 2008.

Walsh, J. & Warren, K. (1979), Selective primary health care: an interim strategy for disease control in developing countries, *New England journal of medicine*, 301: 967-974.

W.H.O (2000), *The World Health Report, 2000, Health Systems: Improving Performance*, World Health Organization.

Worthington, A.C. (2004), Frontier Efficiency Measurement in Health Care: A Review of Empirical Techniques and Selected Applications, *Medical Care Research and Review, 61(2)*, 135-170.

Young, B. & Briscoe, J. (1988), A case-control study of the effect of environmental sanitation on diarrhea morbidity in Malawi, *Journal of epidemiology and community health*, 42: 83-88.

In: Potable Water and Sanitation
Editor: Joel M. McMann

ISBN 978-1-61122-319-4

Chapter 3

HYDROGEOLOGY AND GROUNDWATER RESOURCES OF GHANA: A REVIEW OF THE HYDROGEOLOGY AND HYDROCHEMISTRY OF GHANA

[1]Bruce Banoeng-Yakubo, [1]Sandow Mark Yidana, [2]Joseph Owolabi Ajayi, [1]Yvonne Loh and [1]Daniel Asiedu

[1]Department of Earth Science,
Box LG 58, University of Ghana, Legon-Accra, Ghana
[2]Department of Geology
Obafemi Awolowo University, Ile-Ife, Nigeria

ABSTRACT

This paper surveys the aquifers and the groundwater resources landscape in Ghana to identify the critical resource issues constraining the development and management of groundwater throughout the country. The research is based on a collation of the recent data on the hydrogeological properties of the different geological terrains in the country. This study finds that there are five main hydrogeological provinces distinguished by their well yields, lithology, and groundwater quality. This latest hydrogeological zonation, which is an improvement over previous works consists of the Birimian Province, the Crystalline Basement Granitoid Complex Province, the Voltaian Province, the Pan African Province, and Coastal Sedimentary Province. The hydrogeological properties of each province are discussed. The most prolific aquifers in the country have been identified among the fractured and weathered zones within the Togo Series, Buem Formation, Birimian and Tarkwaian System rocks. Aquifers of these rock formations also offer groundwater of the best quality for most uses in the country.

Hydrochemical data gathered from wells drilled through all the hydrogeological terrains in the country for this study suggest that groundwater hydrochemistry is controlled principally by the weathering of silicate minerals and cation exchange activity. These processes are pervasive throughout the country. In isolated areas, groundwater contamination due to domestic waste discharge and agricultural activities has been noted. Four groundwater types have been identified based on Q-mode hierarchical cluster analysis and conventional graphical methods. In the coastal areas such as the Keta basin, seawater intrusion has been identified as one of the major influences on hydrochemistry of groundwater in the shallow unconfined aquifers. On the basis of the hydrochemistry, groundwater suitability for irrigation purposes is copiously discussed.

The aquifer management problems can be posed as follows. There is inadequate knowledge base of the aquifer geometry and safe yield characteristics of most of the aquifers in the country. There is also inadequate information to address the status and trends in groundwater quality for most areas of the country. The following emerging issues have been identified: anthropogenic effects on groundwater quality; salt-water intrusion in coastal aquifers, groundwater depletion or inadequate supplies from groundwater to meet human needs; uncertainties about recharge and sustainability due to increased abstraction; as well as lack of institutional capacity to respond adequately to sustainable groundwater use. Recommendations to ameliorate these problems include the strengthening of institutional capacity for comprehensive national evaluation of groundwater resources, including aquifer delineation and pro-active strategies to reduce groundwater abstraction to safe yield levels and adoption of aquifer protection strategies for overall protection of groundwater quality.

1. Introduction

Ghana is located along the southern coast of West Africa, within latitudes 4° 40' N and 11° 15' N by longitudes 3° 15' W and 1° 15' E, covering a land area of approximately 239,000 km^2. The country was originally called the Gold Coast on account of vast deposits of gold found within the country. However, there are other economic mineral deposits, including diamond, manganese, bauxite, salt and more recently commercial quantities of oil and gas. The country is bounded on the north by Burkina Faso, on the east by the Republic of Togo, in the south by approximately 500 km of Gulf of Guinea (South Atlantic Ocean) coastline, and on the west by Ivory Coast (Cote D'Ivoire).

The relief varies from 0 m at sea level at the coast to nearly 900 m in the Buem-Togo Range (highest peak is the Mount Afadjato at 885 m, near the Togo border in SE Ghana). There is a divided plateau in the central part of southern Ghana. Ghana is mostly low-lying with 90% of the country having elevation less than 300 m. The Volta River Basin is a major physiographic feature of the country covering approximately 75% of the total land area. The Volta Lake formed behind Akosombo dam has an area of approximately 8,500 km^2 and is the largest man-made lake in the world. The hydro-electric power station at Akosombo dam generates over 75% of the total energy demand of Ghana.

The population of Ghana in 2000 was approximately 18.9 million people with an overall population density of 79.3 persons per km^2. The rural: urban population ratio then was 56.2: 43.8. A large percentage of the rural population engages in farming and livestock rearing, especially in the north. Irrigated agriculture is not very extensive but may be locally significant, especially along river courses.

The historical temperature extremes lie between 18 °C and 40 °C with a smaller average annual temperature range of 24 °C to 30 °C. The general pattern is for temperature to increase with distance from the coast. The average annual rainfall increases from a low of about 800 mm in the northern latitudes to over 2100 mm in the southern latitudes. Near the coast and in most of the southern parst of the country, two rainy seasons are observed (May – July and

September to November), whereas there is only a single rainy season in the north, separated by a rather long dry spell of six months or more (beginning in November and lasting beyond April). The amount of rainfall increases from N-S. The converse holds for potential and actual evapo-transpiration with consequent increase in groundwater recharge from north to south. The climate ranges from hot and humid in the south near the equator (4° N) to hot and dry towards the Sahara (> 10 ° N).

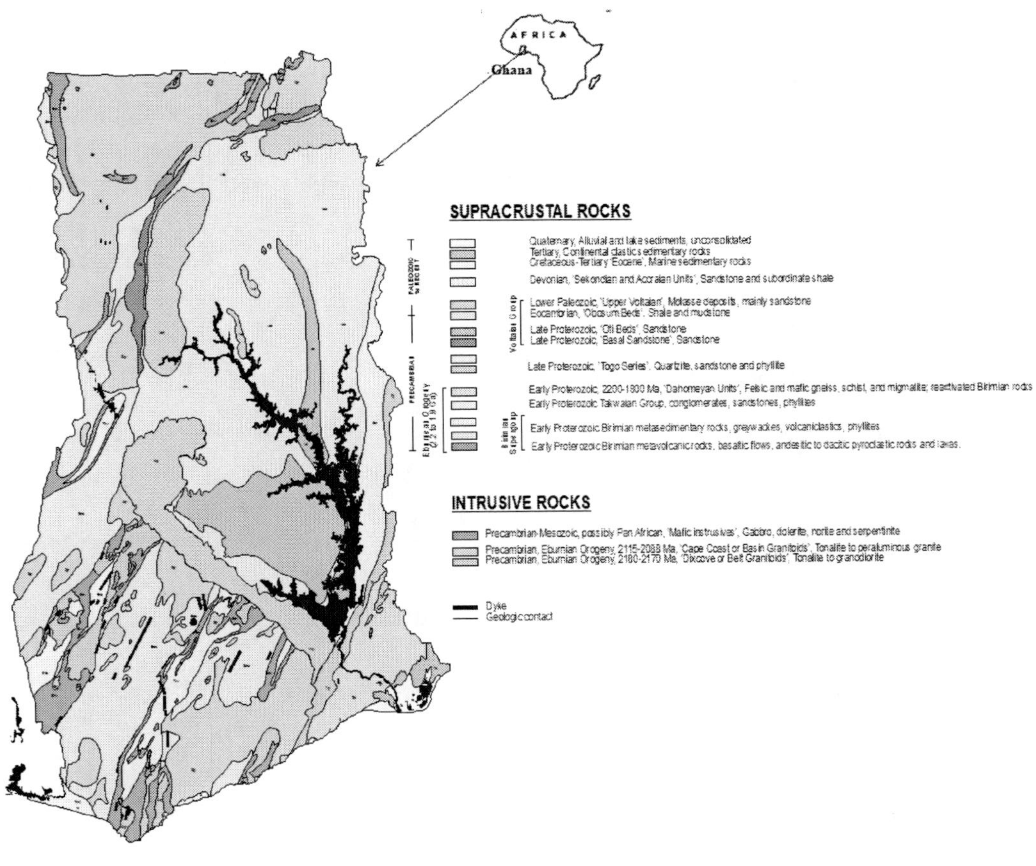

Figure 1. The Geological Map of Ghana.

Vegetation is climate controlled and has the following latitudinal characteristics. With increase in latitude from the coast, the vegetation grades from equatorial rain forest in the southwest through a transition zone to semi-deciduous tropical forest in the central parts to guinea savannah in the northern latitudes and finally to Sudan savannah in the extreme north-eastern part of the country (Dickson and Benneh 1985).

Groundwater is increasingly being phased at various depths in the major hydrogeological provinces for domestic water supply, especially in the rural areas, where the Community Water and Sanitation Agency (CWSA) is mandated to drill prolific wells to satisfy growing rural water needs. As part of the country's millennium development goals, successive governments have tried to project Agriculture as the major pivot around which the poverty reduction strategies in the poverty endemic areas revolve. Discussions on the use of groundwater and available surface water for irrigation of crops all year round to supplement the current rain-fed agricultural practices have been going on for some time now. However, exploitation of groundwater for any commercial agricultural activities should be advised by well balanced information about the properties of aquifers that underlie different parts of the country, and the optimal groundwater extraction rates permissible from these aquifers under current and projected future environmental conditions. There is also the need for an assessment of the sum total of the quality of groundwater from the various aquifer units for irrigation purposes. This is because the use of poor quality groundwater for irrigation purposes has the tendency of destroying agricultural soils, thereby reducing crop productivity in those soils.

This paper provides an update on the general hydrogeology of Ghana, and the quality of groundwater delivered by the different aquifers for various purposes at home and for irrigation purposes. A new hydrogeological zonation of the country is being pursued based on the current knowledge of the rock types, well depths, well yields, and groundwater quality among others.

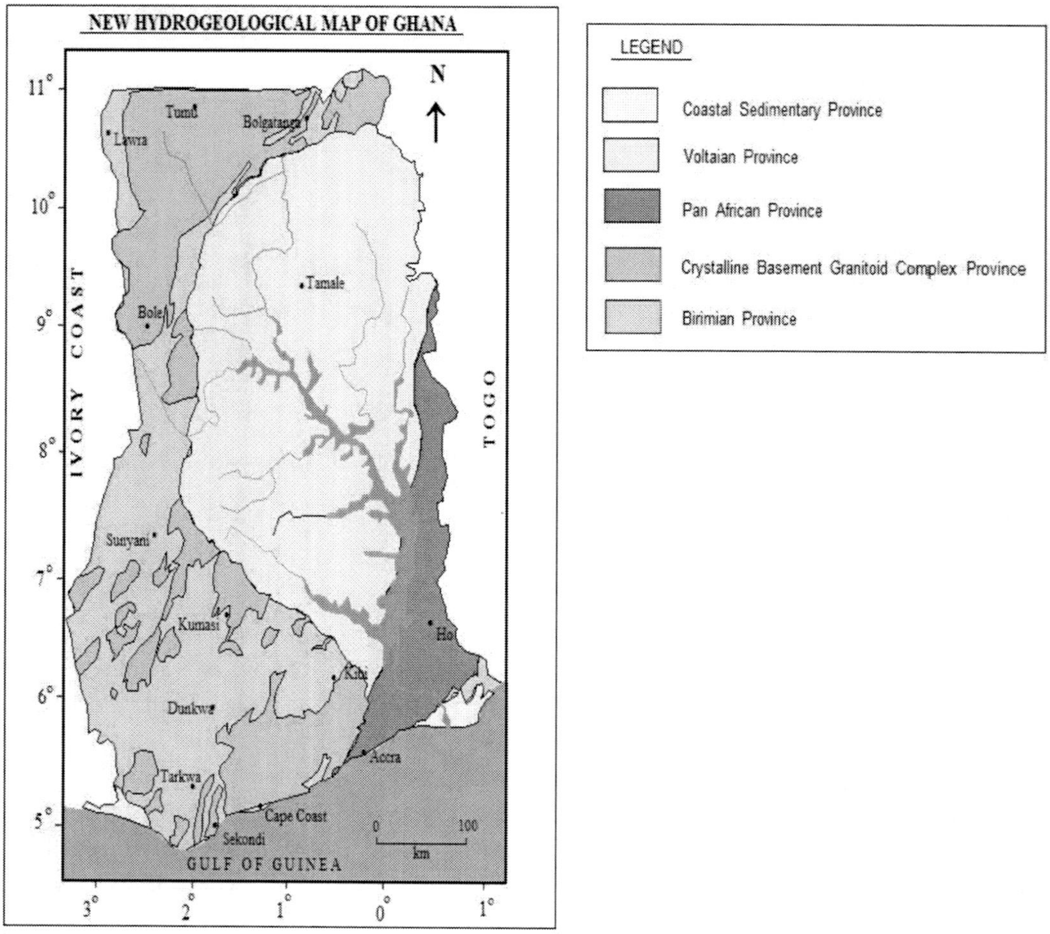

Figure 2. The New Hydrogeological Map of Ghana.

2. General Geology of Ghana

Geologically, Ghana can be divided into several distinct terranes (Figure 1). Rocks of the Birimian System underlie most of southern, western and northern Ghana, and host most of the gold and diamond deposits in the country. The Birimian consists of metamorphosed volcanic and sedimentary rocks which form five subparallel belts of volcanic rocks separated by broad "basins" of sedimentary rocks. Rocks of the Birimian System have been divided into an older "Lower Birimian", consisting of predominantly metasedimentary rocks, and a younger "Upper Birimian", made up of metavolcanic rocks (Junner 1935, 1940; Bates 1955). The rock types present in the Lower Birimian sedimentary belt are greywackes with turbidite features, phyllites, slates, schists, weakly metamorphosed tuffs and sandstones. Some of the phyllites contain pyrite, and finely divided carbonaceous matter is present in most of them. Silicification is common in the phyllites, particularly towards the boundary with the Upper Birimian.

The Upper Birimian volcanic succession consists of lava flows and dyke rocks of basaltic and andesitic composition. Most of these rocks have now been metamorphosed to hornblende-actinolite schists, calcareous chlorite schists and amphibolites (the greenstones) (Kesse 1985). Pillow structures indicating subaqueous eruption of the original basaltic lavas are commonly observed. Available major and trace element chemical data show that these Birimian metabasalts are tholeiitic. However, felsic volcanic rocks also occur in this succession as well as in the predominantly sedimentary sections. The felsic units include dacitic pyroclastic rocks, minor andesitic and rhyolite flows, and undifferentiated volcanogenic sediments. Minor intrusions of mafic and ultramafic rocks cut the volcanics in some places. Mn-rich horizons also occur at stratigraphically lower level in the Upper Birimian and have been found in the uppermost Lower Birimian as well.

A distinctive sequence of clastic sedimentary rocks occurs in elongate troughs developed on top of the Birimian System. These rocks host important paleoplacer gold deposits and are known as the Tarkwaian System. Most workers agree that the Tarkwaian sediments were deposited in intermontane grabens formed by preferential rifting along the axes of the volcanic belts and that there is no evidence that the depositional basins were ever linked. The rocks of the Tarkwaian System represent erosional products of the Birimian and are dominated by coarse clastic sediments (Kesse 1985).

Almost one third of Ghana is covered by sediments of the inland Voltaian Basin which covers an area of about 103,600 km^2. The Voltaian strata are nearly horizontal beds of sandstones, shales, mudstones and conglomerates thought to be of Late Precambrian to Paleozoic age (Kesse 1985). In most places, the flat lying Voltaian strata overlie the Birimian rocks with a marked angular unconformity. Junner and Hirst (1946) subdivided the Voltaian sediments on the basis of lithology and field relationships into Lower, Middle and Upper units.

The Lower Voltaian sediments represent a marine transgression-regression cycle on the craton, whereas the Middle Voltaian records a glacial event followed by prolonged marine incursion and subsidence of the basin. In the eastern part of the basin, the adjacent Togo Belt crops out. The Upper Voltaian, otherwise known as the Obosum Formation, is thickest and coarsest in the southeast. The conglomerates contain pebbles of granite and other igneous rocks, as well as quartzite fragments. Sedimentary structures show the direction of transport to have been from the southeast. The Obosum beds are molasse deposits formed by the erosion of the Togo Series following its uplift in the Pan African event.

The Dahomeyan System is a part of the second major tectono-stratigraphic terrane in Ghana; it underlies eastern and southeastern Ghana. The Dahomeyan is the easternmost rock group in Ghana and differs significantly from other rocks in Ghana in that it is composed of high grade metamorphic rocks. The system consists of four lithologic belts of granitic and mafic gneiss. The mafic gneisses are relatively uniform oligoclase, andesine, hornblende, salite and garnet gneisses of igneous parentage and generally tholeiitic composition (Holm, 1974). The granite gneisses interlayer with the mafic gneiss and are believed to be metamorphosed volcaniclastic and sedimentary rocks. Persistent bands of nepheline gneiss in the system appear to be metamorphosed calc-alkaline igneous rocks (Holm 1974).

A distinctive, but normal, lithology in the Dahomeyan is the "Kpong Conglomerate", a calcareous rock which has been interpreted to be a carbonatite (Mani 1978). Structurally, the granite gneiss is the lowest unit in the system. All of the gneisses have undergone at least two stages of penetrative deformation. The latest deformation is believed to be of Pan-African age (500-600 Ma) and is referred to as a "reactivation" of Birimian crust by Kennedy(1964). Fitches (1970), however, suggests that the metamorphism may be Eburnian (Proterozoic) in age. The gross structure of the Dahomeyan is that of alternating northeast trending lithologic belts with moderate dips to the southeast. Along the western boundary of the belt the gneisses are in fault contact and overthrust onto rocks of the Buem-Togo Belt. The original age of the Dahomeyan protolith is unknown. Rb-Sr and K-Ar dates by Agyei et al. (1987) show Pan African ages for the last metamorphism. Blay (1991) postulates that the Dahomeyan are Birimian rocks. This conclusion is supported by Grant (1969) and Affaton et al. (1980).

The second major lithologic group which makes up the eastern Ghana terrane is the Togo Belt comprising the Buem and Togo Series. This group of rocks comprises three distinctive lithologic assemblages. The rocks which comprise the north to northeast trending Togo Range consist of strongly tectonised phyllite, quartzite and serpentinite. These rocks are variously known as the Togo Series (Kesse 1985) or the Togo Tectonic Unit (Blay 1991). The contacts between the Togo and Dahomeyan to the east and the Buem to the west are thrust faults. The unit grades from east to west from phyllite and chlorite schist upwards into quartzite, micaceous quartzite and sandstone. Serpentinites occur along the western contact and appear to be emplaced along thrust faults (Grant 1969).

West of the Togo Range is a belt of volcanic and sedimentary rocks known as the Buem Series (Kesse 1985; Jones, 1990) or the Buem Tectonic Unit (Blay 1991). The Buem consists of two lithologic assemblages, volcanic and sedimentary. The volcanic assemblage is made up of pillow basalt, agglomerate, hawaiite and trachyte. The sedimentary assemblage, which encloses the volcanics, consists of red shales, feldspathic to quartz arenite, conglomerate, tillite, jasper and minor limestone. Whereas the volcanics were deposited in a submarine environment, the sediments appear to be shallow water to subaerial in origin (Jones 1990). Jones (1990) described the Buem as an eastward dipping homoclinal sequence, whereas Kesse (1985) stated that the rocks are strongly folded. Blay (1991) cited the presence of a basal conglomerate at the Buem-Togo contact as an indication that the Buem is younger than the Togo unit.

Lashmanov (1991) mapped the northern part of the Dahomeyan belt and noted the presence of a sandy mudstone unit between the Buem and the Togo. Lashmanov's mapping supports Blay (1991) in concluding that the Buem is younger than the Togo. Lashmanov's sandy mudstone unit is laterally equivalent to the "Oti Beds" in the Voltaian Basin, a conclusion reached earlier by Grant (1969). Radiometric dating of glauconite in the Middle Voltaian Obosum Beds, which overlie the Oti Beds, gives an Upper Precambrian age of 620 Ma (Bozhko et al. 1971). This would indicate an Upper Proterozoic age for the Togo and a Proterozoic to Lower Cambrian age for the Buem.

Relatively minor outcrops of sedimentary rocks along the coast from Keta and Accra in the east to Half Assini in the west constitute remnants of rocks of the Phanerozoic coastal basins. From east to west, these rocks occur in the Keta, Accraian, Sekondian and Tano basins. Rocks of the Keta basin are of Cretaceous age and consist of sandstones, siltstone, shales, claystone and fossiliferous limestone beds (Kesse 1985). The Accraian is considered mid-Devonian in age and consists predominantly of sandstones and shales. The Sekondian strata are made up of sandstones, shales, silts and beds of chalcedony, sands and pebbly beds and range in age from Devonian to Cretaceous. The Tano basin is located in the extreme southwestern corner of Ghana. It is made up of Cretaceous-Tertiary sediments consisting of limestones, shales and sands which have a large off-shore extension.

Four main types of granitoids are recognized in the Birimian of Ghana. They include Winneba, Cape Coast, Dixcove and Bongo granitoids (Junner 1940; Kesse 1985). The latter three have been recently termed "Basin", "Belt" and "K-rich" granitoids (Leube et al., 1990; Mauer 1990; Hirdes et al. 1993). The Cape Coast and Dixcove type granitoids are widespread in Ghana, the Winneba type is limited to small areas near Winneba, and the Bongo type outcrops in the Bole-Navrongo Belt and in the Banso area.

The Cape Coast granites occur only within the Birimian sedimentary basins. Some of them are two mica granites. This group also includes gneisses, which are especially well developed in the metasedimentary belts. They are typically biotite-bearing. It has been suggested that the Cape Coast granitoids, which appear migmatitic in some localities, might represent an older continental basement on which the Birimian supracrustals were deposited. However, there is no geochronologic support for this theory. Contacts between these granitoids and the metasediments are irregular; rafts of metasediments and relict structures from metasediments rise into the granitoids, and tendrils of granite vein the metasediments (Taylor et al. 1992). Based on the degree of foliation, early workers assumed that the Cape Coast granitoids intruded during regional deformation and that Dixcove granites were emplaced after deformation. (Junner 1940; Kesse 1985). However, later work by Hirdes et al. (1993) demonstrated, in contrast to long held views, that Dixcove granitoids formed at about 2,175 Ma and are about 60 and 90 Ma older than the Cape Coast granitoids. Taylor et al. (1992) suggest that the Cape Coast and Dixcove granitoids are coeval.

Dixcove-type granitoids are metaluminous and typically dioritic to granodioritic in composition. They intrude Birimian the volcanic rocks. They are typically hornblende-bearing and are commonly associated with gold mineralisation where they occur as small plutons within the volcanic belts. The granitoids are massive in outcrop, do not have a compositional banding or foliation, and are thus generally considered post-deformation. However, Dixcove-type granitoids have never been shown to intrude or crosscut Cape Coast granitoids, and some workers (Murray 1960) have recognised Dixcove granitoid clasts in Cape Coast granitoids. The presence or absence of a foliation is not a sufficient criterion to establish timing relationships in granitoids. In particular, amphibole bearing granitoids have been demonstrated to be less likely to develop a foliation during deformation than biotite-rich granitoids (Vernon and Flood 1988). Locally, sheared granitoids were observed in the Sefwi

Belt, and these may have been deformed during regional deformation (Eisenlohr and Hirdes 1992). Dixcove granitoids have a porphyritic texture defined by plagioclase set in a quartz-hornblende-actinolite matrix. The plagioclase is always saussuritised or sericitised and actinolite appears to crosscut the fabric. Such features are typical of granitoids that have undergone metamorphism (Vernon and Flood 1988). The granitoids commonly contain basalt xenoliths, and there appears to be a gradational boundary between finer and coarser grained Dixcove granitoids and basalts (e.g., Hirst 1946). These observations indicate a close association between Birimian basalts and Dixcove granitoids and suggest they may be part of the same igneous event. The above features do not necessarily prove pre-deformation Dixcove granitoid emplacement, but the data contradict the established view that the granitoids intruded after deformation.

The Winneba granitoid occurs at a single locality near the town of Winneba. It is the only rock suite so far encountered in Ghana which shows evidence for an Archean sialic precursor (Sm/Nd model age of about 2.6 Ga (Taylor et al. 1992). The Bongo granitoid is located in northern Ghana where the granites intrude Tarkwaian sediments that overlie the Bole-Navrongo Volcanic Belt. This granitoid is peraluminous and lacks a foliation (Leube et al. 1990). The Rb-Sr whole-rock isochron age of the granitoid is 1,968 ± 49 Ma (Hirdes et al. 1993). A granitoid similar in composition to the Bongo type, the Banso granitoid, outcrops within the Ashanti Belt south of Kumasi. Formerly, this granitoid was thought to be unconformably overlain by Tarkwaian rocks (Woodfield 1966), but recent mapping indicates that the granite crosscuts the tectonised Birimian/Tarkwaian boundary. Contact metamorphic minerals have been observed in Tarkwaian rocks close to the granitoid (Mauer 1986). The latter observations plus petrographic and geochemical similarities described by Mauer (1986) suggest that the Banso granitoid intrudes the Tarkwaian and thus occupies a similar tectonic position to that of the Bongo type granitoid.

3. Current Understanding of the Hydrogeology of Ghana: A New Hydrogeological Zonation of Ghana

3.1. Hydrogeology

The hydrogeology of Ghana is dominated by rocks of the Birimian Supergroup of the Early Proterozoic age and post intrusive granitoids, the Tarkwaian System and the unconsolidated sedimentary rocks. Gill (1969) delineated the hydrogeological Provinces of Ghana from the existing geological map of Ghana and Dapaah-Siakwan and Gyau-Boakye (1998) identified four geohydrological regions from the overall geological map. In the assessment of Dapaah-Siakwan and Gyau-Boakye (2000), the country is dominated by two major hydrogeological provinces: the Basement Complex composed of the Precambrian igneous and metamorphic rocks constituting about 54% of the landmass, and the Paleozoic sedimentary basin which comprises the Voltaian sedimentary basin which underlies about 45% of the total landmass of Ghana. Based on the then knowledge of the hydrogeology of the country, Dapaah-Siakwan and Gyau-Boakye (2000) opined that two minor provinces: the Cenozoic, Mesozoic, and Paleozoic sedimentary strata along the coast and the Quaternary alluvium along major stream courses constitute the remaining 1% of the hydrogeology of

Ghana. These divisions have been used to describe the hydrogeology of Ghana for some time now, although the existing data suggest that the geological formations lumped together as the Basement complex are significantly different in terms of their groundwater occurrence, well yield, and groundwater quality.

In this paper, five hydrogeological provinces have been identified based on the current data from recent wells, geophysical investigations, and hydrochemical analyses (Figure 2). These are the Birimian Province, the Crystalline Basement Granitoid Complex Province, the Voltaian Province, the Pan African Province, and Coastal Sedimentary Province. These provinces follow, to some extent, the lithology of Ghana but do not necessarily connote the grouping of rocks that define the various geological formations. The alluvial deposits form a thin aquifer unit which is worth mentioning but which usually does not contribute significantly to the hydrogeology of Ghana. The hydrogeology of the alluvial deposits has not yet been studied in detail. In areas where these deposits are permeable, relatively thick, and located adjacent to perennial streams, they offer desirable locations for wells, especially for irrigation purposes (Dapaah-Siakwan and Gyau-Boakye 2000).

The Birimian Province is made up of aquifers of the Lower and Upper Birimian rocks, and Tarkwaian aquifers. The Basement Complex Province consists of aquifers of all granioids in Ghana including the Dahomeyan rocks as well as migmatites of the Cape Coast granite type. The Pan African Province is made up of aquifers of the Buem and Togo Formations. The reasons for the re-designation of the hydrogeological provinces of Ghana are manifold. For instance, basement complex rocks are typically granitoid rocks in which the occurrence of groundwater is significantly different from the Birimian and Tarkwaian metasedimentary sequences. In the previous definition of the hydrogeology of Ghana, the granitoids were lumped together with the (meta)sediments of the Birimian and Tarkwaian Systems. Though in places metamorphosed, most of the rocks of the Birimian and Tarkwaian Systems are purely sedimentary in origin and retain their sedimentary nature. In addition, the designation of the Voltaian basin alone as 'Paleozoic' is erroneous because the Togo and Buem were also formed during the Paleozoic as a result of the Pan African Orogeny. On the other hand, the hydrogeology of the Togo and Buem Formations is significantly different from that of the Voltaian System. The Tarkwaian System is a derivative of the Birimian Supergroup and has been affected by the same geodynamic processes in the Eburnean which have defined the structural controls on the hydrogeology of Ghana. The Minor alluvial Provinces can be limited to the well defined Coastal Sedimentary Province, comprising the Cenozoic, Mesozoic and Paleozoic sedimentary system of the Tano and Keta Basins and the Sekondian Group. The Quaternary to Recent alluvial deposits along the various river valleys and valley bottoms have not been properly defined and delineated. Currently their potential as major reservoirs of groundwater for domestic water supply is unknown. It is very important that the valley bottom aquifers are assessed and used for both domestic and dry season agricultural purposes.

3.2. Mode of Occurrence of Groundwater

The occurrence of groundwater in all but the Coastal Sedimentary Province and alluvial, aquifers depends on secondary structures made up of fractures, veins and faults. The development of secondary permeability in these rocks is dependent on the nature, aperture, length, density, thickness and the degree of interconnection of the fractures (Dapaah-Siakwan and Gyau-Boakye 2000) and in-filling of the fractures as well as the extent of weathering. The development of fractures and faults is a direct consequence of the cooling processes and the pressure release due to erosion, compressional and tensional activities during regional tectonic processes. In these rocks, the mode of occurrence of groundwater is basically linked with the nature and thickness of the decomposed zone, rock fracture, joints, quartz veins and pegmatites. Most of the quartz veins and pegmatites are fractured and to some extent brecciated as a result of brittle deformation as a consequence of Proterozooic Eburnean orogenic activities which affected them. The water-bearing and yielding capacities of these rocks depend on the extent of these secondary permeabilities (Banoeng-Yakubo 2000).

Groundwater in the Crystalline Basement Provinces occurs mainly in the saprolite, saprock and in the fractured bedrock. The most productive zones in terms of groundwater in the Birimian Province comprise the lower part of the saprolite and the upper part of the saprock which usually complement each other in terms of permeability and storage (Carrier et al. 2008). The upper less permeable part of the saprolite can act as a semi-confining layer for this productive zone, while the lower, more permeable and usually saturated part of the saprolite is characterized by lower secondary clay content, thus creating a zone of enhanced hydraulic conductivity. Generally, areas underlain by the Birimian and Tarkwaian rocks display deeper seated weathering than areas underlain by the granitoids. Banoeng-Yakubo (1989) identified three types of aquifers in the basement aquifers. These are the weathered rock aquifers, which are fracture related, the fractured quartz-vein aquifers and the fractured unweathered aquifers. In the Crystalline Basement aquifers, the saprolitic zone is a combination of the top soil, the underlying lateritic soil, the highly weathered zone and the moderately weathered zone. The saprock is the poorly weathered zone. Together, the saprolite and the saprock constitute the regolith. In a recent assessment of the hydrogeology of Northern Ghana, a conceptual model was developed similar to what has been outlined (Carrier et al. 2008).

Borehole depths are varied in the Crystalline Basement Province aquifers. This province is characterized by highly fractured sandstones, phyllites, greywackes, greenstones and schists. Many of them are criss-crossed by fractured quartz-veins and prolific aquifers have been encountered at an average depth of 25 m. The productive wells are completed in the moderately to poorly decomposed zones. The depth of weathering is dependent on the nature of the granitoids and the fissility in the rock. For instance, very productive and high-yielding aquifers are encountered in coarse-grained granites especially those traversed by fractured quartz-veins and depths of up to 60 m are common. It must be mentioned that the depths mentioned in this paper do not necessarily indicate the true depths of the aquifers. In Ghana, most boreholes were drilled for rural water supply and these boreholes are usually fitted with hand-pumps and their depths are dictated by their intended purposes. Most groundwater projects terminate drilling when sufficient water is obtained for rural water delivery. Depths of boreholes drilled through rocks of the Birimian and Tarkwaian Systems range between 35 m and 62 m with an average of 42 m. Borehole depths in areas underlain by the granitoids are similar and range between 35 m and 55 m, with an average of 50 m (Carrier et al. 2008). In some areas the regolith is tapped at relatively shallow depths with relatively shallow hand dug wells.

Aquifer transmissivity of the productive zones of the Birimian and Tarkwaian Systems ranges between 0.2 m^2/d and 119 m^2/d, with an average of 7.4 m^2/d. In these aquifers, storativity ranges between 0.003 and 0.008. Transmissivity within the regolith is slightly higher than that observed in the integrated aquifer system, and ranges between 4 m^2/d and 40 m^2/d with an average of 10 m^2/d. For the integrated aquifer systems in the Birimian and Tarkwaian Systems, borehole yields are generally low and range from 0.48 m^3/h to 36.4 m^3/h with a mean yield of 7.6 m^3/h. Differences in the degree of weathering within the granitoids probably account for the lower yields observed in these rocks.

In the Voltaian Province, groundwater occurs in fracture zones and bedding planes since the primary porosity of some of these rocks was reduced due to consolidation and cementation. The hydrogeological properties of the larger Voltaian Province are varied and based on the degree of weathering and fracturing of the otherwise impervious rocks. Available data suggest that borehole yield within the Lower Voltaian sub-province ranges between 1 and 9 m^3/h with an average of 6.2 m^3/h. On the basis of available data from drilling projects in this sub-province, success rate for drilling prolific wells and boreholes is about 55%.

Success rate for drilling successful boreholes in the Middle Voltaian sub-province is about 56% and the average borehole yield is 6.2 m^3/h. Borehole discharge characteristics at some locations within this sub-province indicate artesian conditions at shallow depths (Yidana et al. 2008a). In the Upper Voltaian sub-province, borehole yields in this range between 0.41 and 9.0 m^3/h and average about 8.5 m^3/h. Success rate for drilling successful boreholes in this is 56%.

The average thickness of the regolith over the Voltaian sedimentary rocks is 9 m. However, previous investigations in the Nanumba and West Gonja districts of the Northern region revealed average regolith thickness of 6 m and 11 m respectively. The relatively thin regolith can be explained by the relatively stable clay (shale) and quartz (sandstone) composition or by the fine texture or ductile nature of the sedimentary rocks found in the Voltaian Province (Carrier et al. 2008). Deeper seated weathering may occur in areas underlain by arkose or arkosic sandstones such as the Middle Voltaian and are rich in K-feldspar and low temperature plagioclase feldspars. These minerals weather faster than quartz and clay minerals. Underlying fracture systems are commonly developed at depths greater than 20 m below ground surface but on the average, required yields for rural water supplies are obtained less than 100 m depth.

Borehole data suggest that wells and boreholes in the Voltaian Province are slightly deeper than those in the Birimian. Borehole depths tabulated by Carrier et al. (2008) range between 45 m and 75 m with a mean depth of 55 m. Fractured rock aquifers in the Voltaian aquifers generally have a low to moderate productivity and overall transmissivity will range from 0.3 m^2/d to 267 m^2/d with a mean yield of 11.9 m^2/d.

Rocks of the Pan African Province comprise the Buem and Togo Formations. The rocks are largely impervious but contain openings along joint, bedding, and cleavage planes. Where these openings are extensive, good supplies of groundwater can be developed from boreholes. Springs frequently occur along the flanks of hills where quartzites are in contact with argillaceous rocks of the valleys (Dapaah-Siakwan and Gyau-Boakye 2000). The reported yields of boreholes in the Buem Formation range between 9 and 1260 l/min with an average and median of 100 and 78 l/min respectively. The average depth of boreholes in the area is 60 m. On the basis of available well test data, aquifer transmissivity values range between 0.2

and 11.4 m^2/day. Specific capacities range from 0.04 m^3/h/m to 1.23 m^3/h/m with an average of 0.47 m^3/h/m. Dapaah-Siakwan and Gyau-Boakye (2000) reported an average yield in the Togo Series aquifers to be about 9.2 m^3/h, ranging between 0.72 m^3/h and 24.3 m^3/h. The highest yielding wells and boreholes in the Togo Series tap fracture zones. Rocks of the Pan African Province are among the most prolific aquifers in the country, and can be relied upon to deliver economic quantities of groundwater for various purposes.

The Coastal Sedimentary Province includes the Tano and Keta basin aquifers. The Tano basin comprises Cretaceous to lower Tertiary sedimentary rocks which include a thick section of alternating sand and clay with occasional thin beds of gravel and fossiliferous limestone. The limestones have been reported to have oil and gas potential. Seepages of gas and oil have been reported in boreholes in the area. In the Cretaceous to Lower Tertiary sediments, groundwater is obtained largely from the upper 91.4 m of the section. The average yield of the boreholes in this unit is about 12.6 m^3/h (Dapaah-Siakwan and Gyau-Boakye, 2000). Records of boreholes in the limestone aquifer in the Keta area indicate that the average yield of wells is 13 m^3/h. Boreholes tapping the limestone aquifer in the Keta and surrounding areas have been reported to have a yield of 21.3 m^3/h on the average and range from 4.5 m^3/h to 54 m^3/h.

4. Hydrochemistry of Groundwater in Ghana

Groundwater samples from the major hydrogeological terrains were collected during different sampling events to facilitate the characterization of the general controls on groundwater hydrochemistry in Ghana. Multivariate statistical methods were combined with conventional graphical methods to determine the different water types associated with each of the hydrogeological terrains in the country. In addition a classification scheme based on the suitability of water for irrigation purposes was applied to the hydrochemical data. Factor and cluster analyses were used in determining the principal causes of the hydrochemistry and the spatial groundwater types in Ghana. Multivariate statistical methods have been widely used to facilitate the solution of environmental problems and suggest clues for the understanding of some natural processes. Classical examples are contained in Güler et al. (2002), Farnham et al. (2003), Cloutier et al. (2008), Yidana et al. (2008a-d), Yidana and Yidana (2009). In all these applications, diverse solutions were reached through the application of factor and cluster analyses. For instance Yidana et al. (2008a-d) combined cluster and factor analysis with inverse geochemical modeling to suggest that the hydrochemistry of groundwater in the southern Voltaian aquifers is dominated by silicate mineral weathering and cation exchange. Later, Yidana and Yidana (2009) used factor scores with ArcGIS to determine the spatial distribution of the major sources of variation in the hydrochemistry of groundwater in the Afram Plains portion of the southern Voltaian. Cloutier et al. (2008) used principal components and Q-mode HCA to distinguish groundwater types in confined and unconfined aquifers in Quebec. Through the combined use of Q-mode HCA and factor scores from principal components analyses, they were able to distinguish seven different water types in the sedimentary aquifers of the area. Earlier studies facilitated the identification of groundwater recharge areas from discharge areas in basins (e.g. Farnham et al. 2000; Stetzenbach et al. 2001). It is in the light of these successful applications of the multivariate statistical methods that it was considered apposite to employ these techniques in the hydrochemical characterization of the major hydrogeological provinces in the country. When used successfully with graphical techniques, the multivariate techniques are a robust hydrochemical classification tool.

4.1. Sampling and Analyses of Groundwater

A total of 383 groundwater samples were taken for this study. The samples were collected in 250 mL sterilized polythene bottles. For all samples collected, parameters such as electrical conductivity (EC), temperature (T) and pH were measured in the field. On-site testing was necessary for these parameters since they are likely to change during transport. These variables were measured using a WTW Universal Conductivity Meter MultiLine P4 set and probes. The pH-meter was calibrated against a standard solution of 3M KCl each morning and also during fieldwork if unusual readings were observed. The groundwater samples were analyzed at the Ecological Laboratory (Ecolab) of the Department of Geography and Resource Development at the University of Ghana, Legon. The parameters analyzed include major ions: Ca^{2+}, Mg^{2+}, K^+, Na^+, Cl^-, SO_4^{2-} and minor ions such as F^-, NO_3^-, and PO_4^{3-} in milligram per liter (mg/L) using Dionex DX 120 ion chromatograph (IC). The bicarbonate ion concentration in the water was determined by titration. The stable isotopes (^2H and ^{18}O) analyses were carried out at the Geological Institute of the University of Copenhagen using the Delta E class mass spectrometer. A 1-10 mg portion of the water was reduced with zinc in a zinc-reaction furnace at 400 - 415°C to prepare hydrogen for isotopic analysis based on Florkowski (1985). Measurement of $\delta\,^{18}O$ was based on the Dugan et al. (1985) technique in which guanidine hydrochloric acid was used for the preparation of the samples for ^{18}O analysis. The isotopic compositions of water are reported as the deviation of D/H or $^{18}O/^{16}O$ ratio from that of Vienna Standard Mean Ocean Water (VSMOW) in parts per thousand (‰).

The resulting data for all the parameters were log-normal and had to be log-transformed to meet the requirements of optimal multivariate statistical analyses (Güler et al. 2002. For the Q-mode hierarchical cluster analysis, HCA, the data for each parameter were standardized to a range of –1 to 1. Data standardization is required in HCA due to the fact that parameters with highest natural variances tend to have greater influences on Euclidean distances. This will result in erroneous results and faulty interpretations. Phosphate was not included in the R-mode factor analysis and Q-mode HCA. This is because concentrations of PO_4^{3-} were either below the detection limit or were not measured at most of the locations.

4.2. General Hydrochemistry

Table 1 presents the statistical summary of all the parameters analyzed for this study. The mean and median concentrations of the major ions in the aquifers of the Voltaian Province are within the WHO (1984) guidelines for drinking water. The maximum NO_3^- and F^- concentrations of 120 mg/l and 3.04 mg/l respectively are, however higher than their respective WHO (1984) standards of 50 mg/l and 1.5 mg/l. These are isolated cases suspected contamination from surface point sources such as domestic sewage. These sources of contamination can be checked through monitoring and supervision. The mean and median EC values of 638 μS/cm and 549 μS/cm, which correspond to total dissolved solids concentrations of 382 mg/l and 329 mg/l respectively are below the WHO (1984) guideline value of 1000 mg/l. Maximum concentrations of some of the major ions such as Na^+ and Cl^- are higher than the WHO (1984) recommended values for taste and aesthetic purposes. However, the low mean and median values of the concentrations of these ions imply that most of the locations sampled have concentrations lower than the recommended maximum values for domestic usage. The pH is slightly acidic but is well within the acceptable WHO (1984) range. In the Buem Formation, the ionic concentrations follow a similar trend as that observed in the Voltaian. Except for Na^+ and EC, whose maximum values of 249 mg/l and 1946 μS/cm (equivalent to TDS of 1167 mg/l) are slightly higher than the WHO (1984) recommended values all other parameters have concentrations lower than the standard guideline limits. The aquifers of the Buem formation, which are mostly sedimentary and meta-sedimentary aquifers, therefore produce groundwater of acceptable quality for most uses.

In the aquifers of theTogo Series, concentrations of the major chemical parameters are lower than those from both the Voltaian and Buem Formations. Values of the major chemical parameters are far below the recommended WHO (1984) guideline concentrations. For instance the maximum concentrations of NO_3^- and F^-, which have definitive maximum limits in the WHO (1984) report, are as low as 27 mg/l and 0.35 mg/l respectively. The very low concentrations of F^- (lower than 0.5 mg/l) might be too low for drinking water. However, it is believed that the required F^- will be acquired from diets. The pH is lower (more acidic) than the recommended WHO (1984) range of 6.5 – 8.5. It has been noted in the WHO (1984) guidelines that pH usually does not have a direct bearing on consumers of water. It might, however affect the aesthetic value or lead to the dissolution of heavy metals in the rocks. As yet, there have been no concerns about heavy metal contamination of aquifers in this area. However, in future studies, attention will be focused on heavy metal concentrations in groundwater from this formation.

Table 1. Statistical summary of the hydrochemistry of the terrains sampled

	Parameter	Mean	Median	St_dev	Minimum	Maximum
Voltaian	EC (μ/cm)	638.38	549.00	379.60	41.30	2650.00
	pH	6.87	6.78	0.58	5.60	9.49
	HCO_3^-	295.10	278.00	129.63	29.30	800.00
	SO_4^{2-}	13.62	4.30	29.33	0.73	191.00
	Cl^-	51.23	27.00	72.11	1.21	494.02
	NO_3^-	2.26	0.20	10.15	0.00	120.00
	PO_4^{3-}	0.13	0.07	0.13	0.00	0.64
	F-	0.47	0.40	0.36	0.01	3.04
	Ca^{2+}	44.01	39.30	25.01	3.45	196.00
	Mg^{2+}	17.35	15.50	11.00	0.28	54.90
	Na^+	62.07	42.62	67.14	4.90	489.00
	K^+	1.89	1.40	1.47	0.30	8.90

Table 1. (Continued)

Togo	EC	300.29	291.50	148.48	28.10	510.00
	pH	5.32	5.39	0.82	3.27	6.48
	HCO_3^-	104.29	86.62	81.85	3.66	252.00
	SO_4^{2-}	2.30	0.91	2.72	0.38	9.22
	Cl^-	32.09	29.24	31.50	0.56	109.85
	NO_3^-	5.19	0.19	8.69	0.00	27.25
	PO_4^{3-}	0.04	0.00	0.12	0.00	0.44
	F^-	0.15	0.16	0.09	0.03	0.35
	Ca^{2+}	18.48	18.71	15.09	1.14	61.14
	Mg^{2+}	8.69	9.30	6.21	0.94	20.38
	Na^+	20.28	21.51	9.61	3.14	33.65
	K^+	2.70	1.90	2.53	0.34	10.83
Keta	EC	222.29	205.00	135.17	55.00	484.00
	pH	4.13	3.96	0.93	2.96	6.32
	HCO_3^-	76.45	61.00	43.56	32.94	184.22
	SO_4^{2-}	3.92	0.46	7.22	0.09	21.22
	Cl^-	11.82	3.98	13.59	2.17	58.52
	NO_3^-	8.71	2.24	18.09	0.00	81.85
	PO_4^{3-}	0.06	0.00	0.23	0.00	1.05
	F^-	0.07	0.06	0.06	0.01	0.27
	Ca^{2+}	10.61	8.26	7.73	2.13	28.82
	Mg^{2+}	4.59	4.50	3.25	0.63	11.47
	Na^+	11.11	8.22	6.19	3.15	27.99
	K^+	1.19	0.93	0.89	0.09	4.52
Lower Birimian	EC	2104.78	963.50	2944.25	228.00	16000.00
	pH	7.33	7.31	0.39	6.35	8.48
	HCO_3^-	350.13	263.52	221.08	104.92	1148.02
	SO_4^{2-}	132.53	63.08	212.86	0.05	1450.22
	Cl^-	490.59	104.89	960.16	16.14	5448.40
	NO_3^-	39.01	5.94	79.45	0.04	449.42
	PO_4^{3-}	1.33	0.54	3.38	0.00	20.44
	F^-	2.25	0.34	3.67	0.01	19.84
	Ca^{2+}	114.87	72.21	157.03	19.14	1093.70
	Mg^{2+}	38.03	12.58	61.64	1.78	359.55
	Na^+	326.42	87.16	603.38	10.23	3643.15
	K^+	49.22	25.75	72.68	1.07	565.27

Table 1. (Continued)

	Parameter	Mean	Median	Stdev	minimum	Maximum
Upper Birimian	EC	139.75	129.00	54.32	88.00	213.00
	pH	3.70	3.71	0.22	3.44	3.92
	HCO_3^-	79.30	68.93	31.83	53.68	125.66
	SO_4^{2-}	3.52	2.11	3.97	0.51	9.34
	Cl^-	8.93	7.77	5.81	3.87	16.30
	NO_3^-	6.13	6.63	4.97	0.67	10.59
	PO_4^{3-}	0.00	0.00	0.00	0.00	0.00
	F^-	0.09	0.04	0.12	0.02	0.27
	Ca^{2+}	12.49	9.80	9.66	4.11	26.23
	Mg^{2+}	2.57	2.29	1.88	0.59	5.11
	Na^+	10.20	9.69	4.53	6.26	15.16
	K^+	1.28	1.26	0.28	1.04	1.58
Cape Coast granite	EC	773.56	588.50	464.46	169.00	1999.00
	pH	6.39	6.41	0.90	4.16	8.78
	HCO_3^-	161.85	136.03	120.87	34.16	523.38
	SO_4^{2-}	60.30	25.88	115.76	3.18	508.30
	Cl^-	70.38	64.45	43.38	9.28	144.56
	NO_3^-	32.08	16.06	35.36	0.00	111.79
	PO_4^{3-}	0.13	0.00	0.25	0.00	0.76
	F^-	3.91	0.69	5.98	0.11	18.46
	Ca^{2+}	42.94	29.70	50.85	12.56	234.20
	Mg^{2+}	15.83	13.51	9.16	3.60	38.73
	Na^+	65.42	56.10	53.63	11.75	251.62
	K^+	5.86	2.83	10.89	1.39	49.00
Surface water	EC	114.23	114.40	37.02	78.00	150.10
	pH	6.88	6.82	0.18	6.74	7.12
	HCO_3^-	111.50	95.00	39.44	86.00	170.00
	SO_4^{2-}	1.10	1.03	0.47	0.69	1.67
	Cl^-	3.86	3.88	2.63	1.52	6.15
	NO_3^-	0.00	0.00	0.00	0.00	0.00
	PO_4^{3-}	0.02	0.00	0.03	0.00	0.06
	F^-	0.39	0.30	0.34	0.12	0.87
	Ca^{2+}	4.06	4.38	1.43	2.25	5.24
	Mg^{2+}	3.11	2.82	1.10	2.13	4.68
	Na^+	4.58	4.82	1.40	2.93	5.78
	K^+	2.19	2.30	0.88	1.14	3.00

In the Upper and Lower Birimian aquifers mean and median concentrations of the chemical parameters are within the acceptable levels for most domestic uses. The low pH might render some of the waters aesthetically unsuitable for drinking purposes. It has been noted that some of the rocks in the Upper and Lower Birimian contain pyrite. The low pH might therefore be related to the oxidation of pyrite. This process has been of considerable interest in the mining areas of Ghana where acid mine drainage, influenced by low acidity resulting from pyrite oxidation, is said to facilitate the leaching of heavy metals into surface and groundwater sources. No such concerns have so far been observed in the study area. The maximum NO_3^- concentration of 81 mg/l observed in the aquifers of the Lower Birimian is an isolated case and probably resulted from contamination from local sewage and/or agricultural activities. The acid neutralizing capacity (ANC) of the groundwater samples collected from the Upper and Lower Birimian aquifers range between 0.26 meq/l and 2.09 meq/l, indicating that in spite of the low pH levels, the waters still have the capacity to neutralize acids in the area. A similar observation was made by Kortatsi et al. (2008) in the lower Offin basin, also underlain by rocks of the Birimian System. The Cape Coast granitoids, which are intrusives into the Birimian rocks, have comparatively higher values of the chemical parameters. In spite of this observation, the concentrations of the major chemical players are within acceptable limits for consumption. The high mean F^- concentration of 3.91 mg/l contrasts with the median of 0.69 mg/l and is probably due to the isolated extremely high fluoride concentrations observed in the study area. High fluoride in groundwater in the area has two major sources: rock-water interactions and contamination from domestic wastes. In the Keta basin, salinity is very high and might be associated with the combined effects of seawater intrusion and rock-water interactions in the area. Although the mean F^- of 2.25 mg/l is higher than the 1.5 mg/l recommended by the WHO (1984), the median of 0.34 mg/l suggests that most of the locations have concentrations lower than the recommended levels. Mean and median NO_3^- in the Keta basin are lower than the WHO (1984) values. There are, however, isolated cases of extremely high values, which result predominantly from contamination from domestic sewage and agricultural chemicals in the area. This is particularly of concern in the Keta basin where the aquifers are shallow and the sanitary conditions are unsatisfactory. The surface water samples have the concentrations of all the major chemical players lower than the maximum permissible concentrations. Nitrate concentrations were not analyzed from the samples of surface waters collected for this study.

The results of the factor analysis (with principal components and varimax rotation) are presented in Table 2. Two principal components resulted from the application of the Kaiser criterion (Kaiser 1960), which requires that principal components to be retained in the factor analysis should have a minimum eigenvalue sum of 1.0. Component 1 is highly positively correlated with EC, pH, HCO_3^-, Ca^{2+}, Mg^{2+}, Na^+ and F^- and probably represents salinity enhancements resulting from the weathering of silicate minerals in the area. The positive correlation of EC (as a measure of salinity) implies that this component represents the major processes controlling the salinity of groundwater among all the hydrogeological aquifers sampled in the southeastern part of the country. A plot of $(Mg^{2+} + Ca^{2+})$-$(SO_4^{2-} + HCO_3^-)$ in meq/l against $(Na^+ + K^+)$-Cl^- in meq/l (Figure 3) suggests that cation exchange activity is an important factor in the hydrochemistry of groundwater from all the major terrains, with the exception of the Keta basin, where the samples scatter around the area and do not fall close to the line with a slope of approximately -1. The samples cluster around a line of slope -0.8, which can be approximated to equal -1. The negative slope implies a negative correlation of the parameters on the two coordinate axes, which suggests that when the contributions of carbonate and sulfate minerals are subtracted from the total concentrations of calcium and magnesium, the result is a function of the cation exchange activity between sodium and potassium in the groundwater system. It is also noteworthy that most of the samples plot within the fourth quadrant (positive ordinate and negative abscissa). This implies that Ca^{2+} and Mg^{2+} are enriched relative to Na^+ and K^+ in most of the aquifers. Thus the cation exchange sites preferentially adsorb Na^+ and K^+, and discharge Ca^{2+} and Mg^{2+}. Previous studies by Kortatsi et al. (2008) and Yidana (2008) in the Birimian and Voltaian aquifers respectively suggest that cation exchange and silicate mineral (predominantly feldspar) weathering are the major factors controlling the hydrochemistry of groundwater from aquifers in these major geological terrains in the country. Yidana et al. (2008a) pointed out that as a result of these weathering processes in the aquifers, groundwater in the aquifers of the southern Voltaian System are stable within the montmorillonite field. Drever (1988) suggests that stability within the montmorillonite field indicates restricted flow conditions in which sustained rock-water interactions resulting from long residence times of groundwaters, lead to enough silica enrichment to form montmorillonite. This agrees with the observed flow characteristics there since primary permeability is destroyed and groundwater flow is restricted to fractured and weathered zones. The strong correlation of the HCO_3^- with Mg^{2+} and Ca^{2+} also suggests that carbonate mineral weathering might be playing a major role in the hydrochemistry. This is in fact true of some of the aquifers of the Voltaian sedimentary rocks, which are limestones and conglomerates with some quantities of calcite and dolomite. Yidana and Yidana (2009) indicated that the predominant carbonate mineral whose weathering plays an active role in the hydrochemistry is dolomite, facilitated by carbonic acid from dissolved carbon dioxide in infiltrating rainfall and plant root respiration. Incongruent silicate mineral weathering is also facilitated by the same mechanism of carbonic acid resulting from plant root respiration and dissolved atmospheric carbon dioxide in infiltrating rainfall. Kortatsi et al. (2008) and Anku et al. (2008) suggest similar mechanisms to account for the major causes of ionic enrichment in some aquifers of the Upper and Lower Birimian aquifers elsewhere in the country. In the Keta basin, whereas previous studies (e.g. Helsrup et al. 2007; Jorgensen and Banoeng-Yakubo 2001; Banoeng-Yakubo et al. 2009) suggest that rock-water interactions have a major role in the hydrochemistry of groundwater from some of the aquifers there, no mention has been made of the role of cation exchange. It is obvious that whereas ion exchange cannot be ruled out as a mechanism in the hydrochemistry (Figure 3), it may not play an active role among aquifers of the Keta basin. Appelo and Postma (2005) indicate that in coastal aquifers, freshwater aquifers are dominated by Ca-HCO_3 water types resulting from the weathering of carbonate minerals. In the Keta basin, however, the dominant water types, based on visual inspection of the hydrochemical data are the Na-Cl, and Ca-Cl, although there are Ca-HCO_3 and Na-HCO_3 water types farther inland. Na-HCO_3 water types usually result from the freshening of groundwaters by cation exchange activity in some coastal aquifers (Appelo and Postma 2005). The second component in the factor analysis (Table 2) correlates positively with EC, SO_4^{2-}, Cl^-, NO_3^-, Na^+, and K^+. This is a mixed component, which represents seawater intrusion and anthropogenic pollution. Component 2 therefore illustrates salinity enrichment from seawater and anthropogenic activities. Predominant anthropogenic activities that have been noted as potential sources of pollution in the physical tours of some of the study locations are the improper disposal of domestic wastes, especially among some communities in the Keta basin and Voltaian basin, and the use of chemical fertilizers and organic manures in farms. The aquifers of the Keta basin and the southern

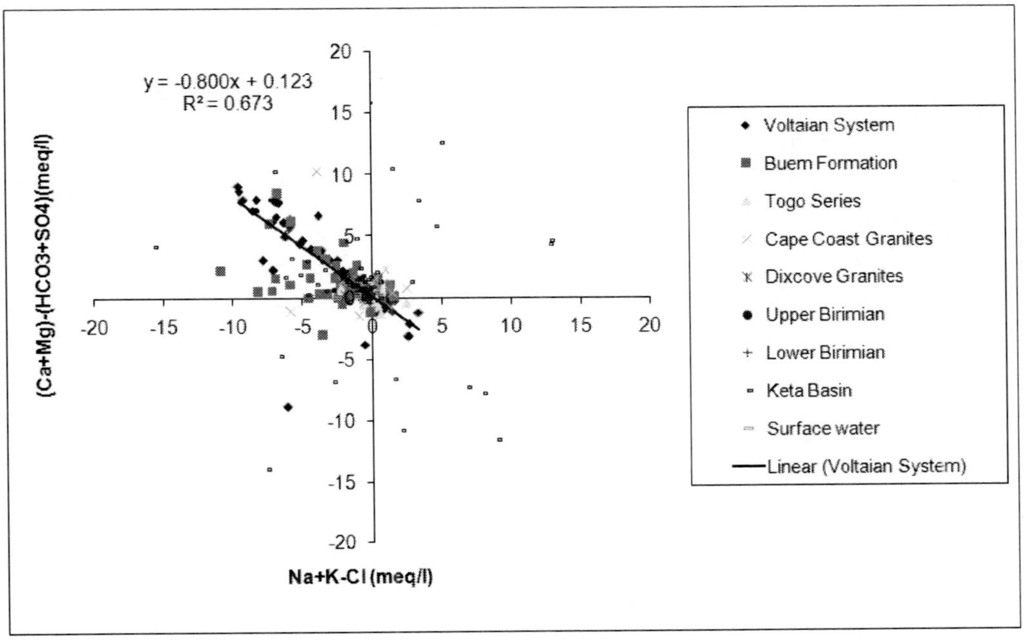

Figure 3. The Linear plot of Ca+Mg-SO_4-HCO_3 against Na+K-Cl.

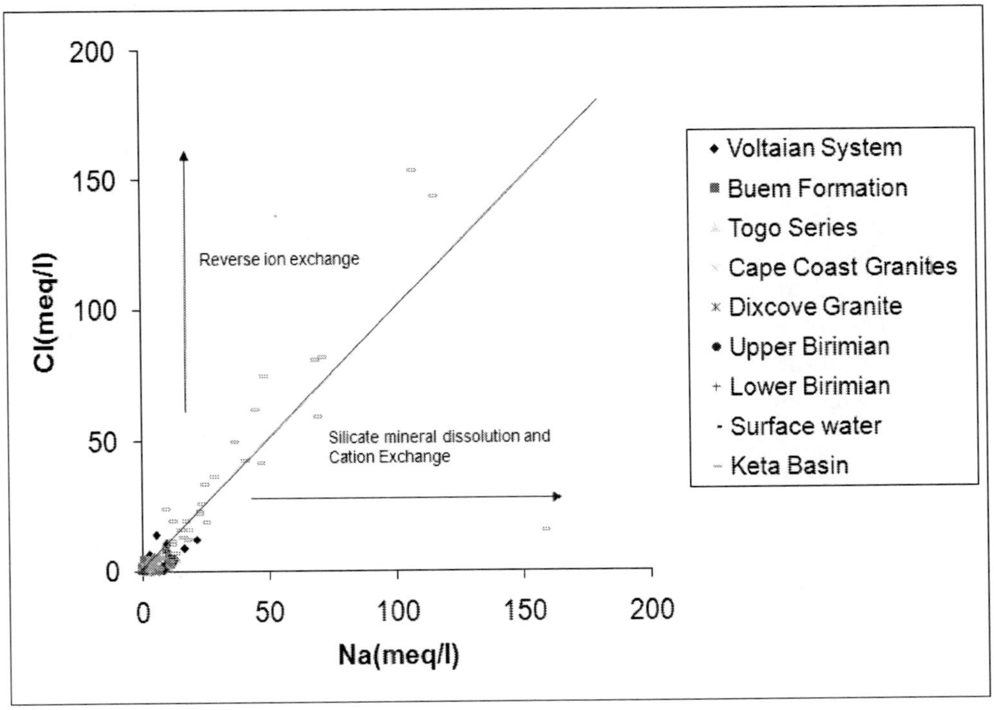

Figure 4. A scatter plot of Na against Cl.

Table 2. Principal component loadings from the factor analysis

Parameter	Component 1	Component 2
LogEC	0.80	0.50
LogpH	0.82	0.00
LogHCO_3^-	0.91	-0.04
LogSO_4^{2-}	0.50	0.70
LogCl^-	0.40	0.80
LogNO_3^-	-0.21	0.85
LogF^-	0.48	0.35
LogCa^{2+}	0.69	0.47
LogMg^{2+}	0.68	0.38
LogNa^+	0.75	0.52
LogK^+	0.29	0.76

Voltaian are particularly vulnerable to these activities since these aquifers have been noted to be shallow. The positive correlation of Na^+ and Cl^- with component 2 suggests seawater intrusion. Although Na^+ correlates with component 2, it has a higher loading with component 1, suggesting that the processes represented by component 1 have a greater effect on the concentration of Na^+ than do those represented by component 2. A plot of Na^+ (in meq/l) in the abscissa against Cl^- (in meq/l) in the ordinate (Figure 4) shows samples from the Keta basin cluster around the 1:1 line, with some falling on the Cl^- side. This suggests the effects of seawater intrusion in these areas. Although some samples from the other geological terrains also fall on the Cl^- side of the diagram, this plot is likely showing the predominant effects of reverse cation exchange activity, rather than seawater intrusion since these areas are rather farther inland and seawater effects have not been noted. Some of the Keta basin groundwater samples also plot on the Na (meq/l) side of the diagram. These samples account for the contribution of silicate mineral weathering.

Principal component scores were calculated from the component loadings and the log-transformed concentrations of the parameters used for this study. Figure 5 illustrates a plot of the component scores for component 1 against those of component 2. As stated above, component 1 is labeled as “salinity increment from mineral weathering”, whilst component 2 is described as “salinity from seawater intrusion and anthropogenic activities”. It is obvious from Figure 5 that the samples are clustered according to the geological/hydrogeological formation. It is obvious that the samples of the Buem and Birimian (Upper and Lower) lean more closely towards component 1 than component 2, suggesting that groundwater from these aquifers are predominantly affected by the effects of mineral weathering. Whereas the effects of component 1 are obviously higher among the samples of the Voltaian, Togo and the granitoids, there is also a significant influence of component 2. This is obviously due to the effects of anthropogenic activities such as chemical fertilizers and domestic sewage. Samples from the Keta basin are very high on the salinity side and have very significant contributions from both components.

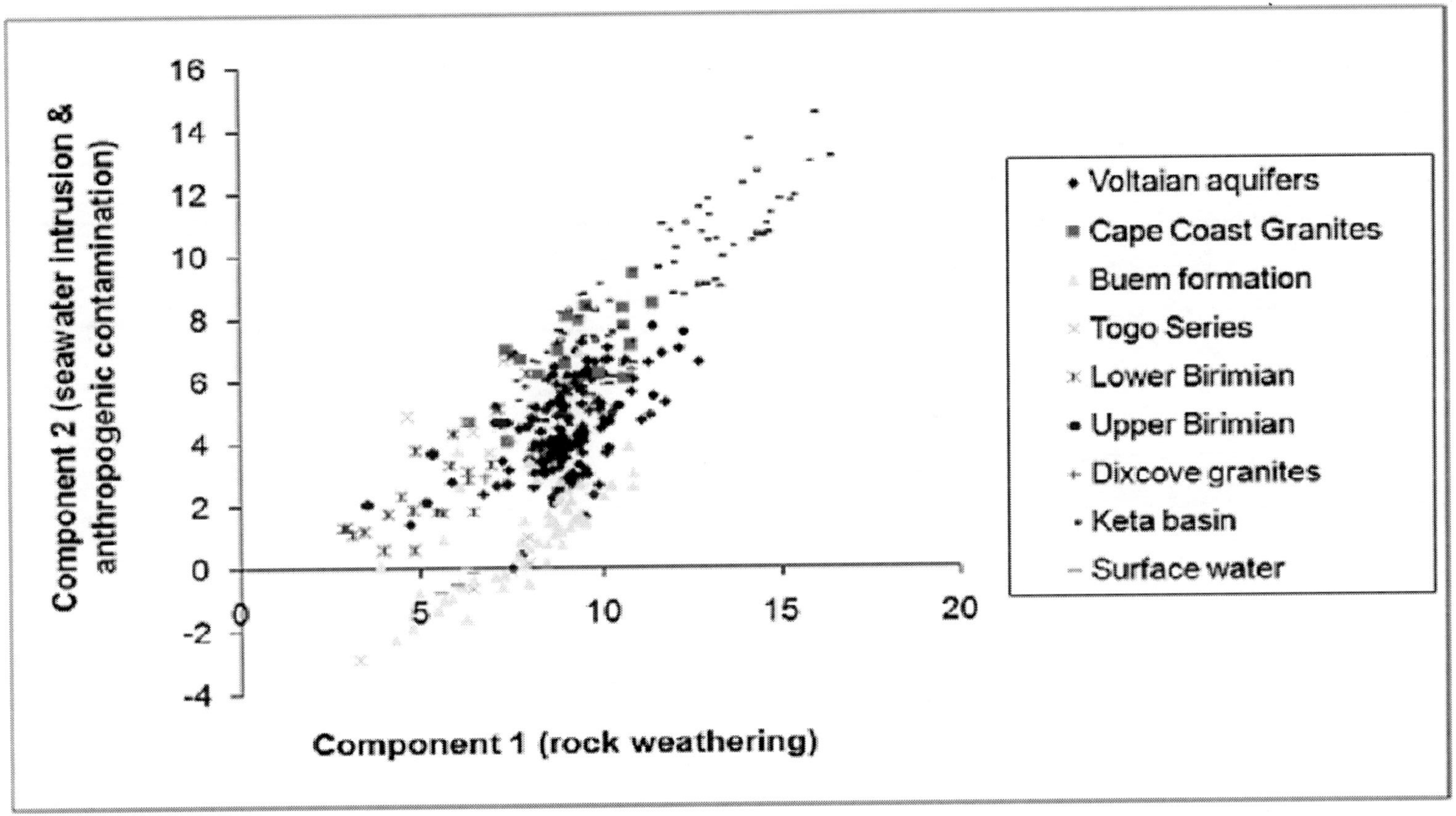

Figure 5. Principal component scores against Principal component 2 scores.

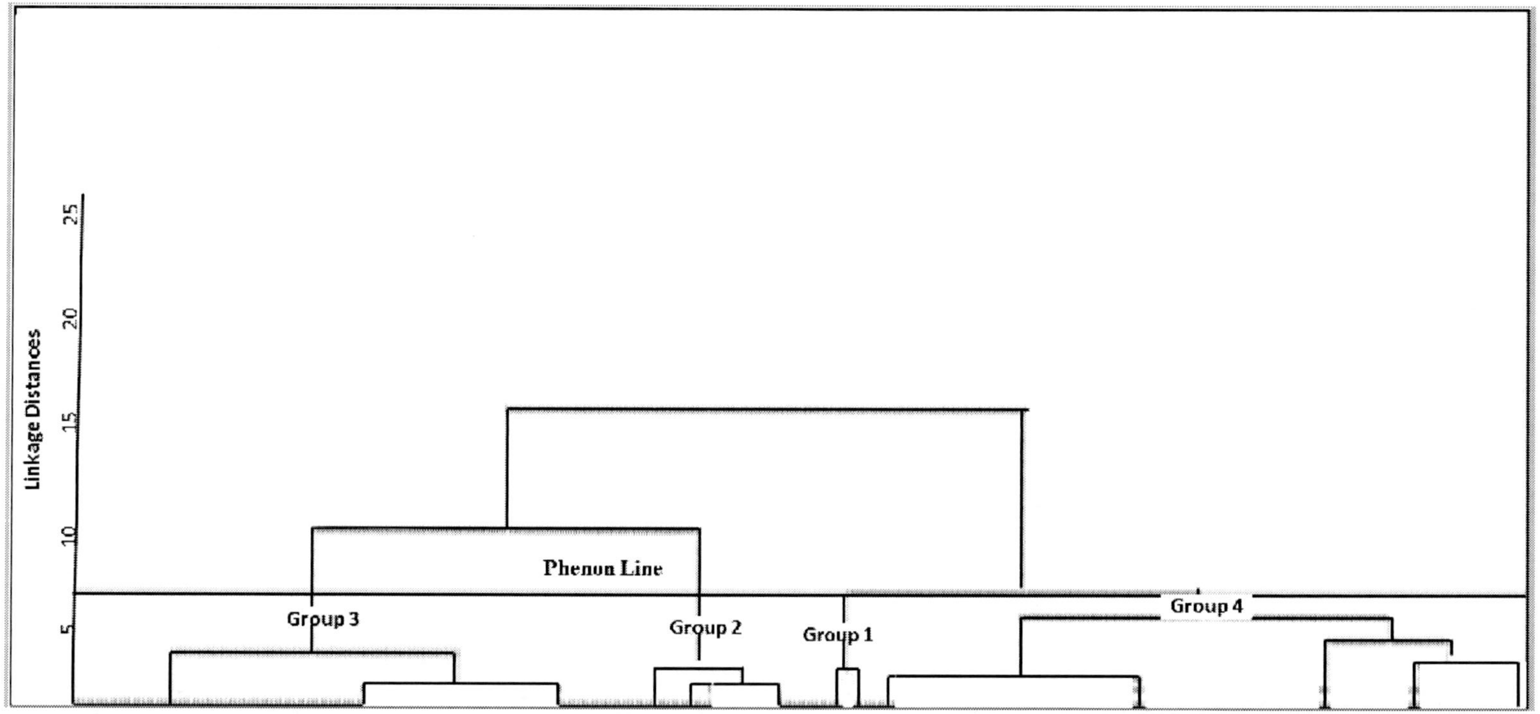

Figure 6. The dendrogram from the Q-mode HCA on the hydrochemical dataset.

This is as expected since there have been concerns of significant marine influence on the quality of groundwater from the shallow aquifers of the Keta basin. In addition to the seawater effects, Yidana et al. (2010) raised concerns about the vulnerability of groundwater in the area to contamination from domestic sewage. Irrigation is regularly practiced in the area, and farmers use animal manure and chemical fertilizers significantly to enhance crop yield. Enhanced nitrate concentrations at some locations in the study area therefore result from the leaching of some nitrate from agricultural activities. Figure 4 shows a general degradation of the quality of groundwater towards the coast, if salinity as measured by the EC is a yardstick to measure water quality. The degradation is a combined effect of seawater intrusion, anthropogenic activities and rock-water interactions, although seawater intrusion is the dominant source of salinization in the coastal areas. This is expected to worsen as groundwater tables fall as high demands lead to heavier groundwater extraction. The implications of this are varied. First, the livelihood of the people in the Keta basin largely hinges on their ability to tap groundwater at shallow depths for irrigation purposes. The use of high salinity waters for irrigation has been determined to affect the quality of soils since it affects their water infiltration qualities. In the second place high salinity waters are not recommended for domestic uses and might cause health problems when used for drinking.

The HCA is one of the most widely applied data clustering methods in Earth Science (Davis 1986; Cloutier et al. 2008) and is often used in the classification of hydrogeochemical datasets (Steinhorst and Williams 1985; Schot and van der Wal 1992; Güler et al. 2002). In this study the Euclidean distance was used as a similarity/dissimilarity measure, whereas the Ward's linkage method was used to link clusters. Many methods exist by which clusters can be created and linked but a combination of the Euclidean distance as a similarity measure, and the Ward's method as a linkage method has been observed as the best combination that results in optimal results in HCA (Güler et al. 2002; Cloutier et al. 2008). The classification of samples in the Q-mode HCA is based on the visual inspection of the resulting dendrogram and in most cases the division of clusters is semi-objective or subjective. However in the interpretation of the results, researchers are guided by their best understanding of the hydrogeology and other factors in the area of study. In this study, the phenon line was drawn across a scaled linkage distance of 10. This means that those clusters and samples under this line are recognized as being under the same cluster. No specific rules govern the choice of the linkage distance through which the phenon line should be drawn and it is influenced by one's best judgment based on a visual inspection of the dendrogram. Four clusters were identified under the phenon line. Mean and median concentrations of the chemical parameters of the samples under each cluster were calculated and the discussions will be based on these values. Figure 6 is the dendrogram showing the four clusters. Group 1 is made predominantly of samples from the Voltaian and Buem aquifers, with few samples from the Cape Coast granitoids. These are largely freshwater aquifers characterized by median and mean EC values of 416 μS/cm and 423 μS/cm respectively (Table 3a). The high standard deviation implies that these EC values are an input from a variety of factors which include natural weathering processes and anthropogenic activities. Group 2 members have the lowest median and mean EC values of the four groups (Table 3b). This group also has the lowest median pH of 4.85 and draws membership from samples of the Birimian and some Togo Series aquifers. Groundwater from members of this group is largely suitable for most uses, except that the low pH might cause some aesthetic effects. Silicate mineral weathering is a buffer against groundwater acidity in the absence of carbonate minerals (Appelo and Postma 2005).

However, silicate mineral weathering is usually very slow and might not be sufficient to erase acidity resulting from contaminating sources. Intense pyrite oxidation is suspected among the Birimian aquifers. Silicate mineral weathering is apparently insufficient to buffer the resulting acidity. In the case of the intrusive granitoids (Cape Coast and Dixcove), the pH is close to neutral and suggests that silicate mineral weathering occurs at a more rapid rate. In addition to the relatively higher pH, the concentrations of the major cations are relatively higher among the granitoid aquifers than the metasediments, implying higher rates of silicate mineral weathering. In addition, where the aquifers contain heavy metals, the low pH can influence the leaching of these metals into the groundwater system, rendering it unsuitable for drinking at elevated concentrations. Group 3 has three members (Table 3c) drawn from the Keta basin alone. Groundwater from this group has intolerably high EC values and is not recommended for domestic use and irrigation. These waters have had significant marine influence and have salty tastes. The F^- and NO_3^- concentrations are also far in excess of the WHO (1984) guideline limits for domestic water. Group 4 (Table 3d) represents mixed water types. It draws members from all with significant input from the Keta basin and the Voltaian.

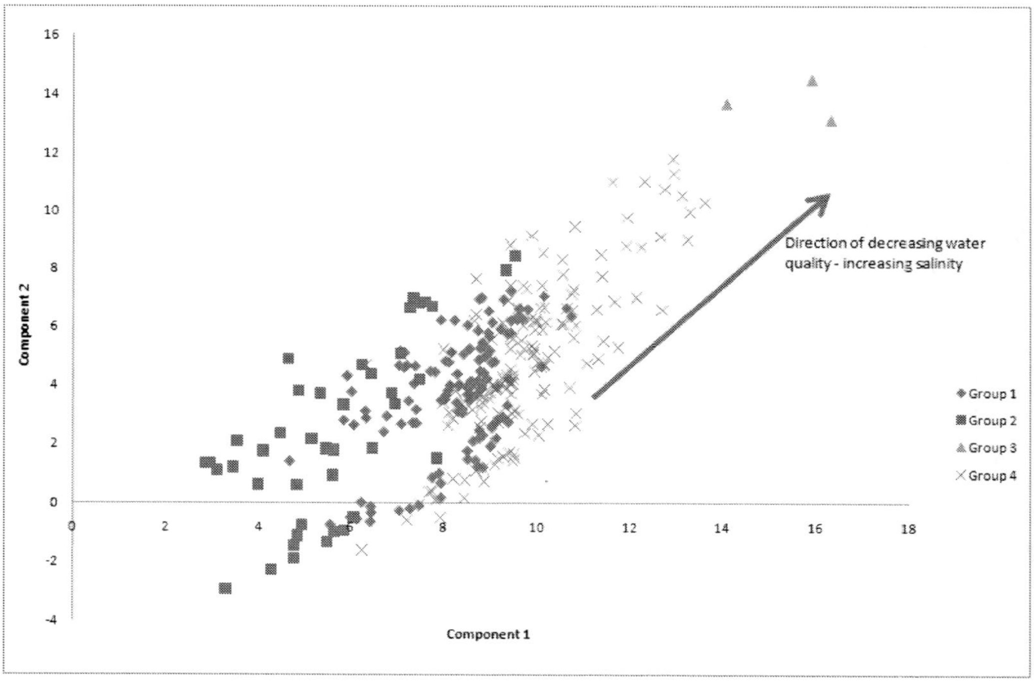

Figure 7. PC1 scores against PC2 scores for the four groundwater types distinguished from the HCA.

Although the median and mean EC values are high, they are within tolerable limits. Median and mean NO_3^- concentrations are also low. The observed maximum NO_3^- concentration of 447 mg/l is an outlier and probably results from a point source contamination. Component scores were calculated for members of all four groups. Figure 7 shows the plot of the scores of component 1 against component 2 for the four groups. Groups 1 and 2 members are mostly influenced by component 1. Most members of group 4 also have considerable influence from factors represented by component 1, whilst the group 3 members have significant contributions from both components. Groups 1 and 2 members serve the best quality for domestic uses are made predominantly of aquifers of the Voltaian, Buem, Togo and Birimian (including the intrusive granitoids) formations. Group 4 members have elevated salinity values but the quality of groundwater they deliver is acceptable for most uses. Groundwater from group 3 members may not be used for any purpose without prior treatment.

Table 3a. Statistical summary of chemical parameters in group 1

Group 1					
ID	**mean**	**standard deviation**	**minimum**	**maximum**	**Median**
EC	423.55	191.61	41.30	1395.00	416.00
pH	6.35	0.66	2.98	7.43	6.48
HCO_3^-	190.12	72.67	18.30	573.00	184.00
SO_4^{2-}	7.86	10.87	0.19	63.08	3.40
Cl^-	29.00	28.47	0.64	155.00	18.00
NO_3^-	2.32	5.37	0.00	34.66	0.30
PO_4^{3-}	0.18	0.38	0.00	3.87	0.07
F^-	0.42	0.27	0.009	1.55	0.40
Ca^{2+}	32.78	17.41	2.247	119.01	32.26
Mg^{2+}	13.81	7.67	1.50	39.30	12.60
Na^+	27.85	18.06	0.00	150.63	25.84
K^+	2.37	2.02	0.40	14.24	1.80

Table 3b. Statistical summary of chemical parameters in group 2

Group 2					
ID	**mean**	**standard deviation**	**minimum**	**maximum**	**Median**
EC	228.71	196.05	28.10	955.00	165.00
pH	4.72	0.87	3.17	6.59	4.85
HCO_3^-	70.33	35.80	3.66	184.22	66.00
SO_4^{2-}	4.62	10.81	0.00	61.04	0.64
Cl^-	22.98	32.91	0.37	144.56	4.73
NO_3^-	15.82	31.01	0.00	120.00	1.12
PO_4^{3-}	0.07	0.18	0.00	1.045	0.00
F^-	0.15	0.18	0.014	0.82	0.09
Ca^{2+}	10.73	7.91	1.14	28.10	8.65
Mg^{2+}	5.46	6.04	0.59	29.51	2.87
Na^+	16.04	16.28	2.65	83.45	8.71
K^+	1.84	1.83	0.09	10.83	1.25

Table 3c. Statistical summary of chemical parameters in group 3

Group 3					
ID	**mean**	**standard deviation**	**minimum**	**maximum**	**Median**
EC	11104.33	8246.82	1583.00	16000.00	15730.00
pH	7.50	0.16	7.35	7.67	7.47
HCO_3^-	692.96	299.83	351.36	912.56	814.96
SO_4^{2-}	1019.65	405.00	646.32	1450.22	962.40
Cl^-	3820.28	2521.68	915.57	5448.40	5096.86
NO_3^-	186.45	233.56	3.12	449.42	106.80
PO_4^{3-}	1.00	1.25	0.00	2.40	0.59
F^-	3.41	1.75	1.45	4.80	3.98
Ca^{2+}	406.83	360.04	157.30	819.57	243.62
Mg^{2+}	227.30	62.02	186.18	298.64	197.08
Na^+	1893.63	1173.50	543.16	2665.14	2472.60
K^+	180.79	49.10	133.78	231.74	176.84

Table 3d. Statistical summary of chemical parameters in group 4

Group 4					
ID	**mean**	**standard deviation**	**minimum**	**maximum**	**Median**
EC	954.01	624.25	169.00	3420.00	752.50
pH	7.17	0.52	6.28	9.49	7.10
HCO_3^-	378.48	130.19	61.00	800.00	356.00
SO_4^{2-}	36.41	65.93	0.00	508.30	6.70
Cl^-	91.57	150.64	0.15	847.81	41.00
NO_3^-	10.15	44.50	0.00	447.80	0.16
PO_4^{3-}	0.21	0.40	0.00	2.58	0.04
F^-	1.22	2.86	0.00	18.46	0.30
Ca^{2+}	60.31	51.30	3.28	435.18	52.10
Mg^{2+}	19.63	13.32	0.28	55.17	16.84
Na^+	99.46	96.63	1.22	517.64	61.40
K^+	9.21	20.25	0.13	104.66	1.80

The Chadha diagram (Chadha 1999) was used to define the major water types in the country. It is a modified version of the Piper (1944) diagram and plots the equivalent percentages of the (Ca^{2+} + Mg^{2+}) - (Na^+ + K^+) of the samples against their corresponding equivalent percentages in terms of (CO_3^{2-} + HCO_3^-) – (SO_4^{2-} + Cl^-). Figure 8 is the Chadha diagram for samples from this study. The predominant groundwater types in the southeastern aquifers are the Ca-Mg-HCO_3 types. These water types have three major sources: carbonate mineral weathering and the incongruent dissolution of high temperature silicate minerals and cation exchange activity. These waters are also typically hard waters and may leave scales in hot water kettles. About 20 % of the samples from the Voltaian aquifers are Na-K-HCO_3 water types, which usually result from the incongruent weathering of low temperature feldspars and cation exchange. These are "soft" waters, which have no tendency to leave scales in hot water kettles. Over 50 % of the Keta basin samples fall within the Na-K-Cl-SO_4 water types. This is as expected because of the significant marine influence. About 45 % of the samples from the Cape Coast granitoids plot close to the origin and represent mixed water types. Over 90 % each of the Togo, Buem and Birimian samples plot within the Ca-Mg-HCO_3 section. It is very likely that these are waters, which have evolved from Na-HCO_3 waters due to reverse cation exchange since the aquifers do not have significant carbonate minerals to generate predominantly Ca-Mg-HCO_3 waters. Groups 1 and 2 samples (Figure 6) predominantly fall within the Ca-Mg-HCO_3 category in the Chadha (1999) diagram (Figure 8). Group 3 members fall within the Na-K-Cl and Ca-Mg-Cl categories whilst group 4 members largely represent all water types with Ca - Mg – HCO_3^- and Na-K-HCO_3 predominating.

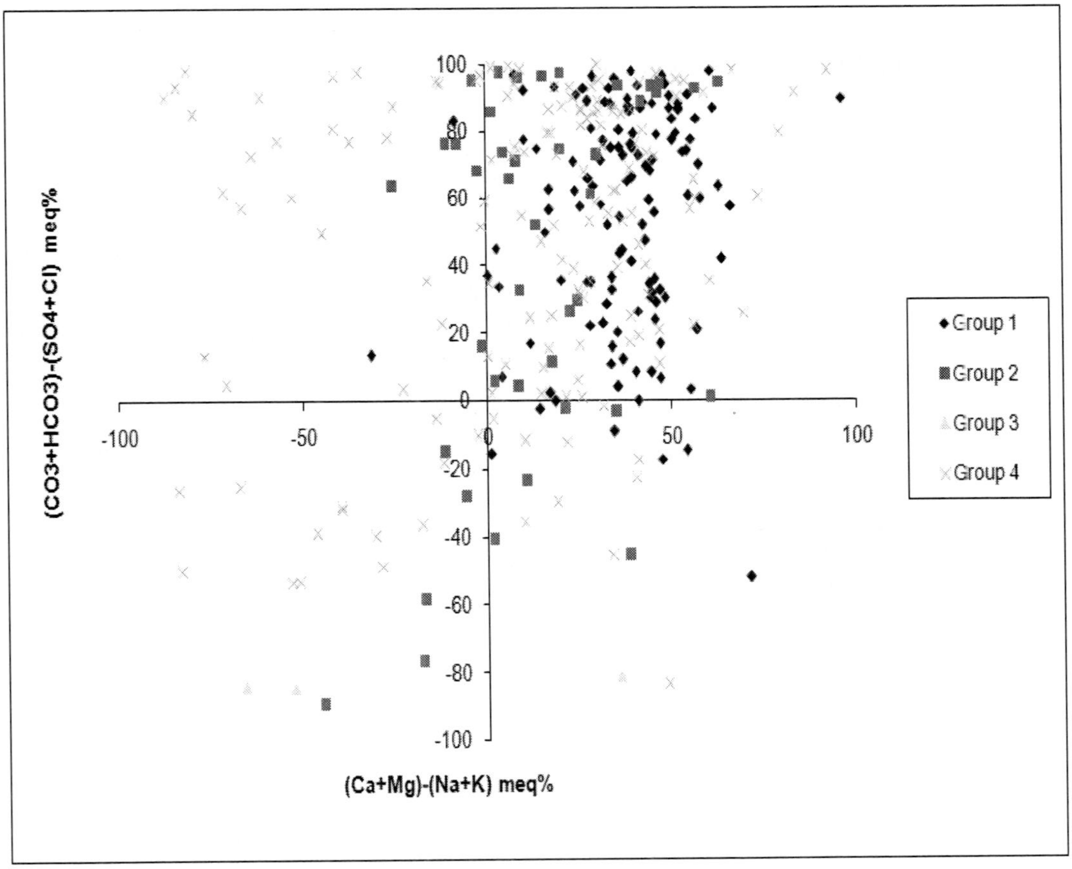

Figure 8. Chadha diagram for the four major groundwater groups distinguished from the HCA.

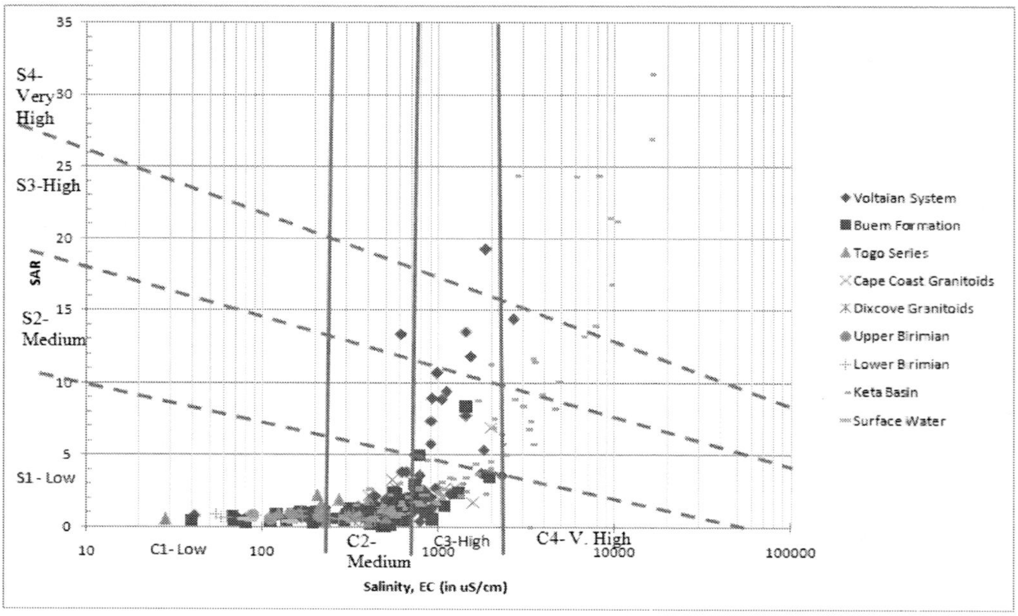

Figure 9. The USSL (1954) plot for assessing the quality of groundwater from the major geological formations in Ghana for irrigation purposes.

4.3. Suitability of Groundwater for Irrigation

The suitability of water for irrigation purposes is measured by many indices. In this study, the sodium adsorption ratio, SAR, which measures the proportion of sodium relative to the combined concentrations of calcium and magnesium, and the salinity as measured by the electrical conductivity, EC, were used to construct the United States Salinity Laboratory's diagram (USSL 1954) for groundwater from the aquifers of southeastern Ghana. Soils with high sodium content are notorious for having a poor soil structure (Appelo and Postma 2005). Problems with soil structure arise when Na^+ forms about 15% of the exchangeable cations and the ionic strength is less than 0.015 (Appelo and Postma 2005). The Exchangeable Sodium Exchange, *ESR*, is function of the activity of sodium divided by the square root of the sum of Ca^{2+} and Mg^{2+} activities (Appelo and Postma 2005). The ESR is related to the sodium adsorption ratio, SAR, by Equation 1:

$$ESR = 0.0158 * SAR \quad (1)$$

where

$$SAR = \frac{m_{Na}}{\sqrt{\left(m_{Ca^{2+}} + m_{Mg^{2+}}\right)}} \quad (2)$$

with all the ions measured in mmol/l.

When waters of high sodium relative to the concentrations of magnesium and calcium are used for irrigation, they tend to impart this property to the soils, thus affecting their agricultural properties. The USSL (1954) diagram relates the SAR to the EC of groundwater and places the samples into irrigation water categories based on the combination of the two parameters. Figure 9 is the USSL (1954) diagram for samples from the study area. About 75% of the samples are within the S1, low SAR category and range in salinity between C1, low salinity to C3, high salinity. These groundwaters are generally acceptable for irrigation, without the risk of causing sodium permeability effects resulting from high exchangeable sodium percentage. Groundwater from the Togo Series, Birimian (including the granitoid intrusive), and Buem Formations are generally within this category. About 15% of the samples from the Voltaian aquifers have SAR values beyond the S2 (medium SAR) categorization, and salinities in the C3 (High salinity) category. These groundwaters may not be used for irrigation without prior dilution with low salinity waters. It is obvious that the highest salinity and SAR waters are those from the Keta basin aquifers. This is mainly caused by seawater intrusion. Although groundwater from most of the wells sampled from the Keta basin aquifers are of acceptable quality for irrigation purposes (Figure 9), some of the wells have extremely high salinities and are likely to lead to the accumulation of huge quantities of ions in soils when used as irrigation waters.

4.4. Origin of Groundwater in Ghana

Oxygen-18 and Deuterium (^{2}H) data were used to assess the genesis and evolution of groundwater from the different geologic terrains sampled. The relationship between δ^2H and δ^{18}O for the major hydrogeological terrains, rainwater and the Volta Lake is presented in Figure 10. The equation derived for each terrain was compared with the equation for the Global Meteoric Water Line (GMWL) (Craig 1961) (Equation 3), which is the standard for recent meteoric water. Groundwater recharged during cool events such as the Pleistocene is depleted with respect to both isotopes. In addition, precipitation in the higher colder latitudes is depleted with respect to the stable isotopes of oxygen and hydrogen. The converse is true for groundwater recharged in warm low latitude regions. These isotopes can therefore be used to determine the approximate location and period of recharge of groundwater.

The data from the Voltaian aquifers fall rightly on the GMWL, suggesting that groundwater from these aquifers are probably meteoric. The data from the adjacent Volta Lake are quite enriched with respect to these isotopes. This is probably due to evaporative enrichment as the Lake is open to the intense tropical heat. It also suggests that groundwater recharge into the Voltaian aquifers does not come from the Volta Lake. Surprisingly, the rainfall data are more enriched with respect to these isotopes than do groundwater from some of the aquifers in the area. This is probably due to the fact that the rainfall samples were collected in the dry season with a relative humidity of less than 100%, when raindrops are enriched because of evaporative effects. Evaporation of raindrops under a relative humidity less than 100% will result in a slope of 5±2 on a δ^2H-δ^{18}O plot (Gonfiantini 1986). Furthermore, Mathieu and Bariac (1996) suggest that during the beginning or the end of the major monsoon or rainy season, the rainfall is light and relatively enriched with respect to the heavier isotopes.

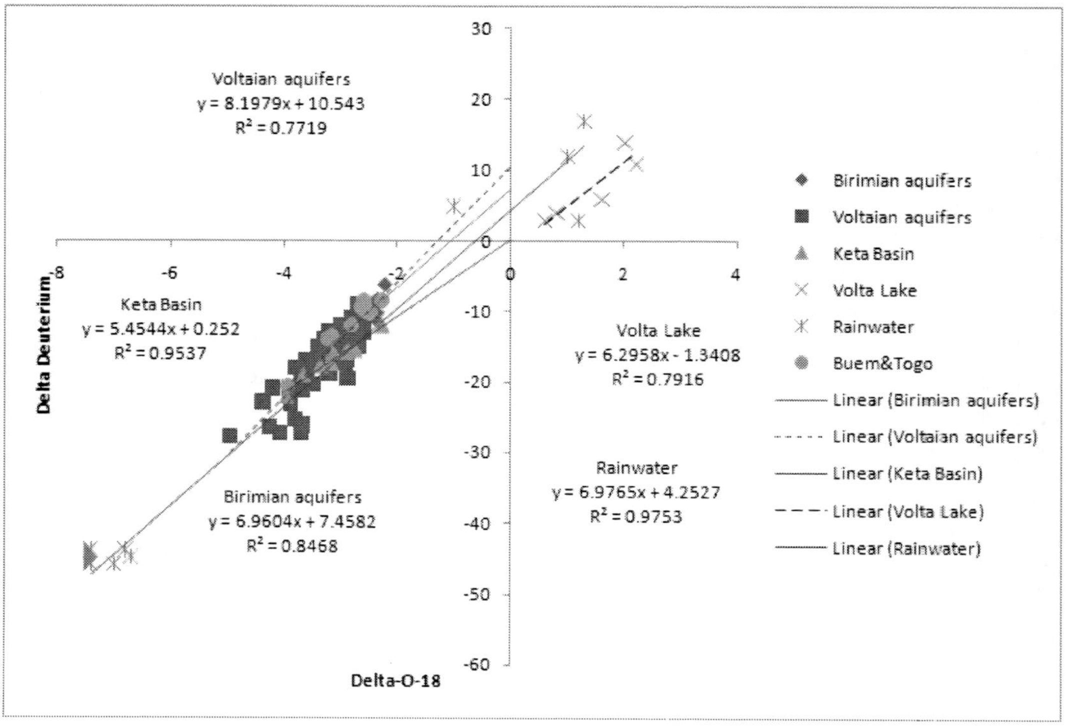

Figure 10. $\delta^{18}O$ against δ^2H for groundwater from some geological formations in Ghana.

$$\delta^2H=8\delta^{18}O+10 \tag{3}$$

The rainfall samples were taken at the end of the major rainy season, when the rainfall was light and relatively enriched in terms of the heavier isotopes. The observed relationship shown in Figure 10 is therefore within the range expected under these conditions. The slope for the data from the Keta basin is 5.5, which implies that groundwater from the Keta area is also enriched with respect to the GMWL (Craig 1961). This is within the range determined for evaporative effects. The aquifers in the Keta basin are quite shallow and prone to evaporative effects. In addition, Gonfiantini (1986) reports that a relative humidity of about 70%, is sufficient to lead to this margin of evaporative enrichment. Dickson and Benneh (1995) report a relative humidity of 65-75% in the area. Another possible contributory factor to the enrichment is the seawater intrusion which is suspected for the elevated salinity levels in groundwater in the area, but seawater samples were not analyzed for this study. However, previous work in the region by Jorgensen and Banoeng-Yakubo (2001) revealed that seawater in the study area ranges between -3.2 and 0.2 for δ^2H and -1.01 to -0.02 for $\delta^{18}O$, which is slightly depleted with respect to normal ocean water. The groundwater samples therefore lie between the local seawater values and the GMWL. Groundwater from the Birimian, Togo and Buem Formations is described by the same equation which has a slope of approximately 7. This represents a slight enrichment with respect to the GMWL, and is attributed to a recharge component from tributaries of the Volta Lake and the Densu River which drain these areas. Silicate mineral weathering under 50°C usually does not affect the groundwater isotopic composition (Drever 1988; Fetter 2001). Therefore the observed isotopic signatures are largely due to the climatic factors as well as the sources of recharge of the various hydrogeological formations sampled.

5. Ground Water Use in Ghana

In arid and semi-arid Africa, the majority of countries suffer from imbalance between the constantly increasing demand of water and the available natural water resources. As the second largest sources of fresh water, groundwater is an important natural resource, which is rudimentary to human life and economic development. In fact groundwater does not only serve as the most cost effective potable water supply but is also becoming an increasingly accepted resource because of the relative ease and flexibility with which it can be tapped. It has been and will remain a key determinant of population growth and distribution, social and political organization and the quality of life, especially in developing countries.

Modern developments and population growth have greatly increased water demands. As water crisis is forecasted in the future, the welfare of the world's population is closely tied to a sustainable exploitation of groundwater, surface and coastal water resources in order to prevent their depletion and contamination (Bear et al. 2000). Groundwater is thus considered not only feasible but also the most economic source of rural potable water supply. Gyau-Boakye and Dapaah-Siakwan (2000) reported that 52 percent of the rural population in Ghana was dependent on groundwater tapped by boreholes with hand pumps and open wells with and without pumps. Groundwater thus plays an essential role in sustaining rural households.

Currently, groundwater use in Ghana has increased tremendously and constitutes the main source of water supply in all the regions, most especially in deprived communities in the country. About 84% of this resource is used for domestic purposes particularly, drinking-water for people who do not receive public-supply of water from treated surface water in some urban centers and most rural communities (Kortatsi, 1994). Few private organizations and very few individuals in the urban cities have drilled boreholes for their private use due to the erratic supply of treated surface water resulting from the growth of population and a high demand for water. There are high potentials of spring and mineral water production in the Volta Region and the Accra Plains. At present, Voltic brand mineral water is produced at Medie located within the Accra Plains which serves as drinking water for most people in Ghana and beyond. Other individual use groundwater in the production of bottled and sachet water on small scale. Industries and mining facilities also use a lot of groundwater for various purposes.

The role of groundwater use in agriculture cannot be overemphasized, and it seems also clear that the absolute role of groundwater in Ghana's agriculture is much smaller to be related more to a limited effective accessibility to (as opposed to availability of) groundwater stemming from a variety of factors including geology, geography/spatial variation, tradition and political economy. Unfortunately, documentary evidence of the socio-ecology of actual agricultural water use in the country in general, and groundwater use in particular, is poor at best. There are hardly any data on small-holder irrigation outside larger schemes. The extensive use of groundwater for the purpose of irrigation is limited to the Accra plains, southern Volta region and the Upper Regions of the country. Kankam-Yeboah (1987) in a report on groundwater utilization in the Accra Plains reported that 33% of the boreholes drilled in the area are used for irrigation of vegetables including cabbages, onion, tomatoes and carrot. Groundwater is also used in the irrigation of pineapples on large scale and for fish farming on a small scale in the area. The people of Keta in the southern Volta Region use groundwater extensively for domestic, industrial and irrigation. Almost every home and farm along this stretch has hand-dug wells. Majority of these wells located on the farms are used for the irrigation of shallots, vegetables and other crops produced all year round in the area. Where the farm is large that irrigation by water cans and buckets becomes difficult, low powered irrigation pumps are fitted on some of these wells. Groundwater in the Northern and Upper Rregions is basically used for livestock and poultry watering and to a lesser extent for the irrigation of crops.

6. Critical Issues Constraining Groundwater Development and Management in Ghana

The efficient development of groundwater resources is very critical to the development of the irrigation industry in Ghana since the effects of climate change have led to the drying out of the major streams and rivers which hitherto fed the growing irrigation industry. In order to protect the groundwater resource base of the country, a set of regulations (L.I. 1827) were put in place by the Water Resources Commission (WRC) to regulate drilling and groundwater exploitation in the country. These regulations require among other things that a qualified well driller should have passed through a qualification procedure prior to the acquisition of a drilling license. It is prohibited for a person or group of persons to be contracted to drill without such a licence. In addition to these regulations, there are other legal documents which govern the acquisition of water rights in the country. The WRC is responsible for the allocation of water rights (including groundwater rights) on behalf of the President of Ghana. The major problem in the groundwater sector is not the absence of regulatory framework but the lack of enforcement of such regulations efficiently. Wells and boreholes are being drilled and constructed without recourse to the known regulations. For instance it is not uncommon for wells and boreholes to be drilled without the supervision of a qualified hydrogeologist or geologist with the requisite qualification. Wells are being dug without the required exploratory and environmental investigations, leading to contamination of groundwater resources by domestic sewage and other pollutants. In some cases, the required sampling and analyses of groundwater prior to use are not properly done. In areas where the required sampling is done, the required sampling and analytical protocols are generally not adhered to, leading to errors in the analytical results or their interpretation.

The Community Water and Sanitation Agency (CWSA), Water Research Institute (WRI), and the WRC have well maintained databases on the hydrogeological properties and hydrochemistry of groundwater from all the major and minor hydrogeological provinces in the country. Usually these databases contain information from sponsored drilling projects which are properly supervised. However, access to these databases is very much limited and the data are not readily released for research. The current knowledge of the hydrogeology of Ghana is therefore limited to the quantity of such data released for research. The research institutions such as the Universities are usually constrained with resources to be able to carry out such detailed hydrogeological investigations. As a result, there is inadequate knowledge base of the aquifer geometry and safe yield characteristics of most of the aquifers in the country. There is also inadequate information to address the status and trends in groundwater quality for most areas of the country. Non-governmental organizations (NGOs) such as the World Vision International, churches, and private individuals have also been involved in well drilling in the country. Apart from the World Vision which employs qualified personnel to oversee well tests and groundwater sampling, the other NGOs cannot be trusted to supply veritable hydrogeological and hydrochemical data. This is because the activities of these NGOs are targeted towards the provision of water of adequate quantity to satisfy the yearnings of populations without due regard to the environmental effects of groundwater over extraction. Well drilling in such cases is not usually guided by thorough exploratory activities, and drilling is not supervised by qualified professionals. Thus reliable well logging is not done. This practice constitutes a grave threat to management of groundwater resources and aquifers in the country. The current proliferation of well drilling companies and NGOs exploiting groundwater without the required personnel is alarming and must be checked by the regulatory agencies.

The following emerging issues in groundwater development in Ghana have therefore been identified: anthropogenic effects on groundwater quality; salt-water intrusion in coastal aquifers, groundwater depletion or inadequate supplies from groundwater to meet human needs; uncertainties about recharge and sustainability due to increased abstraction; as well as lack of institutional capacity to respond adequately to sustainable groundwater use.

Recommendations to ameliorate these problems include the strengthening of institutional capacity for comprehensive national evaluation of groundwater resources, including aquifer delineation and pro-active strategies to reduce groundwater abstraction to safe yield levels and adoption of aquifer protection strategies for overall protection of groundwater quality. The current regulations on extraction of groundwater resources must be strictly enforced. The WRC should be strengthened and adequately resourced to enforce these regulations to the letter. Collaborative work among the research institutions on groundwater resources evaluation and management should be encouraged.

The proper management of groundwater resources of the country will require the use of groundwater management and optimization models which will require detailed well monitoring data and hydrostratigraphic data from all the major hydrogeological provinces of the country. Currently, the hydrostratigraphy of the different groundwater basins in the country are not well understood. This will require adequate funding which the Government of Ghana must provide in order to ensure sustainable groundwater resources management towards the achievement of the millennium development goals.

7. Summary and Conclusion

Aquifers underlie almost everywhere in the country. A new hydrogeological map of Ghana has been developed based on the current data on the hydrogeology of the country. This is an improvement over the version of Gill (1969) and Dapaah-Siakwan and Gyau-Boakye (2000). The new hydrogeological map contains five major hydrogeological provinces distinguished on the basis of groundwater potential and lithology. These are the Birimian Province, the Crystalline Basement Granitoid Complex Province, the Voltaian Province, the Pan African Province, and the Coastal Sedimentary Province. Most of the prolific aquifers have are associated with areas where deep seated weathering and fracturing have created enough secondary permeabilities to facilitate groundwater recharge and storage.

The hydrochemistry of groundwater from all aquifers in the country is controlled by the weathering of silicate minerals and cation exchange. However, in a few areas where the aquifers are relatively shallow and unconfined, the effects of anthropogenic activities such as agricultural activities and unregulated domestic wastes, have been noted to have serious effects on groundwater hydrochemistry. In the coastal areas such as the Keta basin, seawater effects have been noted. Stable isotope data suggest that groundwater recharge is predominantly from precipitation.

Except in a few cases of high fluoride content due to the weathering of fluorite, and high salinity resulting from the intrusion of salty water from the sea, the aquifers generally offer groundwater of acceptable quality for most domestic uses. Groundwater is currently being assessed for possible uses in irrigation on a large scale, although it is being used locally in some areas for irrigation of vegetables.

The major problem facing groundwater development and aquifer management in the country is the lack of enforcement of the regulations governing groundwater exploration and well drilling in the country. Consequently, wells are being drilled without recourse to the environmental regulations. In addition the current databases on aquifer properties are not readily accessible for detailed hydrogeological research. As a result, there is the lack of proper understanding of the characteristics of the hydrogeological basins in the country. The way forward will require the strengthening of the institutional capacity of the enforcement agency, the Water Resources Commission, to be more pro-active in regulatory enforcement. In addition, the national will for a comprehensive assessment and evaluation of the groundwater resources of the country must be done without delay. This will require strengthening the collaborative works among the major research institutions of the country in the area of hydrogeology.

REFERENCES

Affaton P, Sougy J, and Trompette R (1980) The tectono-stratigraphic relationships between the Upper Precambrian and Lower Paleozoic Volta basin and the Pan-African Dahomeyide orogenic belt (West Africa): *American Journal of Science,* v. 280, pp. 224-248.

Agyei EK, Van Landewijk JM, Armstrong RL, Harakai JE, and Scott KL (1987) Rb-Sr and K-Ar Geochronometry of Southeastern Ghana: *Journal of African Earth Sciences,* v. 6, no. 2, pp. 153-161.

Anku SY, Banoeng-Yakubo B, Asiedu D, Yidana SM (2008) Water quality analysis of groundwater in crystalline rocks, Northern Ghana. *Environmental Geology, DOI* 10.1007/s00254-0081578-4.

Appelo CAJ, Postma D (2005) *Geochemistry, groundwater and pollution,* 2nd Edition. Balkema, Amsterdam, 649p.

Banoeng-Yakubo BK (1989) Occurrence of groundwater in basement complex rocks of the Upper Regions of Ghana. Unpublished M.Sc. Thesis, Geology Department, Obafemi-Awolowo University.

Banoeng-Yakubo BK (2000) The application of remote sensing and geographical information systems to hydrogeological studies in the Upper West Region, Ghana. Unpublished Ph.D. Thesis, Geology Department, University of Ghana.

Banoeng-Yakubo B, Yidana SM, Nti E (2009) Hydrochemical modeling of groundwater using multivariate statistical methods – The Volta Region, Ghana. KSCE Journal of Civil Engineering 13(1): 55 - 63.

Bates DA (1955) *Geological map of Ghana: Ghana Geological Survey Department,* scale 1:1,000,000. Reprinted by the Geological Survey Department in 1966.

Bear JA, Cheng D, Sorek S, Ouazar D, and Herrera I (2000) Seawater Intrusion in Coastal Aquifers-Concepts, Methods and Practices.p.625. Kluwer Academic Publishers, The Netherlands. Gyau-Boakye and Dapaah-Siakwan (1998). Groundwater: Solution to Ghana's rural water supply industry? In *Advances in Groundwater exploration and development in Ghana.* Banoeng-Yakubo, B.K. and Asiedu, D.K. (eds.) pp.53-63.

Blay PK (1991) Applying subduction tectonics to the evolution of the Pan-African Dahomeyide deformed belt of Ghana, West Africa: *Proceedings of the First Local Conference on Mineral Exploration and Development and their Impact on the Economy of Ghana, Accra, 7th December 1990*, Minerals Commission Publication, pp. 52-75.

Bozhko NA, Kozakov GA, Trofimov DM, Knorre KG, and Gatinsky YA (1971) A new absolute dating of West African glauconites: Academia Naukrainia SSR Doklady, R.

Carrier, MA, Lefebvre, R., Racicot, J., Asare, EB (2008) Northern Ghana hydrogeological assessment project. *33rd WEDC International Conference, Accra, Ghana.*

Chadha DK (1999) A proposed new diagram for geochemical classification of natural waters and interpretation of chemical data. *Hydrogeology Journal* 7: 431-439.

Cloutier V, Lefebvre R, Therrien R, Savard MM (2008) Multivariate statistical analysis of geochemical data as indicative of the hydrogeochemical evolution of groundwater in a sedimentary rock aquifer system. *Journal of Hydrology* 353: 294-313.

Craig H (1961) Isotopic variations in meteoric water. Science 133: 1702-1703 Dapaah-Siakwan S, Gyau-Boakye P (2000) Hydrogeologic framework and borehole yields in Ghana. *Hydrogeology Journal* 8: 405-416

Dapaah-Siakwan S, Gyau-Boakye, P (2000) Hydrogeologic framework and borehole yields in Ghana. Hydrogeology 8, 405–416.

Davis JC (1986) *Statistics and Data Analysis in Geology.* John Wiley & Sons Inc., New York.

Dickson KB, Benneh G (1995) *A New Geography of Ghana*, 3rd revised edition. Longman, Malaysia.

Drever JI (1988) *The geochemistry of natural waters*, 2nd Edition. Prentice Hall, New Jersey, 437p.

Dugan JP, Borthwick J, Harmon RS, Gagnier MA, Glahn JE, Kinsel EP, MacLeod S, Viglino JA, Hess JW (1985) Guanidine hydrochloride method for determination of water oxygen isotope ratios and the oxygen-18 fractionation between carbon dioxide and water at 25°C. *Anal. Chem.* 57: 1734-1736.

Eisenlohr BN, and Hirdes W (1992) The structural development of early Proterozoic Birimian and Tarkwaian rocks in southwest Ghana, West Africa: *Journal of African Earth Science.,* v.14, no.3, pp. 313-325.

FAO (1967) Land and Water Survey of the Upper and Northern Regions of Ghana. Fetter CW (2001) *Applied Hydrogeology*, 4th Edition. Prentice Hall, New Jersey.

Farnham IM, Stetzenbach KJ, Singh K, Johannesson KH (2000) Deciphering groundwater flow systems in Oasis Valley, Nevada, using trace element chemistry, multivariate statistics, and geographical information system. Math. Geol. 32, 943–968.

Farnham IM, Johannesson KH, Singh, AK, Hodge VF, Stetzenbach, KJ (2003) Factor analytical approaches for evaluating groundwater trace element chemistry data. Anal. Chim. Acta 490, 123–138.

Fitches WR (1970) "Pan African Orogeny" in the coastal region of Ghana: *Nature*, v. 226, p. 744-746.

Florkowsky T (1985) Sample preparation for hydrogen isotope analysis by mass spectrometry. *Int. J. Appl. Radiat. Isot.* 36: 991-992.

Gill HE (1969) A ground-water reconnaissance of the Republic of Ghana, with a description of geohydrologic provinces. *US Geological Survey Water-Supply paper* 1757-K.

Gonfiantini R (1986) Environmental isotopes in lake studies. In: P. Fritz and J.Ch. Fontes (eds). Handbook of environmental isotope geochemistry, 113-168. Elsevier, Amsterdam.

Grant NK (1969) The late Precambrian to early Paleozoic orogeny in Ghana, Togo, Dahomey and Nigeria: *Geological Society of America Bulletin,* v. 80, no.1 pp. 45-55.

Güler C, Thyne G, McCray JE, Turner AK (2002) Evaluation and graphical and multivariate statistical methods for classification of water chemistry data. *Hydrogeology Journal* 10: 455-474.

Helstrup T, Jørgensen NO, Banoeng-Yakubo B (2007) Investigation of hydrochemical characteristics of groundwater from Cretaceous – Eocene limestone aquifers in southern Ghana and Togo using hierarchical cluster analysis. *Hydrogeology Journal* 15: 977-989.

Hirdes W, Senger R, Adjei J, Efa E, Loh G, and Tetty A (1993) Explanatory Notes for the Geological Map of Southwest Ghana 1:100,000, sheet Wiawso (0603D), Asafo (0603C), Kukuom (0603B), Goaso (0603A), Sunyani (0703D) and Berekum (0703C): Geologisches Jahrbuch, Reine B, Heft 83, Hannover, Germany, 139 pp., 32 Fig., 18 tables, 1 map (6 sheets).

Hirst T (1946) *Reports on the Bibiani goldfield, Part I: The geology of the field: Gold Coast Geological Survey Department Memoir* No.9, Accra, pp 1-16.

Holm RF (1974) Petrology of alkalic gneiss in the Dahomeyan of Ghana: *Geological Society of America Bulletin,* v.85, pp. 1441-1448.

Jones WB (1990) The Buem volcanic and associated sedimentary rocks, Ghana: a field and geochemical investigation: Journal of African Earth Sciences, v. 11, no. 3/4, pp. 373-383.

Jørgensen NO, Banoeng-Yakubo BK (2001) Environmental isotopes (^{18}O, ^{2}H, and $^{87}Sr/^{86}Sr$) as a tool in groundwater investigations in the Keta Basin, Ghana. *Hydrogeology Journal* 9: 190-201.

Junner NR (1935) *Gold in the Gold Coast: Gold Coast Geological Survey Memoir No. 4,* Accra, 67 pp.

Junner NR (1940) Geology of the Gold Coast and Western Togoland: *Gold Coast Geological Survey Bulletin No. 11*, Accra, 40 pp.

Junner NR, Hirst T (1946) The geology and hydrogeology of the Voltaian basin. Gold Coast Geological Survey Mem 8. *Gold Coast Geological Survey*, The Gold Coast.

Junner NR, Service H (1936) Geological notes on the Volta River District and Togoland under British mandate. *Annual Report on the Geological Survey by the Director*, 1935-36.

Kankam-Yeboah K (1987) Status report on the pilot scheme on groundwater utilisation in the Accra Plains. *Unpublished technical report,* Water Resources Research Institute, Accra, Ghana.

Kaiser HF (1960) The application of electronic computers to factor analysis. *Educ. Psychol Meas.* 20: 141-151.

Kennedy WQ (1964) The structural differentiation of Africa in the Pan-African (±500 m.y.) tectonic episode: Leeds University Institute for African Geology, *Annual Report,* Scientific Results (abstracts), pp 48-49.

Kesse GO (1985) The Mineral and Rocks Resources of Ghana. A.A. Balkema Publishers. Netherlands-Rotterdam. Pp 39-50.

Kortatsi BK, Tay CK, Anornu G, Hayford E, Dartey G (2008) Hydrogeochemical evaluation of groundwater in the lower Offin basin, Ghana. *Environmental Geology* 53: 1651-1662.

Kortatsi, BK (1994) *Groundwater utilization in Ghana,* IAHS Publication No. 222 pp. 149-156.

Lashmanov V (1991) Geology and Minerals of the Area of the Nkwanta (Eastern Part) and Dutukpene Field Sheets: *Ghana Geological Survey Department and Minerals Commission Archive Report No. 58*, Accra, 124 pp.

Leube A, Hirdes W, Mauer R and Kesse G (1990) The Early Proterozoic Birimian Supergroup of Ghana and some aspects of its associated gold mineralization: *Precambrian Research,* v. 46, pp. 139-165.

Mani R (1978) The geology of the Dahomeyan of Ghana: Geology of Ghana Project. *Ghana Geological Survey Department,* Accra, (unpublished).

Mathieu R, Bariac T (1996) An isotopic study (^{2}H and ^{18}O) of water movements in clayey soils under semiarid climate. *Water Resources Research* 32(4): 779-789.

Mauer R (1990) Petrographische und geochemische Untersuchungen der prakambrischen (Birimian) Granitoide Ghanas: *Disertation,* Techn. Univ., Berlin, 202 pp.

Murray RJ (1960) The geology of the "Zuarungu" 1/2° Field Sheet: *Ghana Geological Survey Department Bulletin No.25,* Accra, 118 pp.

Piper A M (1944) A graphic procedure in the geochemical interpretation of water samples. Am Geophys.

Quist LG, Bannerman RR, Owusu S (1988) Groundwater in rural water supply in Ghana. In: Groundwater in rural water supply. Report of the West African Sub-Regional Workshop, Accra, Ghana, 20–24 October 1986, *UNESCO Technical Documents in Hydrology,* UNESCO, Paris, pp 101–126.

Schot PP, van der Wal J (1992) Human impact on regional groundwater composition through intervention in natural flow patterns and changes in land use. *Journal of Hydrology 134:* 297-313.

Steinhorst RK, Williams RE (1985) Discrimination of groundwater sources using cluster analysis, MANOVA, canonical analysis and discriminant analysis. *Water Resources Research* 21: 1149-1156.

Stetzenbach KJ, Hodge VF, Guo C, Farnham IM, Johannesson KH (2001) Geochemical and statistical evidence of deep carbonate groundwater within overlying volcanic rock aquifers/aquitards of southern Nevada, USA. J. Hydrol. 243, 254–271.

Taylor PN, Moorbath S, Leube A and Hirdes W (1992) *Early Proterozoic crustal evolution in the Birimian of Ghana—Constraints from geochronology and isotope geology: Precambrian Research* v. 56, pp. 97-111.

USSL (1954) Diagnosis and improvement of saline and alkali soils. USDA, Handbook 60, p.147.

Vernon RH and Flood RH (1988) *Contrasting deformation of S- and I-type granitoids in the Lachlan Fold Belt, eastern Australia: Tectonophysics* v. 147, p.127-143.

WHO (World Health Organization) (1984) *Guidelines for drinking water quality,* vol 1. Recommendations. Geneva, Switzerland.

Woodfield PR (1966) The geology of the ¼° field sheet 91, Fumso N.W.: *Ghana Geological Survey Department Bulletin* No. 30, Accra, 66 pp.

Yidana SM (2008) Groundwater resources management for productive uses in the Afram Plains area, Ghana. *Doctoral Dissertation*, Montclair State University, USA.

Yidana SM (2009) Hydrochemical framework of surface water basins in southern Ghana. Journal of Environmental Geology. 57(4):789-796, DOI 10.1007/s00254-008-1357-2

Yidana SM, Banoeng-Yakubo BK, Akabzaa T (2010) Analysis of groundwater quality using multivariate and spatial analyses in the Keta Basin, Ghana. Journal of African Earth Sciences, 58, 220-234, DOI: 10.1016/j.jafrearsci.2010.03.003

Yidana SM, Ophori D, Banoeng-Yakubo B (2008a) Hydrogeological and hydrochemical characterization of the Voltaian Basin: the Afram Plains area, Ghana. *Environmental Geology* 53: 1213-1223.

Yidana SM, Ophori D, Banoeng-Yakubo B (2008b) Hydrochemical evaluation of the Voltaian system – The Afram Plains area, Ghana. *Journal of Environmental Management* 88: 697-707

Yidana SM, Ophori D, Banoeng-Yakubo B (2008c) Groundwater availability in the shallow aquifers of the southern Voltaian system: a simulation and chemical analysis. *Environmental Geology* 55: 1647-1657.

Yidana SM, Ophori D, Banoeng-Yakubo B (2008d) Groundwater quality evaluation for productive uses – The Afram Plains area, Ghana. *Journal of Irrigation and Drainage Engineering* 134(2): 222-227.

Yidana SM, Yidana A (2009) Assessing groundwater quality using water quality index and multivariate statistical analysis – the Voltaian basin, Ghana. Environmental Earth Science, DOI 10.1007/s12665-009-0132-3.

In: Potable Water and Sanitation
Editor: Joel M. McMann

ISBN 978-1-61122-319-4

Chapter 4

CHARACTERIZATION OF THE QUALITY OF GROUNDWATER FROM THE MAJOR AQUIFERS IN GHANA

Sandow Mark Yidana*

Department of Earth Science, Box LG 58, University of Ghana, Legon, Accra, Ghana

ABSTRACT

In Ghana, groundwater plays a very crucial role in the overall water delivery system, especially in the rural areas. This resource is being assessed for possible exploitation for irrigation projects since the over reliance on rain-fed agriculture does not auger well for national development in the face of dwindling and equally erratic rainfall patterns in the country. In this study, Q-mode hierarchical cluster analysis (HCA) and R-mode factor analysis were used jointly with Geographic Information Systems (GIS) to determine the spatial groundwater quality types in Ghana, and the spatial variation of the major sources of variation in the hydrochemistry of groundwater in the country. In addition to assisting in the understanding of the different groundwater types in the country, the study demonstrates the applicability of multivariate statistical methods in groundwater quality classification in Ghana. This study finds that two major groundwater types exist in the country: the Na-Cl groundwater types which dominate the coastal aquifers where groundwater salinity is high and attributed partly to seawater intrusion and Na-K-HCO_3 groundwater types which dominate aquifers farther inland. It has also been determined that the effects of seawater intrusion on groundwater quality in Ghana is significant along the coast where over extraction of the resource has led to seawater up-coning. Farther inland, groundwater hydrochemistry is controlled principally be mineral weathering processes, among which silicate mineral weathering has been noted to be the predominant. The effects of chemical fertilizers, organic manure, and domestic sewage are significant in areas where the aquifers are relatively unconfined and shallow or where the well head is poorly protected and where agricultural activities are high. High nitrate

* Corresponding author: Email: smyidana@ug.edu.gh, yidanas117@gmail.com; Tel. +233203143351, +233265683724.

concentrations have been noted in groundwater sample from the northern parts of the country, some parts of the forested areas and the Keta strip of the Volta region of Ghana. The oxidation of organic carbon is a minor factor because its effects are relatively insignificant when compared to mineral weathering and anthropogenic contamination. Interpolation maps have been generated to highlight the intensity of each of the major processes in the country. Groundwater quality for irrigation purposes is discussed in relation to the major hydrogeological basins in the country.

1. INTRODUCTION

Groundwater is the mainstay of the rural water economy of Ghana. The resource has been tapped at various levels in the hydrogeological basins to supply potable water resources to communities. The demand for groundwater resources is expected to rise in the very near future in the wake of rapidly dwindling surface water resources, and the need to expand the scope of agricultural practice to include irrigation in the country. Such a practice is a vital part of the policy of the Government of Ghana. Sustainable irrigation in the country would require groundwater of acceptable hydrochemical and biological content to boost crop yield whilst maintaining acceptable soil quality for future generations. In the assessment of water quality for irrigation, salinity and sodicity are critical factors which are usually considered (e.g. Jalali, 2007; Arumugam and Elangovan, 2009), although the assessment of specific ion toxicities in irrigation waters is commonly done. A complete assessment of the causes of variation and the spatial variability of groundwater quality in the country is therefore critically important in assessing the irrigation water needs of the country. Such an assessment will identify areas of critical groundwater quality problems for remediation prior to usage. This is particularly important in the coastal areas where some aquifers have been contaminated by suspected seawater intrusion attending high levels of groundwater extraction.

Fortunately, hydrogeologists, governmental and non-governmental organizations have collated data on the hydrochemistry of groundwater from the major aquifers in the country. In addition, there have been studies on the major factors controlling the hydrochemistry of groundwater at the local basin level for most of the major groundwater basins. Recent notable examples include Acheampong and Hess (1998, 2000), Kortatsi et al. (2008), Yidana et al. (2008a-d), Banoeng-Yakubo et al. (2009), and Yidana (2010). In each of these studies, the researchers applied mass balance and statistical methods to determine the major sources of variation in the hydrochemistry of groundwater at the respective basins, and therefore set the tone for a discussion of the groundwater control parameters at the national scale. The natural recharge regimes and the sources of variation in the quality in Ghana are typical of tropical aquifers in western Africa. The groundwater quality types in Ghana are therefore typical of what pertains in much of western Africa.

One of the weaknesses of the existing literature on groundwater hydrochemistry in Ghana is the inability to identify the spatial variations in the major control factors. In addition, except in a few cases such as the Voltaian (Yidana et al., 2008b), Buem formation (Banoeng-Yakubo et al., 2009), and southeastern Ghana (Yidana, 2010), no attempt has been made to rank the different groundwater control factors in order of importance as most of the other studies have concentrated on the role of silicate mineral weathering in groundwater hydrochemistry. Moreover, there has not been any study that provides information on the spatial variation of the major processes controlling groundwater hydrochemistry on the national scale. Such a study is critically important as the country prepares towards accessing groundwater in large quantities for commercial purposes including irrigation.

This research therefore studies the processes of relevance in the general hydrochemistry of groundwater in Ghana, and the spatial variation of the intensities of these processes on a national scale. It tests the effectiveness of the combined use of multivariate statistical methods and geographical information systems in groundwater quality studies on a national scale. Sequential cluster analysis and factor analysis are being used together with robust spatial interpolation procedures to study the variations in groundwater quality on a nationwide scale.

Factor and cluster analyses have been copiously applied in groundwater quality studies all over the world, and have been noted to provide helpful information on groundwater quality trends. For instance Viero et al. (2009) applied factor and cluster analysis to the study of the sources of high fluoride in a granitic aquifer in Porto Alegre in southern Brazil. Their study identified three distinct groundwater groups, principally distinguished on the basis of the concentrations of fluoride and the major ions. Earlier, Farnham et al. (2003) applied multivariate statistical methods to hydrochemical data to attribute the sources of variation in the hydrochemistry of groundwater to rock-water interactions. The literature also contains examples of the successful application of multivariate statistical methods to identify groundwater flow lines in a basin (e.g. Farnham et al., 2000; Stetzenbach et al., 2001). Cloutier et al. (2008) combined cluster analysis, principal component analysis and conventional graphical methods to classify groundwater into seven (7) clusters representing different groundwater types. They successfully defined a groundwater hydrochemical evolution path which is controlled by the geology and mineralogy, the degree of confinement of the aquifer, and the geological history. On the basis of factor analysis and graphical methods, they concluded that the high levels of fluoride in groundwater in the area were attributed to the dissolution of interstitial fluorite in the matrix of the aquifers in the area.

Multivariate methods have also been applied to hydrochemical data analyses in some groundwater basins in Ghana. These tools have been used to study hydrochemical processes in surface water basins (e.g. Yidana et al., 2008a, Yidana, 2009), groundwater in the Voltaian (e.g. Yidana et al. (2008b, c), groundwater from the Togo and Buem aquifers (e.g. Banoeng-Yakubo et al., 2009), aquifers of southeastern Ghana (e.g. Yidana, 2010). The current study conducts a holistic assessment of the general causes of variation in the hydrochemistry on a national scale. It also conducts a comparative assessment of the suitability of groundwater from the major aquifers in the country for various uses.

2. Geography, Geology and Hydrogeology of Ghana

2.1. Geography

Ghana is located between latitudes 4° 40' N and 11° 15' N and longitudes 3° 15' W and 1° 15' E, covering a land area of approximately 239,000 km^2. The variations in the principal elements of temperature, rainfall, and humidity that govern the climate are influenced by the movement and interaction of the dry tropical continental air mass (the harmattan), which blows from across the Sahara in the northeast, and the opposing tropical maritime or moist equatorial system which blows from the southwest.

In the summer months in the northern hemisphere, a warm and moist maritime air mass intensifies and pushes northward across the country. At this time, a low-pressure belt in the air mass brings warm air, rain, and prevailing winds from the southwest. The harmattan sets in when the sun returns south across the equator.

Climatic and weather conditions across the country are not uniform. The Kwahu Plateau marks the northernmost extent of the forest area and also serves as an important climatic divide. To the north of this plateau, two distinct seasons occur: the harmattan season with its dry, hot days and relatively cool nights from November to late March or April, followed by a wet period that reaches its peak in late August or September. The southern and southwestern parts of the Kwahu Plateau are characterized by four separate seasons and the annual mean rainfall ranges from 1,250 millimeters to 2,150 millimeters. Normally, heavy rains fall from about April through late June. This pattern has shifted dramatically in recent times as an apparent testimony to the effects of climate change in the country. The heaviest rains are sometimes observed in July and August. The rains are sometimes very heavy within short periods, leading to the flooding of some settlements.

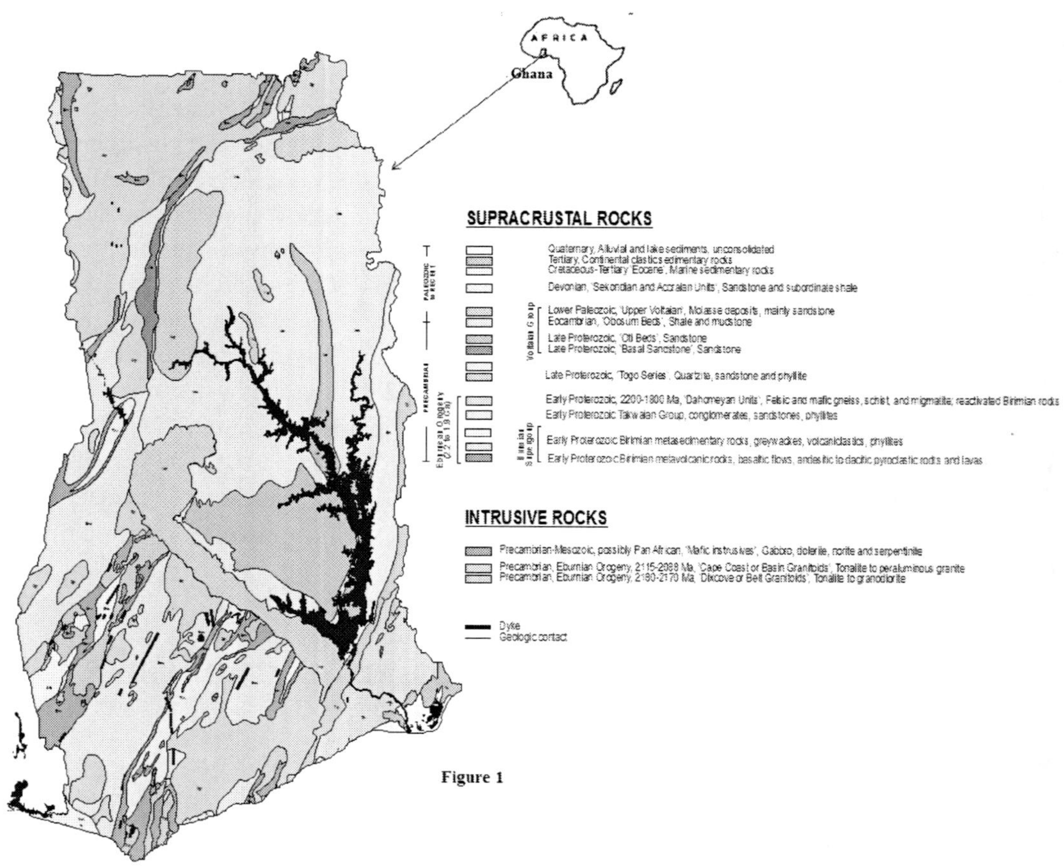

Figure 1. The Geological map of Ghana.

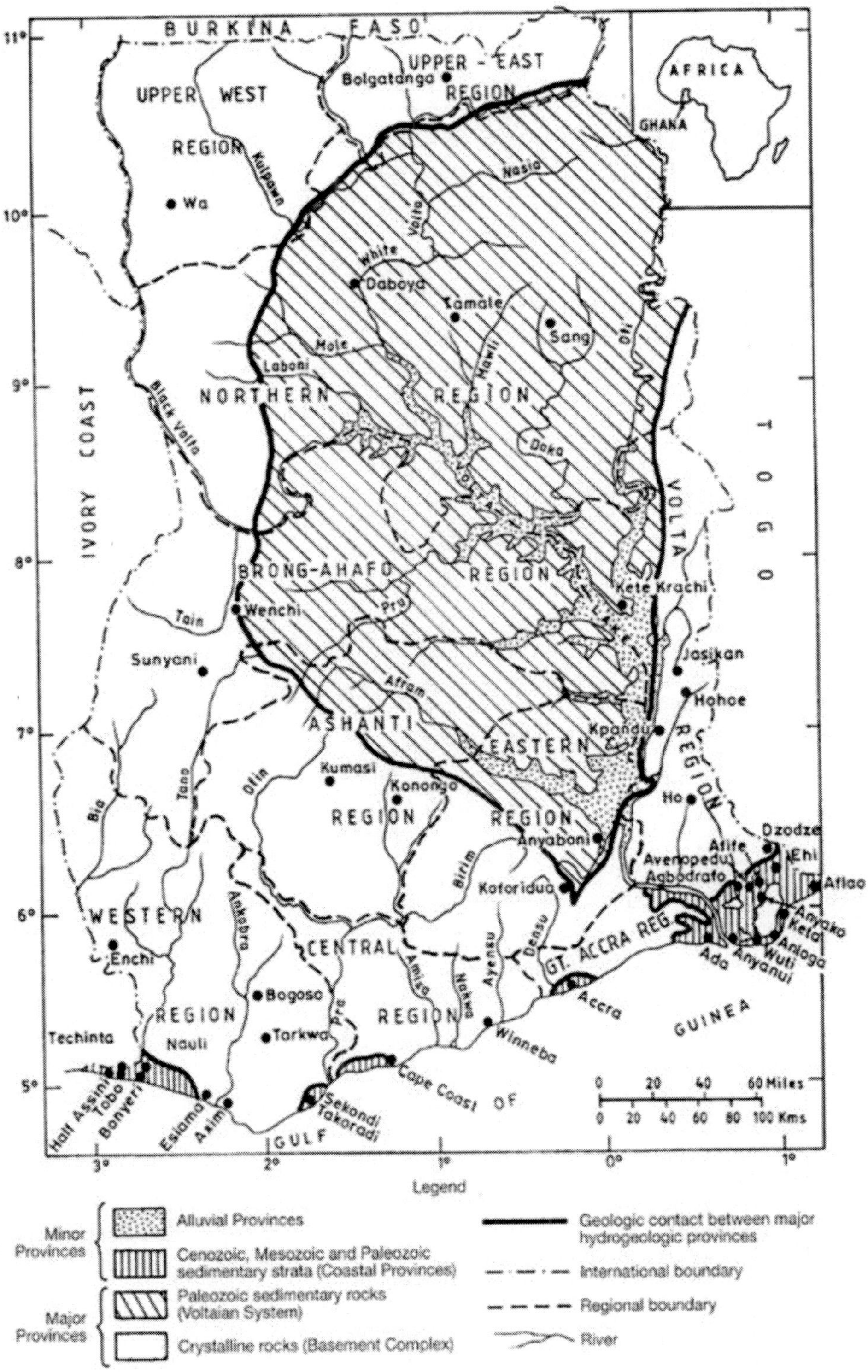

Figure 2. The Hydrogeological map of Ghana showing the major groundwater basins in the country (after Dapaah-Siakwan and Gyau-Boakye 2000).

The intensity and duration of dry and wet spells vary across the country. The heaviest rains occur in the Axim area in the southwest corner of Ghana. Kumasi and its neighborhood in the Ashanti region receive an average annual rainfall of about 1,400 millimeters, whilst Tamale in the drier northern savanna receives rainfall of approximately 1,000 millimeters per year. From Takoradi in the west to the Accra Plains farther east, including the lower Volta region, rainfall averages between 750 millimeters to 1,000 millimeters a year.

The historical temperature extremes are in the range of 18 ^{o}C to 40 ^{o}C with a smaller average annual temperature range of 24 ^{o}C to 30 ^{o}C. Generally, temperature increases with distance from the coast. In the far north, temperature highs of 31°C and higher, are common. The southern part of the country is relatively humid especially at night, when humidities in the range of 95 to 100 % humidity are common. In the harmattan season humidity drops as low as 25 percent in the north although high humidities are observed during the rainy season.

Vegetation is climate controlled and has the following latitudinal characteristics. With increase in latitude from the coast, the vegetation grades from equatorial rain forest in the southwest through a transition zone to semi-deciduous tropical forest in the central parts to guinea savannah in the northern latitudes and finally to Sudan savannah in the extreme north-eastern part of the country (Dickson and Benneh, 1995).

2.2. Geology

Geologically, Ghana can be divided into several distinct terrains (Figure 1). Rocks of the Birimian System underlie most of southern, western and northern Ghana, and host most of the gold and diamond deposits in the country. The Birimian consists of metamorphosed volcanic and sedimentary rocks which form five sub-parallel belts of volcanic rock separated by broad basins of sedimentary rocks. Rocks of the Birimian System have been divided into an older Lower Birimian, consisting of predominantly metasedimentary rocks, and a younger Upper Birimian, made up of metavolcanic rocks (Junner, 1935, 1940; Bates, 1955). The rock types present in the Lower Birimian sedimentary belt are greywackes with turbidite features, phyllites, slates, schists, weakly metamorphosed tuffs and sandstones. Some of the phyllites contain pyrite, and finely divided carbonaceous matter is present in most of them. Silicification is common in the phyllites, particularly towards the boundary with the Upper Birimian.

A distinctive sequence of clastic sedimentary rocks occurs in elongate troughs developed on top of the Birimian System. These rocks host important paleoplacer gold deposits and are known as the Tarkwaian System. Most workers agree that the Tarkwaian sediments were deposited in intermontane grabens formed by preferential rifting along the axes of the volcanic belts and that there is no evidence that the depositional basins were ever linked. The rocks of the Tarkwaian System represent erosional products of the Birimian and are dominated by coarse clastic sediments (Kesse, 1985).

Almost one third of Ghana is covered by sediments of the inland Voltaian Basin which covers an area of about 103,600 km^2. The Voltaian strata are nearly horizontal beds of sandstones, shales, mudstones and conglomerates thought to be of Late Precambrian to Paleozoic age (Kesse, 1985). In most places, the flat lying Voltaian strata overlie the Birimian rocks with a marked angular unconformity. Junner and Hirst (1946) subdivided the Voltaian sediments on the basis of lithology and field relationships into Lower, Middle and Upper units.

The Dahomeyan System is a part of the second major tectono-stratigraphic terrane in Ghana; it underlies eastern and southeastern Ghana. The Dahomeyan is the easternmost rock group in Ghana and differs significantly from other rocks in Ghana in that it is composed of high grade metamorphic rocks. The system consists of four lithologic belts of granitic and mafic gneiss. The mafic gneisses are relatively uniform oligoclase, andesine, hornblende, salite and garnet gneisses of igneous parentage and generally tholeiitic composition (Holm, 1974). The granite gneisses interlayer with the mafic gneiss and are believed to be metamorphosed volcaniclastic and sedimentary rocks. Persistent bands of nepheline gneiss in the system appear to be metamorphosed calc-alkaline igneous rocks (Holm, 1974).

The second major lithological group which makes up the eastern Ghana terrain is the Togo Belt comprising the Buem and Togo Series. This group of rocks comprises three distinctive lithological assemblages. The rocks which comprise the north to northeast trending Togo Range consist of strongly tectonized phyllite, quartzite and serpentinite. These rocks are variously known as the Togo Series (Kesse, 1985) or the Togo Tectonic Unit (Blay, 1991). The contacts between the Togo and Dahomeyan to the east and the Buem to the west are thrust faults. The unit grades from east to west from phyllite and chlorite schist upwards into quartzite, micaceous quartzite and sandstone. Serpentinites occur along the western contact and appear to be emplaced along thrust faults (Grant, 1969).

West of the Togo Range is a belt of volcanic and sedimentary rocks known as the Buem Series (Kesse 1985; Jones, 1990) or the Buem Tectonic Unit (Blay, 1991). The Buem consists of two lithological assemblages, volcanic and sedimentary. The volcanic assemblage is made up of pillow basalt, agglomerate, hawaiite and trachyte. The sedimentary assemblage, which encloses the volcanics, consists of red shales, feldspathic to quartz arenite, conglomerate, tillite, jasper and minor limestone. Whereas the volcanics were deposited in a submarine environment, the sediments appear to be shallow water to subaerial in origin (Jones, 1990). Jones (1990) describes the Buem as an eastward dipping homoclinal sequence, whereas Kesse (1985) states that the rocks are strongly folded. Blay (1991) cites the presence of a basal conglomerate at the Buem-Togo contact as an indication that the Buem is younger than the Togo unit.

Relatively minor outcrops of sedimentary rocks along the coast from Keta and Accra in the east to Half Assini in the west constitute remnants of rocks of the Phanerozoic coastal basins. From east to west, these rocks occur in the Keta, Accraian, Sekondian and Tano basins. Rocks of the Keta basin are of Cretaceous age and consist of sandstones, siltstone, shales, claystone and fossiliferous limestone beds (Kesse, 1985). The Accraian is considered mid-Devonian in age and consists predominantly of sandstones and shales. The Sekondian strata are made up of sandstones, shales, silts and beds of chalcedony, sands and pebbly beds and range in age from Devonian to Cretaceous. The Tano basin is located in the extreme southwestern corner of Ghana. It is made up of Cretaceous-Tertiary sediments consisting of limestones, shales and sands which have a large off-shore extension.

Four main types of granitoids are recognized in the Birimian of Ghana. They include Winneba, Cape Coast, Dixcove and Bongo granitoids (Junner, 1940; Kesse, 1985). The Cape Coast and Dixcove type granitoids are widespread in Ghana, the Winneba type is limited to small areas near Winneba, and the Bongo type crops out in the Bole-Navrongo Belt and in the Banso area.

2.3. Hydrogeology

The hydrogeology of Ghana is dominated by rocks of the Birimian Supergroup of the Early Proterozoic age and post intrusive granitoids, the Tarkwaian System and the unconsolidated sedimentary rocks. Gill (1969) delineated the hydrogeological Provinces of Ghana from the existing geological map of Ghana and Dapaah-Siakwan and Gyau-Boakye (2000) identified the major geohydrological regions from the overall geological map (Figure 2). In the assessment of Dapaah-Siakwan and Gyau-Boakye (2000), the country is dominated by two major hydrogeological provinces: the Basement Complex composed of the Precambrian igneous and metamorphic rocks constituting about 54% of the landmass, and the Paleozoic sedimentary basin which comprises the Voltaian sedimentary basin which underlies about 45% of the total landmass of Ghana. Based on the then knowledge of the hydrogeology of the country, Dapaah-Siakwan and Gyau-Boakye (2000) opined that two minor provinces: the Cenozoic, Mesozoic, and Paleozoic sedimentary strata along the coast and the Quaternary alluvium along major stream courses constitute the remaining 1% of the hydrogeology of Ghana. These divisions have been used to describe the hydrogeology of Ghana for some time now, although the existing data suggest that the geological formations lumped together as the Basement complex are significantly different in terms of their groundwater occurrence, well yield, and groundwater quality.

Groundwater occurrence in Ghana is based on the occurrence and intensity of secondary structures in the rocks. This is particularly true of the crystalline basement rocks and the Voltaian. In the rocks of these basins, primary permeability is virtually non-existent and the hydrogeological properties of the rocks are based on the degree of fracturing and weathering. Aquifer transmissivity of the productive zones of the Birimian and Tarkwaian Systems ranges between 0.2 m^2/d and 119 m^2/d, with an average of 7.4 m^2/d. In these aquifers, storativity ranges between 0.003 and 0.008. Transmissivity within the regolith is slightly higher than that observed in the integrated aquifer system, and ranges between 4 m^2/d and 40 m^2/d with an average of about 10 m^2/d. For the integrated aquifer systems in the Birimian and Tarkwaian Systems, borehole yields are generally low and range from 0.48 m^3/h to 36.4 m^3/h with a mean yield of 7.6 m^3/h. Differences in the degree of weathering within the granitoids probably account for the lower yields observed in these rocks.

In the Voltaian, groundwater occurs in fracture zones and bedding planes since the primary porosity of some of these rocks was reduced due to consolidation and cementation. The hydrogeological properties of the larger Voltaian Province are varied and based on the degree of weathering and fracturing of the otherwise impervious rocks. Available data suggest that borehole yield within the Lower Voltaian sub-province ranges between 1 and 9 m^3/h with an average of about 6.2 m^3/h. On the basis of available data from drilling projects in this sub province, success rate for drilling prolific wells and boreholes is about 55%.

Success rate for drilling successful boreholes in the Middle Voltaian sub-province is about 56% and the average borehole yield is about 6.2 m^3/h. Borehole discharge characteristics at some locations within this sub province indicates artesian conditions at shallow depths (Yidana et al., 2008a). In the Upper Voltaian sub-province, borehole yields in this range between 0.41 and 9.0 m^3/h and average about 8.5 m^3/h. Success rate for drilling successful boreholes in this is about 56%.

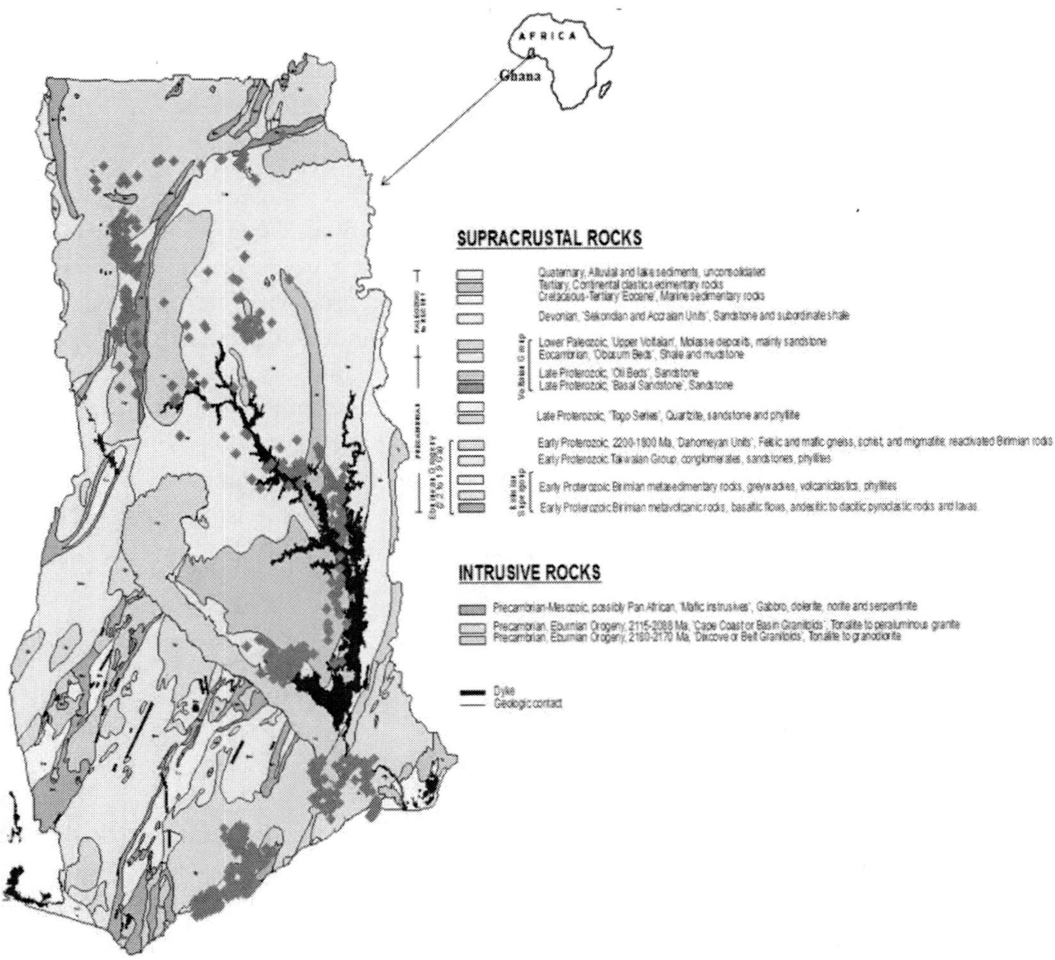

Figure 3. Geological map of Ghana showing the sampling locations.

The average thickness of the regolith over the Voltaian sedimentary rocks is about 9 m. However, previous investigations in the Nanumba and West Gonja districts of the northern region of Ghana revealed average regolith thickness of 6 m and 11 m respectively. The relatively thin regolith can be explained by the relatively stable clay (shale) and quartz (sandstone) composition or by the fine texture or ductile nature of the sedimentary rocks found in the Voltaian Province (Carrier et al., 2008). The yields of boreholes in the Buem formation range between 9 and 1260 l/min with an average and median of 100 and 78 l/min respectively. The average depth of boreholes in the area is 60 m. On the basis of available well test data, aquifer transmissivity values range between 0.2 and 11.4 m^2/day. Specific capacities range from 0.04 m^3/h/m to 1.23 m^3/h/m with an average of 0.47 m^3/h/m. Dapaah-Siakwan and Gyau-Boakye (2000) reported an average yield in the Togo Series aquifers to be about 9.2 m^3/h, ranging between 0.72 m^3/h and 24.3 m^3/h. The highest yielding wells and boreholes in the Togo Series tap fracture zones. Rocks of the Pan African Province are among the most prolific aquifers in the country, and can be relied upon to deliver economic quantities of groundwater for various purposes.

3. METHODOLOGY

3.1. Data Collection and Treatment

Nine-hundred and sixteen (916) data for the major physico-chemical parameters were used for this study. These data included a fair representation of all the major hydrogeological terrains in the country. The data included the concentrations of the major hydrochemical parameters including Electrical conductivity (EC), pH, sodium (Na), calcium (Ca), magnesium (Mg), potassium (K), bicarbonate (HCO_3^-), chloride (Cl^-), sulfate (SO_4^{2-}), nitrate (NO_3^-), fluoride (F^-), and phosphate (PO_4^{3-}). Data was obtained from four major reliable sources: World Vision, Ghana, the Ghana Water Resources Commission, the Community Water and Sanitation Agency (CWSA) and the DANIDA project at the Department of Earth Sciences, University of Ghana. All these are certified reliable sources of data on the hydrochemistry of the groundwater basins in the country. The stable isotope data (δ^2H and $\delta^{18}O$) were generally acquired from the World Vision Ghana, and the DANIDA project. Figure 3 shows the locations of the samples taken for this study.

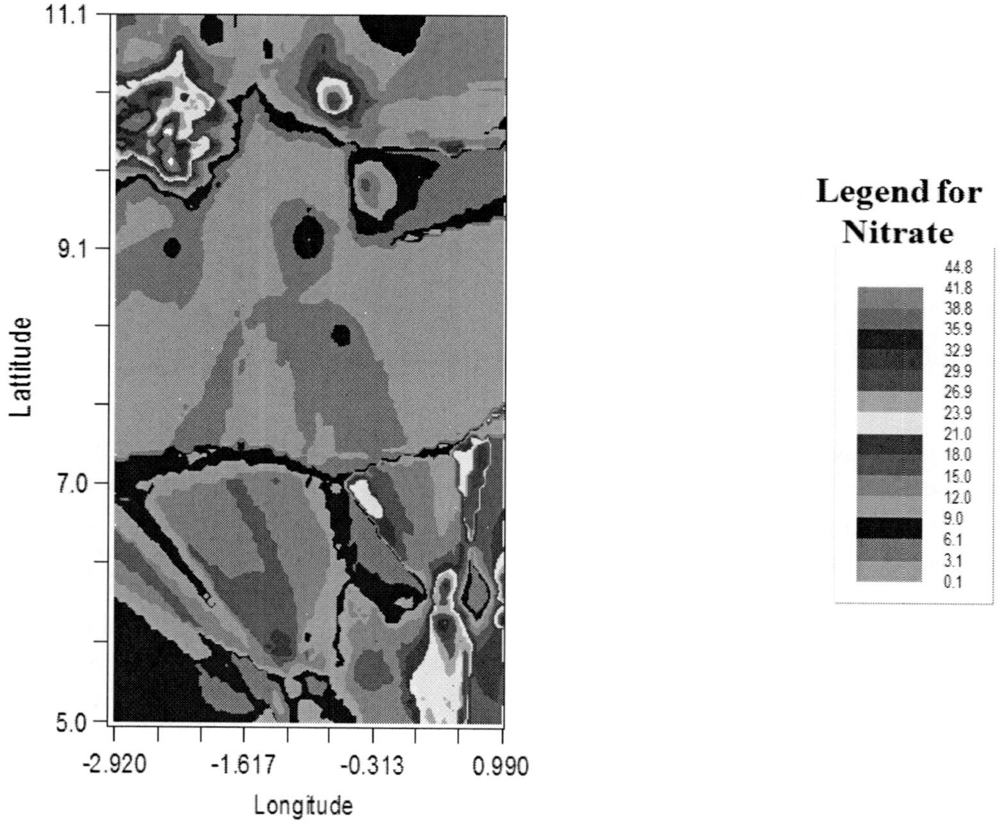

Figure 4. Nitrate distribution map showing locations of high nitrate concentrations in the country.

3.2. Data Processing and Analyses

The data were subjected to internal consistency tests using the charge balance error (CBE) defined by Equation 1. For internal consistency, the CBE for a sample should not exceed 5%. That is, the cations and anions should be within 5% of each other. Most of the samples from the Lake Busumtwi area did not meet this criterion and were dropped. Several reasons account for this: the samples were not adequately preserved in the field and the titration in the laboratory for the determination of the bicarbonate ion concentration amongst the samples. These samples that did not meet the internal consistency criteria were dropped and not used in the next phases of the interpretations of the hydrochemistry of groundwater in the country. Thirty (30) samples were dropped in this way.

$$\text{Charge Balance Error (CBE)} = \frac{\sum Cations - \sum Anions}{\sum Cations + \sum Anions} x100\% \quad (1)$$

where the concentrations of the ions are in milliequivalents per liter (meq/l).

The hydrochemical data for the remaining eight-hundred and eighty-six (886) samples were standardized to their corresponding z-scores using Equation 2. Normal distribution of data is a requirement of optimal multivariate statistical analyses. This is because the analyses make sense if the standard deviations (or variances) are low (very close to 0). Otherwise, the analyses would be influenced mostly by the parameters with the highest variances (Güler et al., 2002; Cloutier et al., 2008). Data standardization helps to achieve the objective of low sample variances among all the parameters analyzed. The standard z-scores were then used for the next phases of the analyses.

$$z = \frac{x - \bar{x}}{s} \quad (2)$$

Where x, $\bar{x}$, s respectively represent the value, mean and standard deviation of the parameter at the sample location in question.

Q-mode hierarchical cluster analysis (HCA) was applied to the data using squared Euclidean distances as a measure of similarity/dissimilarity and ward's linkage as an agglomeration method as a linkage criterion. The combined use of the squared Euclidean distance as a similarity/dissimilarity measure and the Ward's method as a linkage measure among initial clusters has been determined to be the best combination at revealing the most unique set of spatial sample associations in cluster analysis. Cluster analysis is a multivariate statistical method which groups samples or parameters into groups (clusters) based on their relationships in the field. The most similar parameters or samples are clustered together in the same group. Even within the same group, the most similar parameters or samples are clustered at shorter linkage distances. This multivariate statistical analysis has been used extensively in environmental data analyses and interpretations. There are many clustering methods available under cluster analysis and each is suited to a particular purpose. Hierarchical cluster analysis, as the name implies, clusters the samples or variables into hierarchical classes based on similarities/similarities in the field, and is the major clustering method used in Earth Sciences (Davis, 1986). It is often used in hydrogeochemical studies (e.g. Güler et al., 2002; Cloutier et al., 2008; Yidana, 2010) to help explain the causes of variation in groundwater and surface water geochemistry, and assist in the definition of spatial hydrochemical facies based on field associations. One can cluster either the variables (parameters) or the samples in HCA. It is the clustering of the samples that Davis (1986) referred to as Q-mode HCA. The other type is R-mode HCA. In this study, both the Q-mode and R-mode HCA were used in the classification of the hydrochemical dataset. The Q-mode HCA was used to divide the groundwater in Ghana into hydrochemical zones.

The R-mode HCA was applied to the standard z-scores of the parameters to divide the data into clusters based on genesis. This process was intended to be used with the R-mode factor analysis to determine the most probable sources of variation in the hydrochemistry on a national scale. Q-mode clustering was performed on a sequential basis: the first clustering procedure resulted in an initial 17 clusters, which were then clustered into two groups. Squared Euclidean distance and Ward's method were used respectively as similarity/dissimilarity measure and a linkage method for initial cluster for both cases of HCA. The Ward's method is distinct from other clustering method as it uses an analysis of variance approach to evaluate the distances among clusters (Cloutier et al., 2008).

Factor analysis was applied to the z-scores of the parameters for all the data. Principal component analysis (PCA) was used as an extraction method, and the Kaiser criterion (Kaiser 1960) was used to determine the maximum number of factors that can be include in the factor model. Varimax rotation was applied to maximize the variance among factors in order to facilitate interpretation. In the Kaiser criterion, only factors with an eigenvalue of 1 and above would be included in the factor model. This makes practical sense since factor representing unique processes in the hydrochemistry should be able to at least account for themselves. Weighted factor scores were calculated for each of the factors that resulted from the factor analysis with varimax rotation. The weighted factor scores indicate the severity of the processes represented by each of the factors at each sampling location.

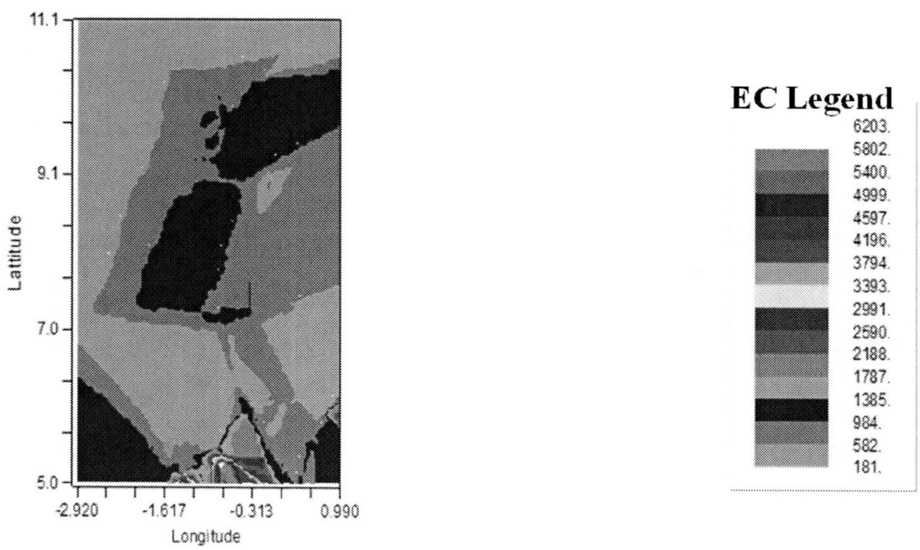

Figure 5. EC distribution map for Ghana. Notice that the high EC values are closest to the coast due to the seawater effects.

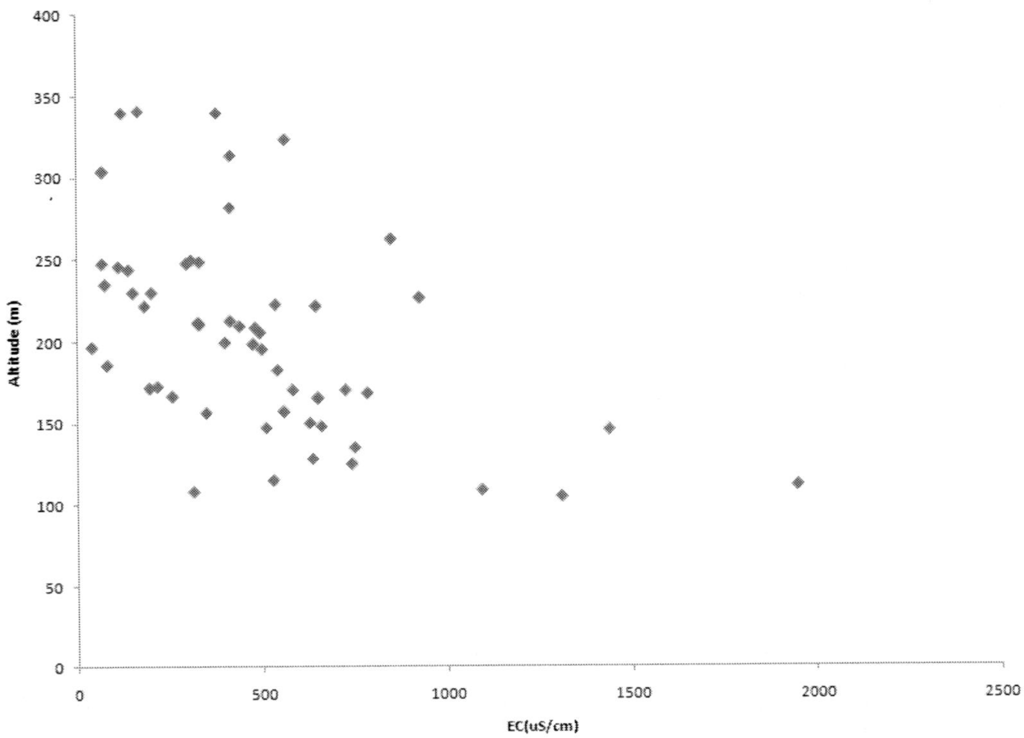

Figure 6. The relationship between EC and altitude in the Buem Formation.

An interpolation procedure was then applied to the factor scores to produce maps which indicate the strength of each of the processes influencing the hydrochemistry throughout the country. Ordinary kriging was used for the interpolation, with the appropriate semivariogram models. Kriging is a sophisticated interpolation method which uses both the proximity measure and the fitness of the semivariogram in predictions. It has an inherent advantage over other interpolation methods such as the inverse distance weighting (IDW) which relies on only the proximity measure.

4. Results and Discussions

4.1. Sources of Variation in the Hydrochemistry and Spatial Groundwater Types

Table 1 shows the statistical summaries of the hydrochemical parameters analyzed for this study. The standard deviations of the parameters are very large, suggesting significant differences among the major hydrochemical parameters in Ghana. Generally, concentrations increase towards the coast. The highest EC value of 43,540 μS/cm was recorded in the Keta basin among the unconfined aquifers. The lowest EC of 28 μS/cm was recorded in the granites. Figures 4 and 5 respectively show the spatial distribution of nitrate and EC in groundwater in Ghana. High nitrate concentrations have been observed in areas where the aquifers are relatively shallow, leading to high impacts of chemical fertilizers, organic manure, and domestic wastes. Concentrations are particularly high in the Keta basin, and some of the shallow aquifers of the Birimian and Voltaian sediments in the mid-sections of the country. These are areas where the well head protection is generally poor and where chemical fertilizers and other chemicals are being heavily used on cocoa farms and vegetable irrigation farms. In the northern sections, there are some patches of high nitrate due to the effects of free range cattle raring and the use of manure for agricultural activities. It has been observed that EC concentrations are general low where the altitude is high. This relationship is illustrated in Figure 6 for the Buem formation in southeastern Ghana. This is probably in line with the general assertions that the high elevations are recharge areas where the concentrations of the ions are generally low, whilst the low elevation areas, which are typically discharge areas have high ionic contents leading to high EC values. This observation is not necessarily the general situation in Ghana, and applies to areas where anthropogenic activities are minimal. High anthropogenic effects and significant marine influence in the Keta basin work to distort this general relationship.

The first phase of the Q-mode HCA resulted in seventeen (17) initial clusters. Table 2 presents the mean and median concentrations of the chemical parameters under each of the 17 initial clusters. The phenon line was drawn across the dendrogram at a linkage distance of 5. The phenon line can be moved up or down the dendrogram to decrease or increase the number of clusters. This aspect of cluster analysis makes it a semi-objective one and is based on the researcher's subjective choice of clusters. The choice of the position of the phenon line and the corresponding number of clusters is advised by due consideration of the researcher's idea about the number of clusters to be expected in terms of the researchers knowledge of the study location. In some cases, the process of selecting the position of the phenon line is repeated several times to arrive at the most optimal position which leads to the most robust hydrochemical model to describe the spatial relationships in the basin.

Table 1. A statistical summary of the parameters analyzed for this study

Parameter	Mean	Standard deviation	Minimum	Maximum
EC(μS/cm)	1245.82	2725.24	28.10	43540.00
pH	6.83	0.90	2.96	10.66
HCO_3^-(mg/l)	228.33	162.58	0.00	1148.02
SO_4^{2-}(mg/l)	134.26	351.20	0.00	5326.95
Cl^-(mg/l)	248.33	1162.59	0.15	22470.98
NO_3^-(mg/l)	11.37	31.48	0.00	449.42
PO_4^{3-}(mg/l)	0.35	3.79	0.00	104.01
F^-(mg/l)	0.76	1.73	0.00	19.84
Ca^{2+}(mg/l)	75.53	145.79	0.00	2219.56
Mg^{2+}(mg/l)	68.55	226.44	0.28	4028.64
Na^+(mg/l)	161.82	656.35	0.00	12493.43
K^+(mg/l)	11.26	29.24	0.09	565.27

Table 2. Mean and Median concentrations of the major chemical parameters in the 17 clusters

Cluster	Cluster 1		Cluster 2		Cluster 3		Cluster 4		Cluster 5	
Para-meter	Mean	Median	Mean	Median	Mean	Median	Mean	Median	Mean	Median
EC	9807.30	8650.00	17968.29	10400.00	2887.19	2100.00	10900.00	12435.00	2494.10	2190.00
pH	7.14	6.95	7.49	7.38	7.31	7.29	7.28	7.27	7.09	7.09
HCO_3^-	226.76	245.42	381.68	282.91	396.44	431.88	798.80	863.76	188.52	167.94
SO_4^{2-}	2098.25	1680.24	1626.35	1789.22	188.67	150.37	519.53	447.26	438.92	416.77
Cl^-	2553.19	2243.29	7665.13	2883.87	602.19	278.47	3492.89	3986.63	522.10	457.77
NO_3^-	5.29	1.47	3.24	2.56	21.22	4.62	29.42	4.51	6.26	2.90
PO_4^{3-}	0.46	0.22	0.16	0.07	0.62	0.12	1.20	1.20	0.19	0.11
F^-	0.70	0.30	2.93	1.00	5.65	4.85	7.69	4.39	0.50	0.41
Ca^{2+}	795.97	529.94	692.21	745.51	113.59	89.97	404.87	200.46	186.35	173.65
Mg^{2+}	1336.54	1311.11	1059.80	744.69	51.28	39.87	254.15	242.41	191.19	167.90
Na^+	982.99	773.65	4285.11	1644.72	360.12	194.22	2590.11	2568.87	273.62	211.09
K^+	13.89	14.41	34.57	16.26	61.82	69.00	297.23	223.41	13.02	13.07

In this research, the best results were obtained when the phenon line was drawn across the dendrogram at a linkage distance of 5. It is apparent in Table 2 that some of the clusters are similar in the distribution of the hydrochemical parameters. For instance clusters 2 and 4 are similar in terms of the EC, although the concentrations of the individual parameters are different in the two groups. These two clusters are the most saline of the 17 clusters, and are not suitable for any purpose. Most of the members of these two clusters are derived from aquifers in the Keta basin, and epitomize seawater effects in the basin. Groups 1, 3, and 5 are characterized by EC values which are higher than 2000 μS/cm, suggesting medium to high salinity groundwater associations. All the others are freshwater groundwater groups whose ionic constitutions are mainly derived from the weathering of minerals. They are differentiated by differences in salinity and pH. The most acidic groups are associated with the Birimian aquifers and represent the effects of the oxidation of sulfides. Groups 9 and 10 are derived from the aquifers of the Birimian metasediments. They have the lowest EC values of all the groundwater associations in the basin. They also present the most acidic groundwater types in the country. Whilst the EC values are generally low, the pH values are quite low and fall outside of the acceptable range for potable water resources (WHO, 1984). The low pH creates the rightful environment for the leaching of heavy and toxic elements into the groundwater system. Heavy metals were not analyzed in this research but there have been indications of acid mine drainage in the Birimian aquifers.

Q-mode HCA was repeated on the mean concentrations of the 17 initial clusters. This resulted in two clusters representing two major groundwater hydrochemical types in the entire country. The dendrogram for the second phase of HCA is presented in Figure 7. The phenon line was drawn across the dendrogram at a linkage distance of 13. Cluster 1 consists of all the fresh groundwater clusters in Table 2, whilst cluster 2 comprises the most saline groundwater types. The Schoeller diagram for clusters 1 and 2 is presented in Figure 8. It is obvious that cluster 1 represents fresh Na-K-HCO_3 water types with relatively low concentrations of the major hydrochemical parameters. These water types have a fair representation from all the major hydrogeological units in the country, especially those located farther inland. Na-K-HCO_3 water types are typical of groundwater from the Obusum beds of the Voltaian in the north of the country, and some portions of the Birimian where cation exchange activity is predominant. Cluster 2 represents a Na-Cl water type which has a significant input from seawater intrusion. As expected, these water types are located along the coast, and have a significant seawater effect. The Chadha (1999) diagram (Figure 9) was used to illustrate the specific water types represented by the seven (17) initial clusters identified from the initial cluster analysis. There is a distinct direction of groundwater hydrochemical evolution from fresh Ca-HCO_3 groundwater types to saline Na- Cl groundwater types from the more inland aquifers towards the coast. The clusters which fall within the freshwater category predominantly belong to the Upper Birimian, Voltaian, Buem, and Togo aquifers, whilst majority of the saline waters belong to the Keta basin. Ca-HCO_3 waters are commonly associated with the weathering of carbonate minerals. However, incongruent weathering of silicate minerals in the presence of carbonic acid can also generate Ca-HCO_3 water types. It is widely accepted that in most of the hydrogeological terrains in Ghana, silicate mineral weathering is one of the major factors affecting groundwater hydrochemistry. Where cation exchange activity is predominant, Na-K-HCO_3 waters are common due to the selective adsorption of Ca-ions onto clay mineral surfaces and the release of Na-ions into solution in the freshwater aquifers of the country. Studies of Kortatsi et al. (2008), Yidana and Yidana (2010) attribute observed hydrochemistry of groundwater from the Birimian and Voltaian aquifers largely to silicate mineral weathering and cation exchange.

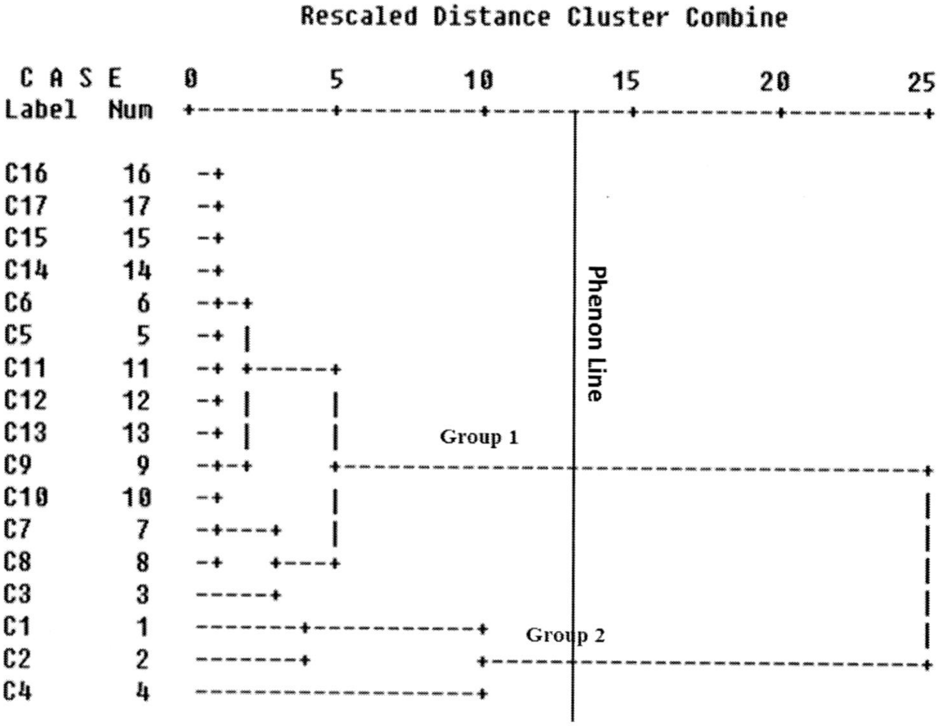

Figure 7. Dendrogram for the second phase of Q-mode HCA.

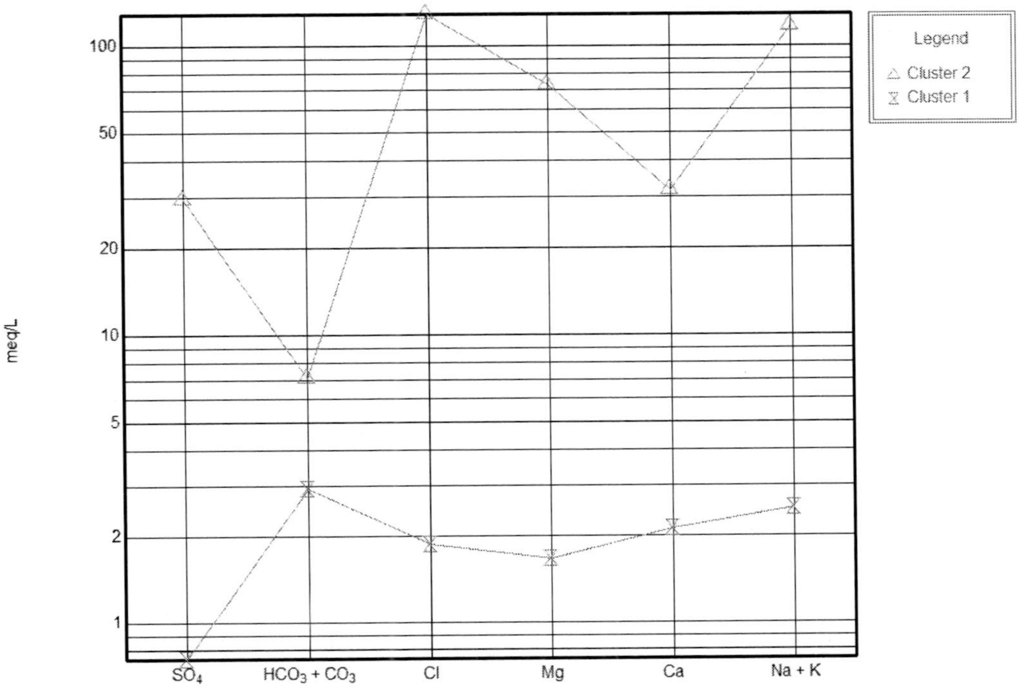

Figure 8. Schoeller diagram for the two clusters resulting from the Q-mode HCA.

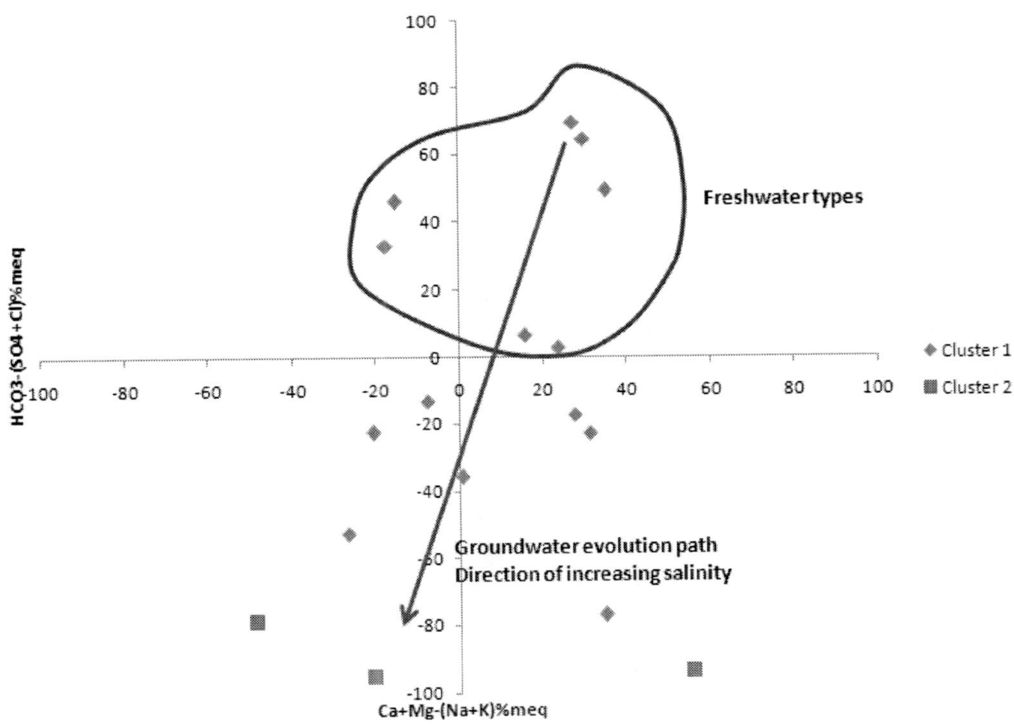

Figure 9. A Chadha diagram for the 17 initial clusters of the Q-mode HCA.

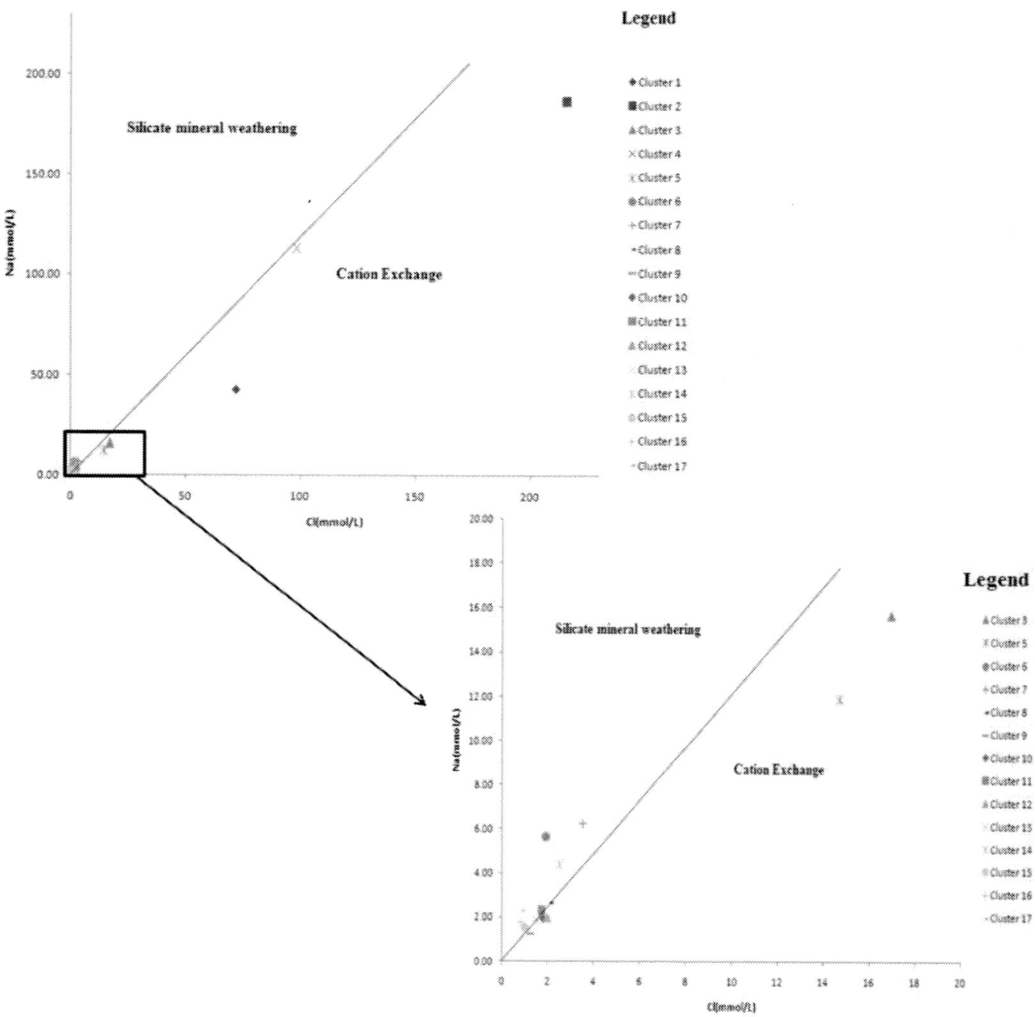

Figure 10. The relationship between Na^+ and Cl^- ions in groundwater in Ghana.

The results of the R-mode factor analysis are presented in Table 3. Table 3 presents the factor loadings for the three most important processes affecting groundwater hydrochemistry in Ghana. Each of the three factors has an eigenvalue of at least 1, and, the three factors together account for about 68% of the variance in the hydrochemistry. It is practically possible to obtain as many factors as there are parameters in factor analysis. However, most of the factors would constitute 'noise', with no significant roles in the hydrochemistry. Factor 1 has high positive loadings for majority of the major ions and EC. It correlates positively with EC, Na^+, Ca^{2+}, SO_4^{2-}, and Cl^-. The Mg^{2+} ion was dropped after the first round of the analysis showed that it had correlated with more than one of the factors, and would therefore not play any role in assessing the different processes influencing the hydrochemistry. Factor 1 is a mixed factor representing the effects of saline water intrusion and mineral weathering processes in both the saturated and unsaturated zone. In the coastal zones, seawater intrusion is significant and is the major factor controlling groundwater salinity in the aquifers, especially in the unconfined aquifers of the Keta basin. Elsewhere in the Obosum beds in the north, the dissolution of halite which is present in high concentrations in the rocks. Seawater effects explain the significantly high salinities in groundwater from the coastal environments. These effects of seawater intrusion are absent farther inland, and the predominant source of variation in the hydrochemistry is mineral weathering, principally silicate mineral weathering. Figure 10 is a linear plot of Na^+ and Cl^- concentrations using the averages of the measures of the parameters for the 17 cluster identified in the first phase of the HCA. Concentrations of both parameters are in mmol/L. The clusters are split between silicate mineral weathering dominance and cation exchange. Yidana et al (2008c) made similar observations in the southern Voltaian basin using inverse geochemical modeling with PHREEQC (Parkhurst and Appelo, 1999). All the initial clusters which were grouped into cluster 2 in the second phase of the HCA show ion exchange dominance, although the most significant factor in the hydrochemistry within this water group is marine water influence. Figure 11 also shows the apparent dominance of cation exchange activity over silicate mineral weathering in the hydrochemistry of groundwater in Ghana to account for the distribution of the bicarbonate ion and calcium in groundwater. Cation exchange reactions may have been fueled by the presence of clay sized minerals which resulted from incongruent silicate mineral weathering in the aquifers. Studies of Yidana et al. (2008b) suggest that montmorillonite is the most stable silicate phase in the hydrochemistry of groundwater in the southern Voltaian aquifers in the Afram Plains area. The same study suggested the transformation of anorthite and albite to their respective clay mineral equivalents along identified groundwater flowpaths in the area. Such a thorough study has not been conducted on a national scale but similar results have been observed in some of the Birimian aquifers in the Ankobra basin.

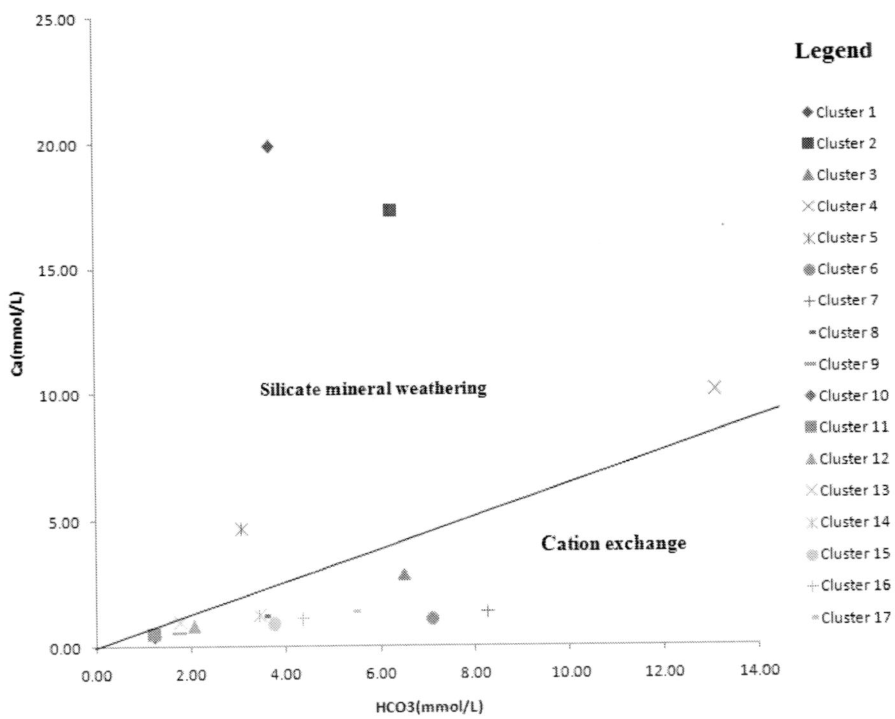

Figure 11. The relationship between Ca^{2+} and HCO_3^- ions in groundwater in Ghana.

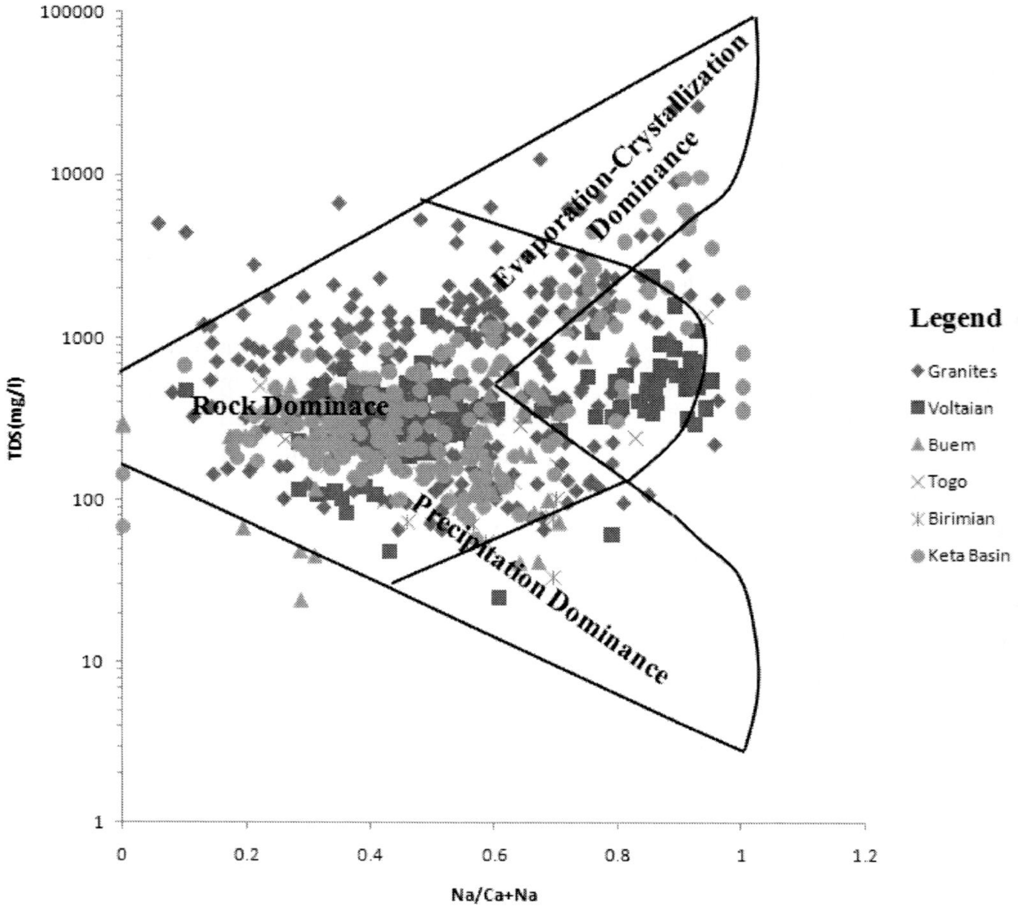

Figure 12. A Gibbs diagram for groundwater in Ghana.

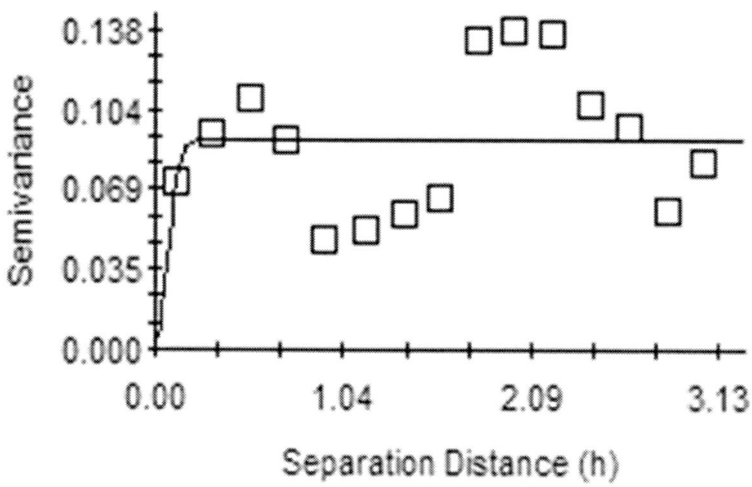

Figure 13a. Semivariogram model for factor 1.

Table 3. Factor loadings for the three factors after Varimax rotation and the application of the Kaiser (1960) criterion

	Component		
	1	2	3
EC	.939	.045	.122
pH	.050	-.041	.740
HCO_3^-	.046	.200	.768
SO_4^{2-}	.810	.067	-.029
Cl^-	.935	-.021	.046
NO_3^-	-.026	.675	-.247
F^-	.055	.743	.238
Ca^{2+}	.791	.209	.008
Na^+	.893	.051	.101
K^+	.210	.754	.207

A Gibbs diagram for data from all aquifer units in Ghana suggests that rock weathering is the predominating process in the hydrochemistry. This buttresses the point that factor 1 represents the mineral weathering and marine water influence. In Figure 12, it is apparent that some of the shallow aquifers with the weathered granites of the Central region have been affected by the evaporative effects of the high dry season temperatures in the area. Some shallow wells in the Keta basin have also been similarly influenced by evaporative effects. Hydrochemistry of groundwater from the Birimian metasediments is predominantly influenced by the weathering of minerals in the aquifers. The Birimian metasediments are among the freshest groundwater aquifers in the country, delivering groundwater of low EC values. Perhaps the major problem with these aquifers is the low pH, probably arising from the oxidation of sulfides to their corresponding sulfates. The Buem formation is arguably one of the very best in the country in terms of groundwater quality. Figure 12 suggest rock weathering as the predominating influence on the hydrochemistry in this formation, and the parameters are within acceptable limits for most purposes in the country. The Voltaian and Togo aquifers are also largely controlled by mineral weathering processes in the aquifers, but some shallow wells in the Afram Plains area of the southern Voltaian have had some evaporative enrichment of the total dissolved solids (TDS).

The second factor accounts for 16.8% of variance in the hydrochemistry and has high positive loadings for NO_3^-, K^+, and F^-. It represents the contribution of agricultural chemicals and domestic waste on the hydrochemistry of groundwater in Ghana. This is the second most important set of processes that influence groundwater quality in Ghana. The effects of factor 2 are particularly high in areas where the aquifers are shallow and unconfined. The impacts are high in the Keta basin, where irrigation activities take place throughout the year, and chemical fertilizers and organic manures are being used. In most locations within the basin, the depth to water is less than 5 m especially in the rainy season when the water table is quite high. Recharge is predominantly from rainfall which leaches down surface contaminants into the groundwater reservoir. In some parts of the northern and upper regions which are underlain by the Voltaian and Birimian rocks respectively, high impacts of agricultural activities are also observed.

Factor 3 has high positive loadings for pH and the HCO_3^-, and accounts for about 11.5% of the total variance in the hydrochemistry. It represents the oxidation of organic matter in the unsaturated zone. Factor 3 is also observed to load negatively (-0.2) with nitrate, suggesting that the oxidation of organic matter is facilitated by the reduction of nitrate in the unsaturated zone. This process is accompanied by an increase in the pH of the medium.

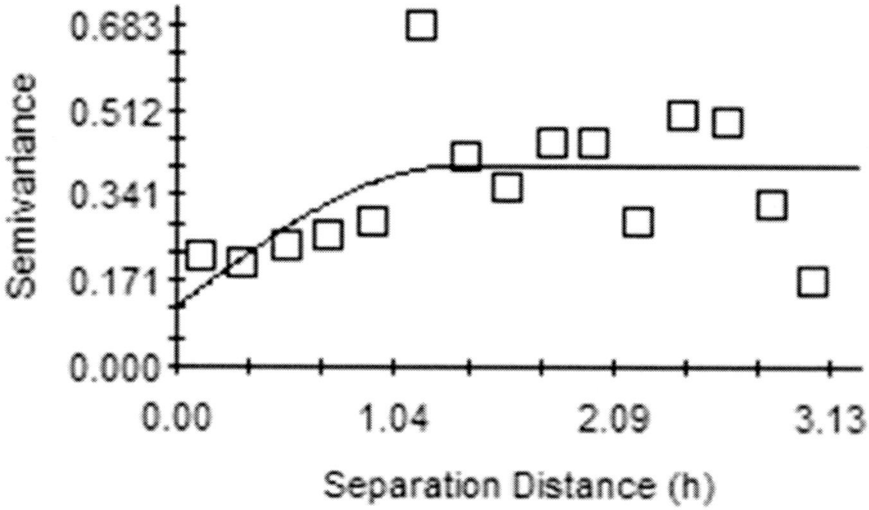

Figure 13b. Semivariogram model for factor 2.

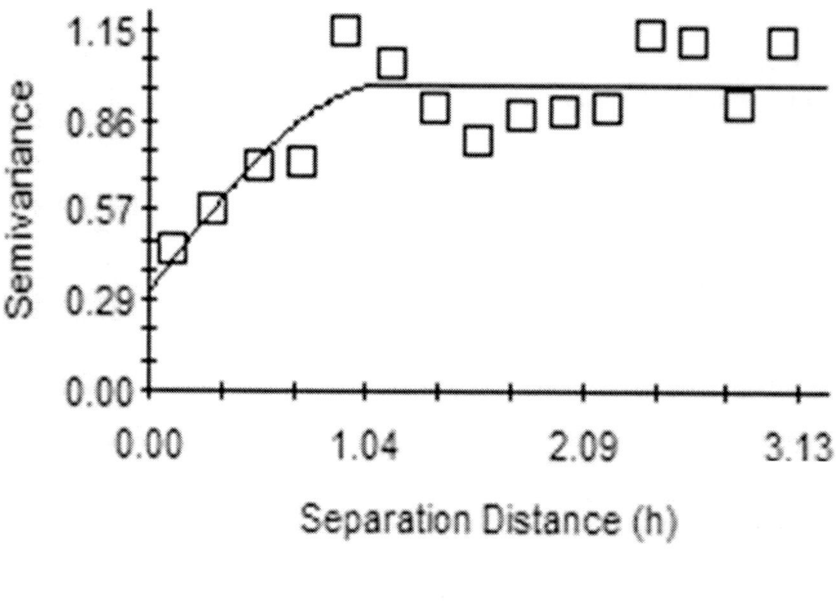

Figure 13c. Semivariogram model for factor 3.

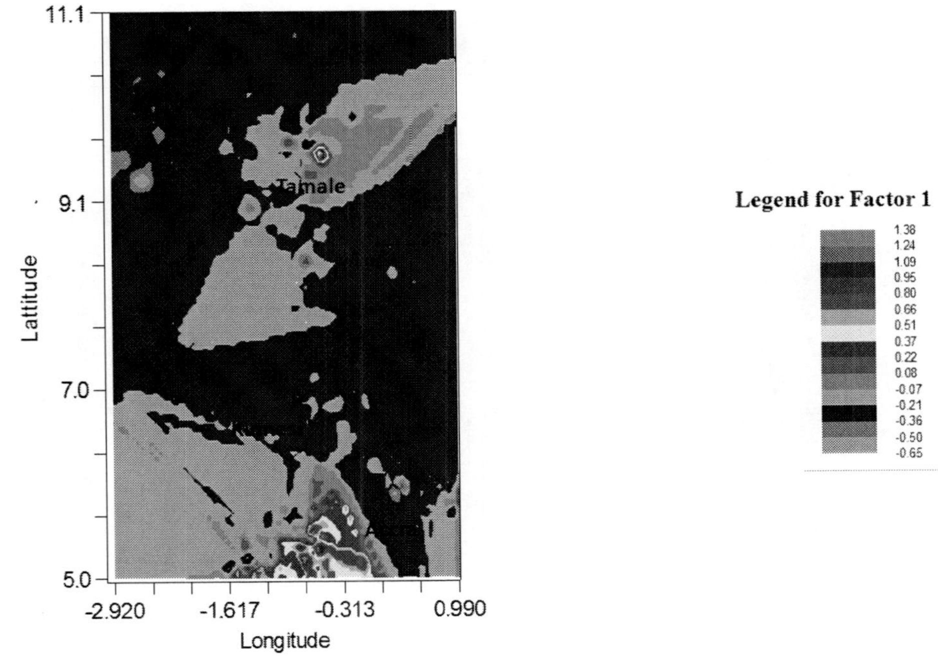

Figure 14a. The variation in the intensity of factor 1 in Ghana.

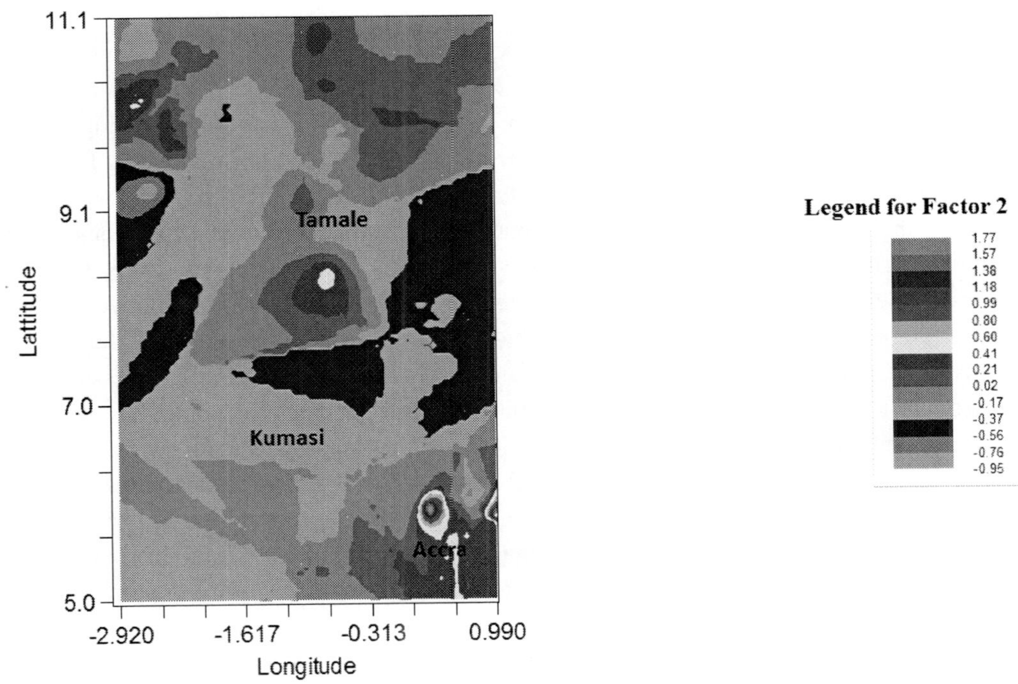

Figure 14b. Variation in the distribution of factor 2 processes in Ghana.

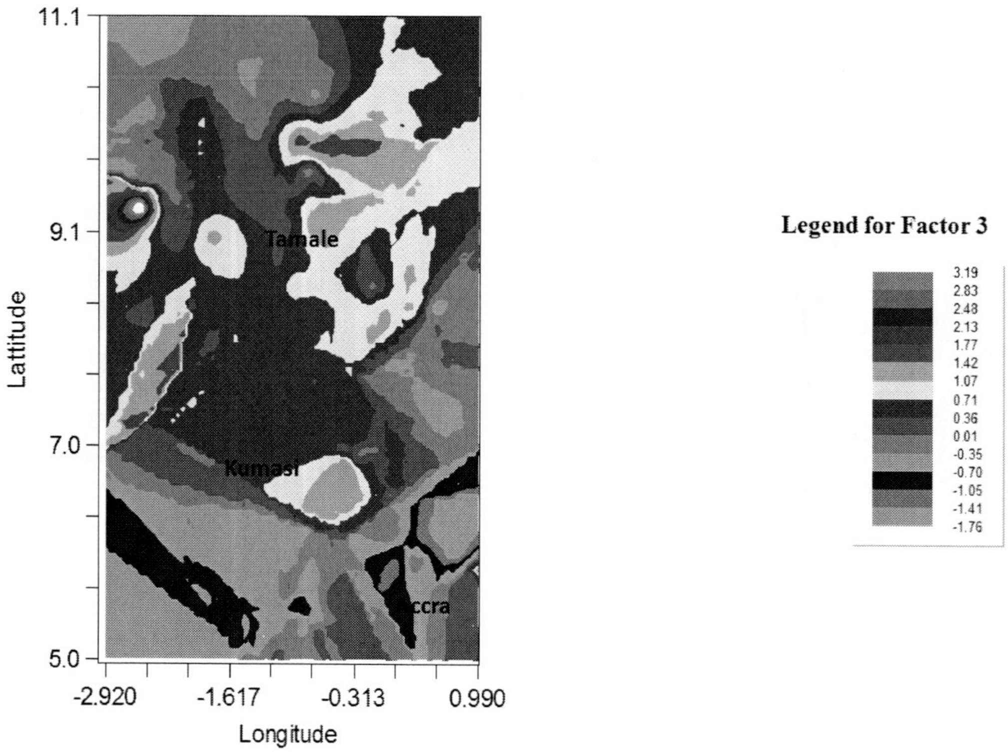

Figure 14c. Variation in the intensity of factor 3 processes throughout Ghana.

Factor scores for all three factors were used to generate prediction maps showing the strength of each of the set of processes represented by the factors on a national scale. Positively high factor scores suggest high intensity of the processes represented by the factor, whereas the converse is true for negative factor scores. Ordinary kriging was used to interpolate the factor scores which were computed from SPSS version 17. A Gaussian model was fitted for factor 1 whilst spherical models were fitted for both factors 2 and 3. Figure 13 presents the semivariogram models for all three factors. These semivariograms were used to produce interpolation maps (Figure 14) for each of the factors in the entire country. It is obvious that factor 1 is highest along the coast because of significant marine influences. In the northern and midsections, the observed patterns are largely attributed to mineral weathering in both the saturated and unsaturated zones. The weathered granite aquifers in the central regions, the Voltaian in the north and midsections of the country have indicated high influences of mineral weathering on their groundwater hydrochemistry. All the hydrogeological basins in the country have significant contributions of mineral weathering to their hydrochemistries. Perhaps one would have expected uniformly positive factor scores for factor 1 throughout the country. However, this expectation is complicated by the fact that factor 1 also represents seawater intrusion. Highly positive factor scores are associated with areas where both seawater effects and mineral weathering are significant such as in the coastal belt. Where mineral weathering is the predominant influence, the factor scores are in the range of -0.20 to 0.60, as observed in the northern and midsections. Aquifers of the Togo series and Buem formation in the Volta region show similar patterns in terms of the contribution of mineral weathering to their hydrochemistry. The contribution of mineral weathering (predominantly silicate mineral weathering) is mild, and probably explains the low EC values associated with groundwater from these aquifers. Silicate mineral weathering is characteristically slow and takes millions of years to register significant effects (Drever, 1988; Appelo and Postma, 2005). A similar trend is observed in the Birimian, where the hydrochemistry is also dominated by silicate mineral weathering (Kortatsi et al., 2008). The degree of influence of silicate mineral weathering is therefore a function of the residence time of groundwater in the aquifers.

The effects of the processes in factor 2 are highest (Figure 14a) in locations where the aquifers are relatively shallow and where agricultural chemicals are heavily used. In addition, high scores for factor 2 (Figure 14b) are observed in areas where domestic waste is not properly disposed, leading to the leaching of nitrate rich chemicals into the groundwater system. The connectivity between surface activities and the aquifers have led to the leaching of the nitrate from anthropogenic activities into the aquifers especially the shallow ones. In some portions of the northern and upper regions of the country, the high values are associated with cattle ranging and intense application of nitrate rich manures and chemical fertilizers. Extensive agricultural activities take place in the northern and Brong Ahafo regions of the country in which food crops are produced. These regions have sometimes been referred to as the food baskets of the country due to the intensity of agricultural activities there. Poor sanitary conditions around some wellheads have led to a connection between aquifers and domestic sewerage and agricultural chemicals. Tube wells and manually drilled wells in the north are usually not protected or covered from surface flows during rains. For this reason, all manner of solid and liquid wastes find their way into wells after heavy downpours. The hydraulic connection between the wells and aquifers gradually lead to the contamination of whole aquifers as nitrate rich surface waters drain into exposed wells. Most villages have no place of convenience for the disposal of human excreta. The populations adopt the practice that is popularly referred free range, whereby they defecate in the open field. In Zangu-vuga, Yakura, Wungu, and other communities in the West Mamprusi district of Northern region, this practice is very rampant. In the rainy season, fecal matter laden runoff drain into shallow wells which they at their farms without any wellheads or proper covering.

In Accra and some portions of the Keta basin, poor sanitary conditions and the very shallow and unconfined conditions of the aquifers in the Keta area partly account for the high scores for factor 2. In the case of Accra where increases in population has placed an extra burden on the conventional water sources, landlords and landladies have resorted to drilling wells and boreholes without the required initial investigations and licensure. This has led to the proliferation of wells and boreholes. High septic tank density without due recourse to proper sanitary practices is another major contributor to the high nitrate levels in groundwater in Accra and neighborhood. In some cases, the septic tanks are deep enough to reach the groundwater table, and as a result of lack of proper lining, the contents of these tanks leak into the surrounding aquifers.

The high effects of factor 3 are localized and probably take the same trend as observed in factor 2 (Figure 14c). In the Ashanti and Western regions, the high effects of factor 3 are associated with the thick vegetation, whereas the intense agricultural activities are responsible for high loadings in the Northern and Brong Ahafo regions of Ghana. The effects are particularly high in the north because of the availability of nitrate from agricultural activities to oxidize organic carbon.

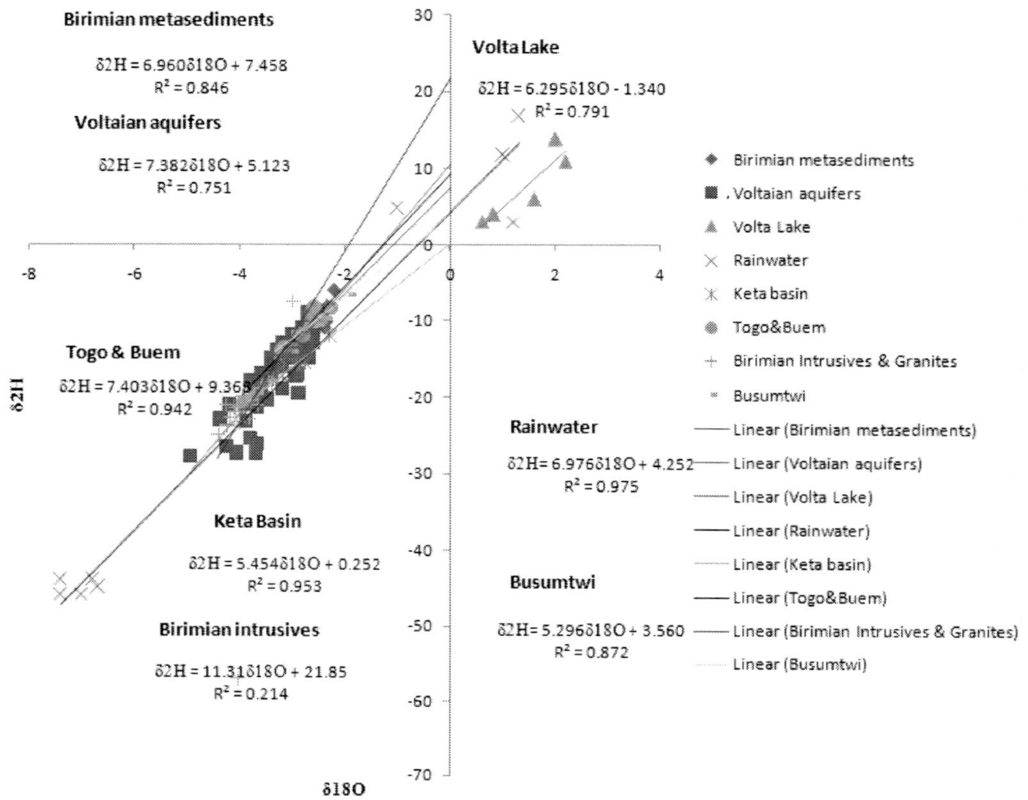

Figure 15. The relationship between $\delta^{18}O$ and δ^2H in groundwater in Ghana.

4.2. The Recharge Regime of Groundwater in Ghana

Stable isotopes of oxygen and hydrogen have been widely used to determine the probable sources of groundwater recharge in many groundwater studies throughout the world. Groundwater recharged during cool events such as the Pleistocene is relatively depleted with respect to the heavier isotopes of both elements (^{18}O and ^2H) (Ophori, 1999). In addition, groundwater recharged in the higher latitudes is relatively depleted with respect to the heavier isotopes. Neither isotope is affected by silicate mineral weathering under the ordinary temperature conditions that prevail in most aquifers (temperatures lower than 50ºC). These characteristics of the heavier isotopes are therefore useful in determining the sources of recharge of groundwater.

The relationship between the distributions of both isotopes in a hydrogeological terrain is frequently compared with the Global Meteoric Water Line (GMWL) (Craig, 1961), which is the standard for determining meteoric recharge. Recharge from precipitation which has suffered evaporative effects under a relative humidity less than 100% can lead to slope of the δ^2H-$\delta^{18}O$ plot of 5±2 (Gonfiantini, 1986). The Volta Lake is very much enriched with respect to the heavier isotopes of both elements, with $\delta^{18}O$ and δ^2H ranging between 0.60 to 2.20 and 3.00 to 14.00 respectively. This is clearly the effects of high evaporative enrichment attending high dry season temperatures and low humidities (<70%). This isotopic enrichment of the surface water of the Volta Lake starkly contrasts what pertains in the isotopic signature of groundwater from the Voltaian basin, which are relatively depleted with values ranging from -4.95 to -2.50 and -27.80 to -9.00 respectively for $\delta^{18}O$ and $\delta^{2}H$. This dispels earlier suggestions that aquifers of the Voltaian sedimentary basin receive recharge from the Volta Lake. This range of values for both isotopes is in keeping with recharge from meteoric waters which have undergone some degrees of evaporative enrichment prior to recharge. The data also plot below the GMWL with a relatively shallower slope and *d*-excess value, suggesting that groundwater in the basin is relatively much more enriched compared to the average global meteoric water (Figure 15). The isotopic signature of the Voltaian aquifers is also relatively enriched compared to recent precipitation in the area. $\delta^{18}O$ and $\delta^{2}H$ data for recent rainfall samples taken from the Afram Plains area fall within the range of -7.40 to 1.30 and -46.00 to 17.00 respectively. The slight enrichment observed is probably attributed to evaporative enrichment of rain drops due to high temperatures and low humidities at the time rainfall samples were taken. The rainwater samples were taken at the end of the major rainy season in the area. Mathieu and Bariac (1996) suggest that during the beginning or the end of the major monsoon or rainy season, the rainfall is light and relatively enriched with respect to the heavier isotopes.

In the Keta basin, heavier isotopes are relatively enriched. Values range between -3.96 to -2.28 and -22.10 to -12.00 for $\delta^{18}O$ and $\delta^{2}H$ respectively. The data also plots slightly below the GMWL with a shallower slope and relatively lower *d*-excess value (Figure 14). This observed enrichment is owed to two possible factors: seawater effects and evaporative enrichment of the meteorically recharged shallow aquifers of the Keta basin. Water samples from the Gulf of Guinea collected and analyzed by Jorgensen and Banoeng-Yakubo (2001) indicate values of both isotopes close to 0 which is the isotopic signature of ocean water. Evaporative enrichment has a significant effect upon the isotopic distribution in the Keta basin since the aquifers sampled are very shallow and relatively humidity in the area ranges between 65 and 75% (Dickson and Benneh, 1995).

The Birimian metasediments are more enriched in the heavier isotopes of both elements than the intrusives and weathered granites. The data for the Birimian metasediments range between -3.36 to -2.22 and -17.00 to -6.10 for $\delta^{18}O$ and $\delta^{2}H$ respectively, whilst the distribution for the intrusives and weathered granites range from -4.42 to -2.99 and -57.10 to -7.50 for $\delta^{18}O$ and $\delta^{2}H$ respectively. The difference is probably attributable to the fact that aquifers of the Birimian metasediments are generally shallow and much more prone to evaporative effects than the intrusive granitoids. The relationship between $\delta^{18}O$ and $\delta^{2}H$ in the intrusives is very weak ($R^2=0.214$), but the slope is steeper than that of the GMWL and the Local Meteoric Water Line (LMWL) of Akiti (1987) and a higher *d*-excess. The steeper slope and higher *d*-excess are attributed to the relatively stronger depletion of the heavier oxygen isotope relative to meteoric water than deuterium. A number of factors could account for this observed disparity. $\Delta^{2}H$ is relatively unaffected by rock-water interactions. On the other hand, $\delta^{18}O$ values can be affected by reactions with carbonate minerals in the unsaturated zone. In both cases, the isotopic signatures are within ranges that can be ascribed to meteoric recharge with slight enrichment in the case of the metasediments.

Perhaps the closest to the GMWL and the LMWL is the distribution of the isotopic data of the Buem and Togo aquifers (Figure 15). Meteoric recharge with slight evaporative enrichment is suspected of the aquifers of both hydrogeological terrains.

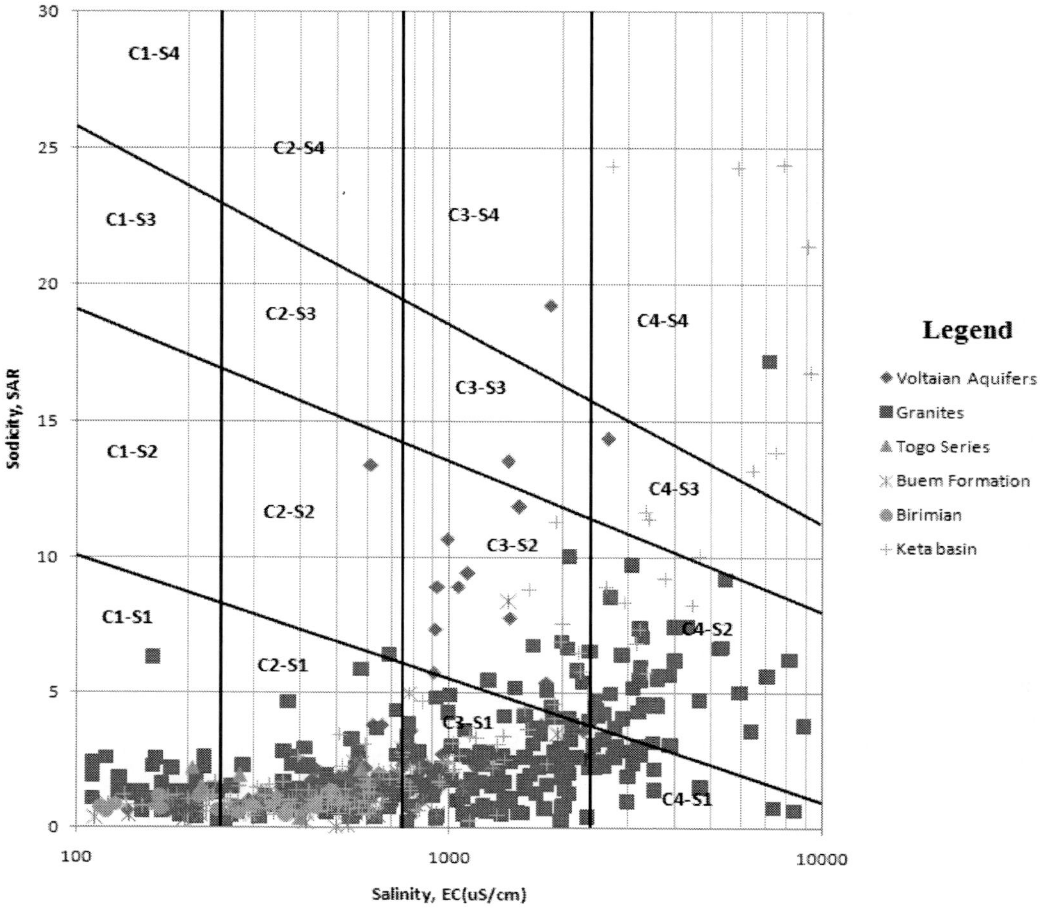

Figure 16. Assessment of groundwater quality for irrigation using the USSL (1954) diagram.

The isotopic signature of groundwater from the Lake Busumtwi suggests recharge from a source which has undergone enrichment of the heavier stable isotopes of oxygen and hydrogen (^{18}O and ^{2}H). $\Delta^{18}O$ and $\delta^{2}H$ values for groundwater from Lake Busumtwi area respectively range between -3.35 and -1.93, and -15.00 and -6.60. A plot of the data in Figure 14 shows that the Lake Busumtwi data plot below the GMWL of Craig (1961), which suggests an enrichment of meteoric water prior to recharge. The slope of the $\delta^{18}O$ - $\delta^{2}H$ relationship for the Lake Busumtwi is 5.3, which is within the range of values that result from evaporative enrichment (Gonfiantini 1986). The Busumtwi data are the most enriched of all the groundwater basins in the country, suggesting that there could be a hydraulic link between the Lake, which is a natural meteoritic crater, and the aquifers that underlie it. The enrichment could also be attributed to Paleorecharge which could attribute older regimes of groundwater recharge in the Lake Busumtwi than what pertains in the other areas of the country. The aquifers of the Busumtwi area belong to rocks of the Birimian and Tarkwaian Systems. The general isotopic signature of groundwater from the Birimian rocks suggests a general a general meteoric recharge with some evaporative effects. Wells sampled for this study are relatively shallow, for which reason Paleorecharge is probably not the reason for the enrichment. The unique enrichment of the Busumtwi groundwater is most likely the result of a hydraulic connection between surface water of the Lake and the underlying aquifers in which the aquifers receive groundwater recharge from the Lake.

4.2. Assessment of the Quality of Groundwater for Irrigation

Groundwater quality for irrigation was assessed on the basis of the sodium adsorption ratio, SAR as a measure of the sodicity, electrical conductivity, EC as a measure of the salinity, and permeability index, PI. Sodium adsorption ratio and PI are calculated from Equations 3 and 4.

$$SAR = \frac{Na^{+}}{\sqrt{\left(Mg^{2+} + Ca^{2+}\right)}} \tag{3}$$

where the concentrations are in mmol/L.

$$PI = \frac{Na^{+} + \sqrt{HCO_3^{-}}}{\left(Na^{+} + Ca^{2+} + Mg^{2+}\right)} x100\% \tag{4}$$

where the concentrations are in meq/L.

The suitability of water for irrigation is based on the presence and concentrations of some specific ions in it. For instance excessive concentrations of sodium and bicarbonate ions in irrigation water can cause long term damage to agricultural soils and thereby reduce productivity. High concentrations of these ions can affect plants both physically and chemically. They reduce the osmotic pressure of plant structural cells, and thereby prevent water from reaching the leaves and branches. Chemically, high concentrations of these ions can affect plant metabolism over the long haul. Excessively high salinity waters also have the capacity to reduce the osmotic pressure of plants, and thereby affect the ability of the plants to absorb nutrients and water (Saleh et al., 1999).

Groundwater in Ghana was assessed using the United States Salinity Laboratory's (USSL, 1954) salinity diagram which compares the SAR with EC values and categorizes water based on its suitability for irrigation purposes. SAR, as indicated in Equation 3, measures the content of sodium relative to the sum of the other major ions (calcium and magnesium). Irrigation waters with high SAR and salinity (EC) have the tendency to cause soil flocculation (Yidana et al., 2007). On the other hand, where the SAR is extremely low, and the salinity is high in irrigation waters, soil dispersion can occur whereby the hydraulic properties of soils are severely affected.

Figure 16 is the USSL (1954) plot for the samples collected for this study in Ghana. Most of the samples plot within the low sodicity (S1) field, with variable salinity. The C1-S1 and C2-S1 groundwaters are acceptable for irrigation of most crops in the country since the sodium content is not so high, and the salinity is within tolerable levels. Ninety percent of the samples from the Buem and Togo aquifers plot within this category. Sixty percent of the samples from the Voltaian aquifers, and about 40% of the samples from the granitic intrusives into the Birimian, plot within the C1-S1 and C2-S1 field. Where the salinity is primarily controlled by natural mineral weathering, the quality is within acceptable limits for most uses. It has been suggested from studies of the major controls on groundwater hydrochemistry in Ghana that the principal factor influencing groundwater hydrochemistry in the Buem and Togo aquifers is the weathering of silicate minerals in the aquifers. On account of this, the electrical conductivity values are low in groundwater from these terrains. All the samples of the Birimian metasediments plot within the C1-S1, and C2-S1 range. Some of the lowest salinities in groundwater have been noted among aquifers of the Birimian metasedimentary rocks. The C3-S1 groundwaters may be useful on well drained soils. About 30% of the samples from the granites plot within this region, whilst 10% and 15% respectively plot within the C4-S1, and C4-S2 categories. These are high salinity groundwaters whose eligibility for irrigation will depend on the soil type and the crop. The water is suited for irrigation of salt tolerant crops such as wheat and barley, which are not cultivated in Ghana. The water type is also acceptable for the irrigation of cotton. Cotton is currently cultivated in the northern parts of the country under rain-fed agriculture. These water types are not good irrigation waters in clayey soils where the drainage is poor. In the Keta basin, about 40% of the samples plot within the C1-S1, and C2-S1 categories whilst 20%, fall within the C3-S1 region. The remaining samples from the Keta basin plot largely within the C4-S2 and C4-S4 categories.

This view is amply and more clearly shown on the Wilcox (1955) diagram in Figure 17. It is obvious most of the data plot within acceptable ranges for irrigation on the basis of the sodium effect and salinity. Those points that fall within the "unsuitable" category are mainly high salinity groundwater samples from the Keta basin and the granites. Whilst 3% of the samples from the Voltaian aquifers plot within the marginal irrigation quality category (Doubtful to unsuitable), about 20% and 15% respectively of the samples from the granites and Keta basin plot within this category. All the other terrains are largely within the acceptable ranges for irrigation, largely on account of low salinities and low to moderate sodicities.

The forgoing analysis suggests that groundwater from the Buem, Togo, and Birimian aquifers are the best suited for irrigation in Ghana on the basis of the salinity-sodicity relationships. Groundwater from the Voltaian is also largely acceptable for irrigation of most of the crops in Ghana. The suitability of groundwater from the shallow aquifers of the Keta basin for irrigation depends on the degree of salinization from seawater and the Keta Lagoon.

The Doneen's chart for all the major aquifers in the country is presented in Figure 18. Three irrigation water categories are clear on the Doneen's chart. The class I waters have the best balance of PI and the total ionic constitution of the water. They are therefore the best water types which have no prospects of causing permeability problems when used for irrigation. Class II waters are acceptable for irrigation but class III types are unacceptable on grounds of high PI values. It can be observed that whereas samples from the Togo and Buem aquifers were adjudged suitable on the basis of the salinity-sodicity relationships in Figure 16, they plot within the class III water types on the Doneen's chart. This results of the relatively high bicarbonate ion content, and suggests that groundwater from these two terrains might reduce the hydraulic properties of irrigation soils and thereby reduce infiltration when used for irrigation. Permeability indices of some samples are as high as 450, whilst the total ion content is low. All the samples from the granite aquifers are within the class I water types and are the best waters with the best combination of PI and total ion content. About 90% of the data from the Keta basin plot within the classes I and II water categories on the Doneen's chart. It therefore goes without saying that despite the high salinities associated with some of the wells in the Keta basin, there appears to be a balance in the bicarbonate ion and the major cations in the groundwater system. However, since the PI is not the only determinant of the suitability of groundwater for productive uses in irrigation, it is still prudent to thread with caution when considering groundwater from the Keta basin for irrigation. About 50% of the data from the Voltaian aquifers are within the class III category, whilst the remaining 50% fall within the classes I and II categories. Most of the Birimian data also plot within the class III category on the Doneen's chart.

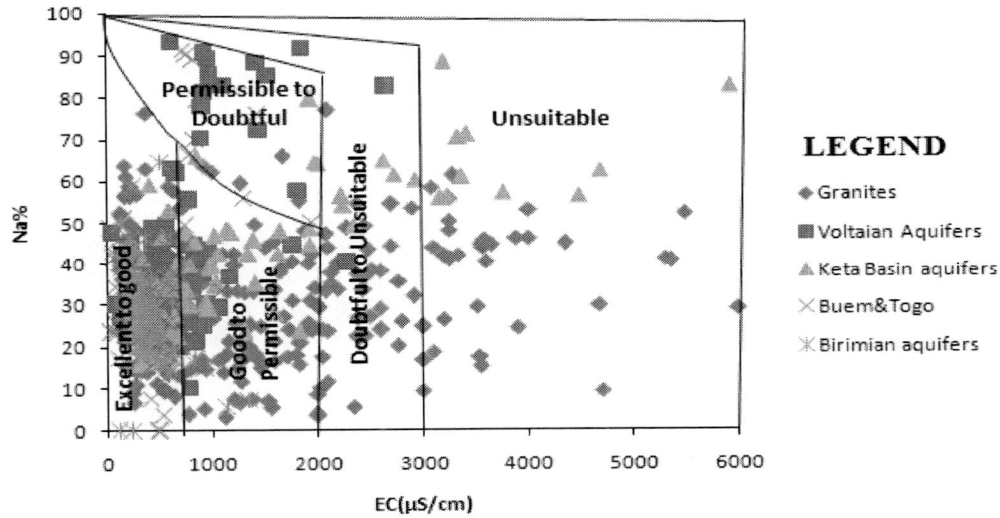

Figure 17. Wilcox (1955) diagram for groundwater from the aquifers in Ghana.

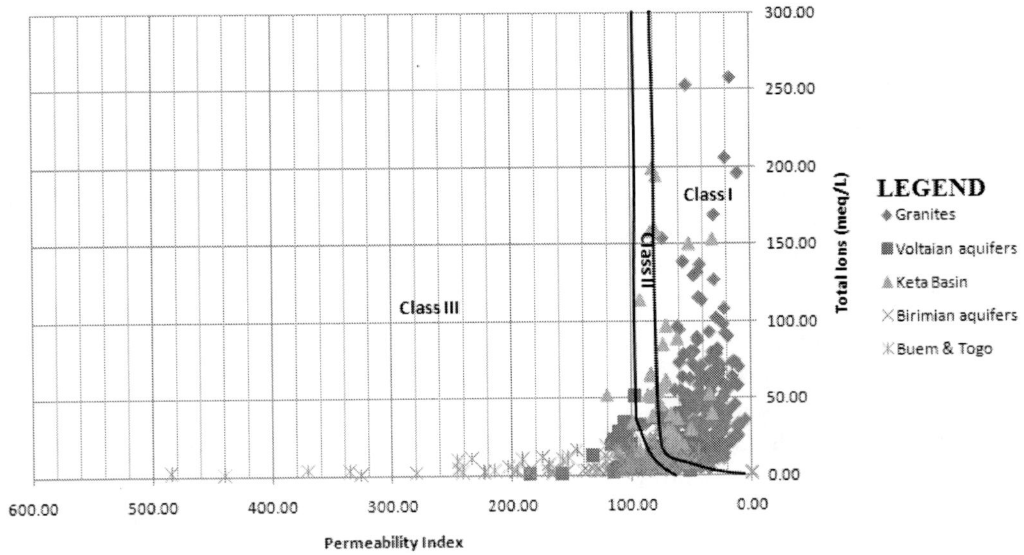

Figure 18. A Doneen's chart showing the permeability indices of groundwater from the various hydrogeological terrains in Ghana.

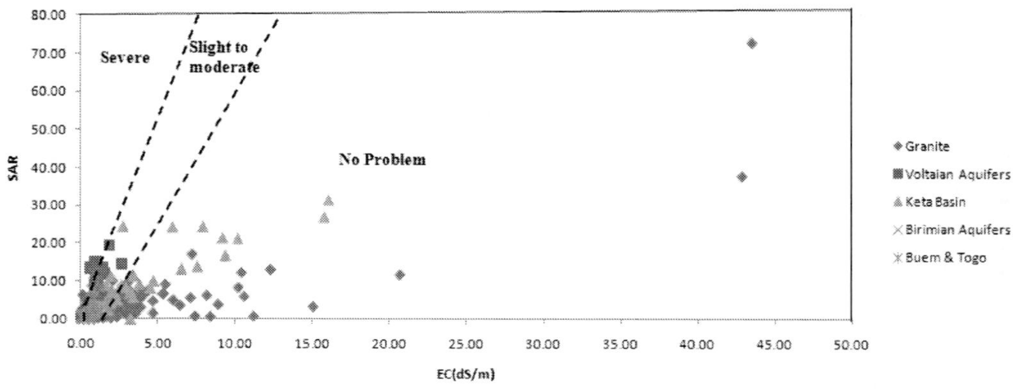

Figure 19. An assessment of the effects of groundwater in Ghana on the infiltration properties of irrigation soils.

Figure 19 assesses the prospective effects of the various groundwaters on the infiltration properties of irrigation soils. It is similar in principle, to the Doneen's chart and ascribes good irrigation water quality to much of the groundwater resource of the country, although one half of the samples plot within the "slight to moderate" category.

Conclusion

Groundwater hydrochemistry in Ghana is dominated by mineral weathering which has significance farther away from the coast, and seawater intrusion which is pervasive along the coast. This study also finds that two other processes: chemical fertilizers, organic manure, and domestic wastes on the one hand, and the oxidation of organic carbon by nitrate on the other hand are respectively the second and third sets of processes influencing groundwater hydrochemistry in Ghana on a national scale. The effects of agricultural chemicals are high and pervasive in areas where the aquifers are shallow and the manures and chemical fertilizers are heavily used to influence crop yields. High effects are observed in the Keta basin and the northern parts of the country. The third set of processes weakly takes the same spatial distribution as the second, and registers high effects in the highly vegetated areas and locations where nitrate is available for the oxidation of organic matter. The spatial distribution maps for these three major processes in the hydrochemistry have been produced in this study using ordinary kriging. Sequential Q-mode HCA identified two major groundwater types in Ghana. These are the predominantly Na - Cl water types identified mostly along the coast, and Na-K-HCO_3 freshwater types identified farther inland. There is a general progression from Ca-HCO_3 water types through to Na-K-HCO_3 water types. Groundwater recharge has been determined to be meteoric with some cases of evaporative effects prior to recharge. This study also stipulates that the aquifers underlying Lake Busumtwi are probably in hydraulic connection with the surface water of the Lake, and that these aquifers probably receive recharge from the Lake. On the other hand, there appears to be no hydraulic connection between aquifers of the Voltaian System and the surface water from the Volta Lake. Groundwater from the major aquifers in the country is largely acceptable for irrigation on the basis of sodicity and salinity. There are a few cases of high salinities such as the case of the

Keta basin and some of the granitic aquifers in the eastern and central regions. The Birimian, Togo, and Buem aquifers deliver groundwater of high permeability indices, rendering them unsuitable for irrigation purposes because of the high prospects of causing infiltration problems on irrigation soils.

ACKNOWLEDGMENTS

The author is very grateful to Mr. Jarvis Ayamsegna and the World Vision Ghana for providing the hydrochemical data for the Voltaian Basin. He is also grateful to the zonal hydrogeologist of the three northern regions of Ghana for hydrochemical data from the northern parts of Ghana. The author also expresses gratitude to Professor Bruce Banoeng-Yakubo for the data from the Keta basin and other parts of the Volta Region of Ghana.

REFERENCE

Acheampong, S.Y., Hess, J.W. (1998). Hydrogeologic and hydrochemical framework of the shallow groundwater system in the southern Voltaian Sedimentary Basin, Ghana. *Hydrogeology Journal* 6, 527-537.

Acheampong, S.Y., Hess, J.W. (2000). Origin of the shallow groundwater system in the southern Voltaian Sedimentary Basin of Ghana: an isotopic approach. *Journal of Hydrology*, 233, 37-53.

Affaton, P., Sougy, J., and Trompette, R. (1980). The tectono-stratigraphic relationships between the Upper Precambrian and Lower Paleozoic Volta basin and the Pan-African Dahomeyide orogenic belt (West Africa). .*American Journal of Science*, 280, 224- 248.

Agyei, E.K., Van Landewijk, J.M., Armstrong, R.L., Harakai, J.E., and Scott, K.L. (1987). Rb Sr and K-Ar Geochronometry of Southeastern Ghana. *Journal of African Earth Sciences*, 6(2), 153-161.

Akiti, T.T. (1987). Environmental isotope study of groundwater in crystalline rocks of the Accra Plains (Ghana). In: Wand U, Strauch G (eds) *Isotopes in nature*. Academy of Sciences of the GDR, Central Institute of Isotope and Radiation Research, Leipzig, pp 107–121

Akpati, B.N. (1975). *Geological structure and evolution of Keta basin*. Rep no. 75/3. Ghana Geological Survey, Ghana.

Appelo, C.A.J., Postma, D. (2005). *Geochemistry, groundwater and pollution*, 2nd Edition. Balkema, Amsterdam, 649p.

Arumugam, K., Elangovan, K., (2009). Hydrochemical characteristics and groundwater quality assessment in Tirupur Region, Coimbatore District, Tamil Nadu, India.Environ. Geol. DOI 10.1007/s00254-008-1652-y.

Banoeng-Yakubo, B., Yidana, S.M., Nti, E. (2009). Hydrochemical modeling of groundwater using multivariate statistical methods – The Volta Region, Ghana. *KSCE Journal of Civil Engineering*, 13(1), 55 - 63.

Bates, D.A. (1955). *Geological map of Ghana*. Ghana Geological Survey Department, scale 1:1,000,000. Reprinted by the Geological Survey Department in 1966.

Blay, P.K. (1991). Applying subduction tectonics to the evolution of the Pan-African Dahomeyide deformed belt of Ghana, West Africa. *Proceedings of the First Local Conference on Mineral Exploration and Development and their Impact on the Economy of Ghana*, Accra, 7th December 1990, Minerals Commission Publication, pp. 52-75.

Carrier, M.A., Lefebvre, R., Racicot, J., Asare, E.B. (2008). Northern Ghana hydrogeological assessment project. *33rd WEDC International Conference, Accra, Ghana.*

Chadha, D.K. (1999). A proposed new diagram for geochemical classification of natural waters and interpretation of chemical data. *Hydrogeology Journal* 7, 431-439.

Cloutier, V., Lefebvre, R., Therrien, R., Savard, M.M. (2008). Multivariate statistical analysis of geochemical data as indicative of the hydrogeochemical evolution of groundwater in a sedimentary rock aquifer system. *Journal of Hydrology* 353, 294-313.

Craig, H. (1961). Isotopic variations in meteoric water. *Science* 133, 1702-1703 Dapaah-Siakwan, S., Gyau-Boakye, P. (2000). Hydrogeologic framework and borehole yields in Ghana. *Hydrogeology Journal* 8, 405-416.

Davis, J.C. (1986). *Statistics and Data Analysis in Geology*. John Wiley & Sons Inc., New York.

Dickson, K.B., Benneh, G. (1995). A New Geography of Ghana, 3rd revised edition. Longman, Malaysia.

Drever, J.I. (1988). *The geochemistry of natural waters*, 2nd Edition. Prentice Hall, New Jersey, 437p.

Dugan, J.P., Borthwick, J., Harmon, R.S., Gagnier, M.A., Glahn, J.E., Kinsel, E.P., MacLeod, S., Viglino, J.A., Hess, J.W. (1985). Guanidine hydrochloride method for determination of water oxygen isotope ratios and the oxygen-18 fractionation between carbon dioxide and water at 25°C. *Anal. Chem.* 57, 1734-1736.

Farnham, I.M., Johannesson, K.H., Singh, A.K., Hodge, V.F., Stetzenbach, K.J. (2003). Factor analytical approaches for evaluating groundwater trace element chemistry data. *Analytical Chimica Acta* 490, 123–138.

Florkowsky, T. (1985). Sample preparation for hydrogen isotope analysis by mass Spectrometry. *Int. J. Appl. Radiat. Isot.* 36, 991-992.

Gill, H.,E. (1969). A ground-water reconnaissance of the Republic of Ghana, with a description of geohydrologic provinces. *U.S. Geological Survey Water-Supply paper* 1757-K.

Gonfiantini, R. (1986). *Environmental isotopes in lake studies*. In: P. Fritz and J.Ch. Fontes (eds). Handbook of environmental isotope geochemistry, 113-168. Elsevier, Amsterdam.

Grant, N.K. (1969). The late Precambrian to early Paleozoic orogeny in Ghana, Togo, Dahomey and Nigeria. *Geological Society of America Bulletin*, 80(1), 45-55.

Güler, C., Thyne, G., McCray, J.E., Turner, A.K. (2002). Evaluation and graphical and multivariate statistical methods for classification of water chemistry data. *Hydrogeology Journal* 10, 455-474.

Holm, R.F. (1974). *Petrology of alkalic gneiss in the Dahomeyan of Ghana*. Geological Society of America Bulletin, 85, 1441-1448.

Jalali, M., (2007). Salinization of groundwater in arid and semi-arid zones: an example from Tajarak, western Iran. Environ. Geol. 52, 1133-1149.

Jones, W.B. (1990). The Buem volcanic and associated sedimentary rocks, Ghana: a field and geochemical investigation. *Journal of African Earth Sciences*, 11 (3/4), 373-383.

Jørgensen, N.O., Banoeng-Yakubo, B.K. (2001). Environmental isotopes (^{18}O, ^{2}H, and $^{87}Sr/^{86}Sr$) as a tool in groundwater investigations in the Keta Basin, Ghana. *Hydrogeology Journal* 9, 190-201.

Junner, N.R. (1935). *Gold in the Gold Coast.* Gold Coast Geological Survey Memoir No. 4, Accra, 67 pp.

Junner, N, R. (1940). *Geology of the Gold Coast and Western Togoland.* Gold Coast Geological Survey Bulletin No. 11, Accra, 40 pp.

Junner, N.R., Hirst, T. (1946). *The geology and hydrogeology of the Voltaian basin.* Gold Coast Geological Survey Mem 8. Gold Coast Geological Survey, The Gold Coast.

Junner, N.R., Service, H. (1936). Geological notes on the Volta River District and Togoland under British mandate. *Annual Report on the Geological Survey by the Director,* 1935-36.

Kaiser, H.F. (1960). The application of electronic computers to factor analysis. *Educ. Psychol Meas.* 20, 141-151.

Kesse, G.O. (1985). *The Mineral and Rocks Resources of Ghana.* A.A. Balkema Publishers. Netherlands-Rotterdam. Pp 39-50.

Kortatsi, B.K., Tay, C.K., Anornu, G., Hayford, E., Dartey, G. (2008). Hydrogeochemical evaluation of groundwater in the lower Offin basin, Ghana. *Environmental Geology,* 53, 1651-1662.

Mathieu, R., Bariac, T. (1996). An isotopic study (^{2}H and ^{18}O) of water movements in clayey soils under semiarid climate. *Water Resources Research* 32(4), 779-789.

Mauer, R. (1990). Petrographische und geochemische Untersuchungen der prakambrischen (Birimian) Granitoide Ghanas". Dissertation, Techn. Univ., Berlin, 202 pp.

Melloul A, Collin M (1992). The 'principal components' statistical method as a complementary approach to geochemical methods in water quality factor identification; application to the Coastal Plain aquifer of Israel. *Journal of Hydrology* 140, 49–73.

Ophori, D.U. (1999). Constraining permeabilities in a large-scale groundwater system through model calibration. *Journal of Hydrology* 224, 1-20.

Parkhurst, D.L., Appelo, C.A.J. (1999). *User's Guide to PHREEQC (Version 2)—A Computer Program for Speciation, Batch-Reaction, One-Dimensional Transport, and Inverse Geochemical Calculations.* United States Geological Survey, Water Resources Investigations Report 99-4259, Washington, DC, p. 326.

Saleh A, Al-Ruwih F, Shehata M (1999) Hydrogeochemical process operating within the main aquifers of Kuwait. *J Arid Environ.* 42, 195–209.

Steinhorst, R.K., Williams, R.E. (1985). Discrimination of groundwater sources using cluster analysis, MANOVA, canonical analysis and discriminant analysis. *Water Resources Research* 21, 1149-1156.

Stetzenbach, K.J., Hodge, V.F., Guo, C., Farnham, I.M., Johannesson, K.H. (2001). Geochemical and statistical evidence of deep carbonate groundwater within averlying volcanic rock aquifers/aquitards of southern Neveda, USA. *Journal of Hydrology* 243, 254–271.

USSL (1954) Diagnosis and improvement of saline and alkali soils.*USDA Agricultural Handbook No. 60*, Washington DC.

Viero, A.P., Roisenberg, C., Roisenberg, A., Vigo, A. (2009). The origin of fluoride in the granitic aquifer of Porto Alegre, Southern Brazil. *Environ Geol* 56, 1707-1719.

Wilcox, I.V., (1955). *Classification and use of irrigation water.* USDA Circ. 696, Washington DC.

World Health Organization, (1984). Guidelines for drinking water quality, vol 1. Recommendations. Geneva, Switzerland.

Yidana, S.M. (2009). Hydrochemical framework of surface water basins in southern Ghana. *Environ Geol*, 57(4), 789-796, DOI 10.1007/s00254-008-1357-2.

Yidana, S.M., Ophori, D., Banoeng-Yakubo, B., (2007). Irrigation water resources management for sustainable agriculture – The Ankobra basin, Ghana. *Journal of Irrigation and Drainage Engineering, 133(6),* 609-615.

Yidana, S.M., Ophori, D., Banoeng-Yakubo, B. (2008a). Hydrogeological and hydrochemical characterization of the Voltaian Basin- The Afram Plains area, Ghana. *Environ Geol* 53, 1213-1223.

Yidana, S.M., Ophori, D., Banoeng-Yakubo, B. (2008b). Hydrochemical evaluation of the Voltaian system – The Afram Plains area, Ghana. *Journal of Environmental Management* 88, 697-707.

Yidana, S.M., Ophori, D., Banoeng-Yakubo, B. (2008c). Groundwater availability in the shallow aquifers of the southern Voltaian system: a simulation and chemical analysis. *Environmental Geology* 55, 1647-1657.

Yidana, S.M., Ophori, D., Banoeng-Yakubo, B. (2008d). Groundwater quality evaluation for productive uses – The Afram Plains area, Ghana. *Journal of Irrigation and Drainage Engineering* 134(2), 222-227.

Yidana, S.M., Yidana, A. (2010). Assessing groundwater quality using water quality index and multivariate statistical analysis – the Voltaian basin, Ghana. *Environmental Earth Science,* 59(7), 1461-1473 DOI 10.1007/s12665-009-0132-3.

Yidana, S.M. (2010). Groundwater classification using multivariate statistical methods: Southern Ghana. *Journal of African Earth Sciences*, DOI: 10.1016/ j.jafrearsci.2009.12.002.

In: Potable Water and Sanitation
Editors: Joel M. McMann

ISBN: 978-1-61122-319-4

Chapter 5

THE CHALLENGE OF SUSTAINABLE WASTE MANAGEMENT IN GHANA – PRACTICES, PROBLEMS AND POSSIBILITIES

Service Opare
Department of Geography, University of Calgary, Canada

ABSTRACT

Adequate sanitation is essential for environmental quality and improved health. Developing countries such as Ghana that experience high mortality and morbidity levels could attain qualitative health improvements through proper waste collection and disposal. However, data from nationally representative surveys conducted in Ghana between 1991/92 and 2005/06 reveal that sanitation infrastructure and services are below optimal levels, households utilize improper waste disposal practices which worsen sanitation conditions and implementation of a number of foreign donor funded water and sanitation schemes has failed to register marked improvements in sanitation coverage. This chapter undertakes an analysis of various sanitation conditions in Ghana and constraints that limit efforts to expand coverage. Based on the findings, strategies whose implementation could ensure sustainable waste management in Ghana such as composting, enforcement of punitive sanctions and increased resource support for acquisition of improved sanitation facilities by households are proposed.

INTRODUCTION

Expanded sanitation coverage is one of the main targets of the Millennium Development Goals (MDGs) formulated by the United Nations and allied international agencies in an effort to draw attention to the critical development needs of the global community. Increasing realization of the importance of water and sanitation has led to renewed focus on expanding coverage to communities (Garange & Dagg, 2005). Despite this, a significant proportion of the world's population continues to lack access to basic sanitation facilities and services of which human excreta disposal and solid waste management are critical. It is estimated that while approximately one-sixth of the world's population lacks access to safe water, a greater proportion of about two-fifths is without adequate sanitation (Tornqvist et al, 2008). Sub-Saharan Africa is considered to have the lowest sanitation coverage of 55% (UNICEF/WHO, 2006, cited in Osumanu, 2007). In Ghana, access to improved sanitation facilities and services remains a daunting problem and confirmed increases in sanitation coverage over the period 1991/92 to 2005/06 have been marginal. Household access to flush toilets, obviously the most appropriate toilet system, recorded an insignificant increase from 7.1% to 10.2% during that period (Ghana Statistical Service, 2007). The proportion of the population without access to adequate sanitation in South Africa is estimated at about 13 million which is slightly more than double the estimated number of those without access to potable water (DWAF, 2008, cited in Manase et al., 2009).

Poor sanitation coverage also appears to be largely a rural phenomenon. Up to 50% of the rural population in Latin America lacks access to basic sanitation (WHO, 2000; PAHO, 2001, cited in Garcia & Brown, 2009). In Sub-Saharan Africa, while urban sanitation coverage stagnated at 74% between 1990 and 2000 (UNICEF/WHO, 2002, cited in Osumanu, 2007), access of the rural population to improved sanitation systems recorded a 2% decline (NEPAD, 2002, cited in Malley et al., 2008). Even though the proportion of urban households with access to flush toilets in Ghana rose from 17.6% to 22.3% between 1991/92 and 2005/06, access of rural households rather registered a decline from a trivial 1.4% to an even lower 1.1% (Ghana Statistical Service, 2007). A similar situation is found in Swaziland where, according to World Bank estimates, 78% of urban residents in Swaziland had access to sanitation facilities compared to only 44% of their rural counterparts (World Bank, 2005, cited in Busari, 2009).

Improved sanitation could make a critical contribution to better living standards in developing countries that are bedeviled with problems of ill health and environmental degradation (Hoko & Hertle, 2006; Mensah & Whitney, 1992; Osumanu, 2007). This is because diseases that emanate from inadequate water and sanitation conditions are estimated to account for over two thirds of common diseases that confront developing countries. Poor water and sanitation conditions are believed to be responsible for an estimated 10 million deaths annually throughout the world (Eyles & Consitt, 2004). These conditions have a disproportionately high impact on children (Freeman, 2010; Shiva, 2004). About 3,900 children below the age of 5 die daily as a result of diseases that emanate from poor water and sanitation conditions (WHO-UNICEF, 2004, cited in Anand, 2007) while the UN puts the figure for all children who die annually from both water and sanitation related diseases as being in excess of 4 million (Kazelli et al., 2003).

Resort to deposition of waste materials in the open is a common household waste disposal practice in developing countries such as Ghana and Botswana (Ghana Statistical Service et al., 2004; Kgathi & Bolaane, 2001). This, together with poor disposal of human excreta and other inappropriate waste disposal practices have been the dominant contributory factors to the prevalence of diseases such as diarrhoea and polio that affect developing countries (Mwendera, 2006). Poor sanitation standards remain daunting problems in South Africa where it is estimated to be responsible for the death of 13,600 people per annum (WHO, 2002, cited in Manase et al., 2009). Satterthwaite (2002:264) argues that

> Many of the diseases and disease vectors (for instance mosquitoes that transmit dengue fever or yellow fever) thrive when provision of water, sanitation, drainage, garbage collection and healthcare is inadequate. Around half the Third World's urban population is suffering from one or more of the diseases associated with inadequate provision for water and sanitation

The declaration of the period 1981-1990 as the International Water and Sanitation Decade by the international community contributed to a renewed focus on expanding access to both water and sanitation as evidenced by implementation of various water and sanitation schemes funded by major international donors such as the World Bank and DANIDA (Danish International Development Agency) in Ghana and other developing countries. Despite the fact that these projects targeted improved waste management, collection and disposal of both solid waste and human excreta remain major bottlenecks in Ghana. These need to be tacked through implementation of strategies that ensure sustainable waste management without which the anticipated benefits of improved health and other socio-economic gains would not accrue to the Ghanaian society. This paper discusses sanitation conditions in Ghana with particular reference to human excreta and solid waste disposal systems which constitute the critical sanitation issues in the country. After the introduction, the next section of the paper provides diverse perspectives on the inter-relationships between sanitation, environment and development that necessitate the institution of sustainable waste management measures. After providing information on data sources and the country setting, the discussions then shift to the sanitation situation in Ghana, its key characteristics and the constraining factors that place limitations on expanding sanitation coverage. Varied measures that could promote sustainable waste management are then explored. The last section highlights key issues and provides some concluding comments.

SANITATION, ENVIRONMENT AND DEVELOPMENT – AN EXPLORATION OF THE RELATIONSHIPS

Sanitation improvement has a direct impact on health conditions in developing countries and is therefore of major relevance to them. This is because better health is a major development challenge that developing countries are struggling to attain. High maternal and child mortality, rising morbidity levels and related health difficulties that are endemic in developing countries contribute to low life expectancy and also affect labor productivity as the working population spends valuable time receiving treatment instead of being engaged in productive work. Failure to dispose of waste materials results in their being drained into surface water bodies that are important sources of water supply to rural and urban poor communities. Direct consumption of water from such sources worsens the disease prevalence levels in these communities. Drainage of contaminating waste items into water bodies especially after heavy rainfall increases treatment costs of water before it become potable enough for consumption. These inter-relationships explain the linked implementation of water and sanitation schemes to ensure maximized health and other impacts.

Apart from its noticeable impacts on health, improved sanitation results in minimized landscape degradation that provides a pleasant setting free of houseflies, rats and mice that tend to dwell in, and feed on, waste items. Organisms such as mosquitoes, pests and other disease causing vectors whose presence adds to the unpleasant scenery, and contributes to odor generation and other inconveniences, would thus be denied the conducive habitat to reproduce, grow and spread. Waste accumulation mars the beauty of the landscape generally, and also dents the aesthetic value that the natural beauty of parks, gardens and other open places generate. It further limits the possibility of their use for recreation, tourism, camping and other leisure events (Boadi & Kuitunen, 2002). Excessive concentration of heavy metals in rivers preclude their use for swimming and other recreational activities due to their potential toxic effects on humans (Adomako et al., 2008).

Poor sanitary conditions adversely affect the successful pursuit of a number of vital activities. For example, successful home care and other support services for persons infected with HIV/AIDS disease are doubtful without a clean home environment and maintenance of hygienic conditions. Areas that are free of bacteria proliferation and pest infestation, and access to adequate toilets, are critical variables for home based care and sound management of opportunistic infections associated with the HIV disease (Ngwenya & Kgathi, 2006). Moreover, availability of improved sanitation facilities has additional benefits of "greater privacy, convenience, safety, and dignity – qualities that are particularly important for women" (Lane, 2006: 271). Increasing global disasters are somewhat linked to poor sanitation and environmental degradation. Waste materials from households and industrial ventures that drain into and choke gutters, apart from being environmentally degrading, are a major cause of flooding in cities including Accra, Ghana's capital city, during prolonged rainfall since they impede the free flow of runoff. Such floods occasionally lead to massive property destruction and loss of life. As noted by Boadi and Kuitunen (2002:307), "the devastation which occurred to lives and property in Accra, in 1995 due to floods, were partly attributed to the accumulation of garbage which blocked the city's drainage channels, particularly in the Korle Lagoon which serves as the major outlet to the sea". Floods that occur as a result of poor drainage also affect agricultural production through denudation of the rich topsoil, destruction of crops and breakdown of irrigation systems.

The maintenance of appropriate sanitation standards is critical for international trade which in turn impacts on the economic growth of developing countries. Strict compliance with a range of specified sanitary standards is one key requirement for exporting products to the European Union (EU) (Newell, 2005). The consumer protection objective of the EurepGap protocol aims at ensuring that established health standards in Europe are not unduly undermined by the entry of contaminated goods produced in unhealthy environments into the EU market (Mausch et al., 2006). Developing countries whose production activities are pursued under sanitation standards that fall below acceptable levels are, as a result, unable to take advantage of existing trading opportunities to export agricultural products such as fruits, vegetables and flowers to the EU. This limits their foreign exchange earning opportunities, which in turn affect their ability to import scarce goods needed for accelerated economic growth. In view of the foregoing, socio-economic development of Third World countries such as Ghana could be affected negatively unless attention is focused on improving waste disposal through the pursuit of effective waste management strategies. These linkages between sanitation, environment and socio-economic development call for the pursuit of policies and strategies that would ensure sustainable waste management. Sanitation improvement measures, in order to be sustainable, should aim not only at elimination of huge waste accumulations but should seek a widespread adoption of appropriate waste disposal attitudes, target institutional restructuring schemes and pursue sectoral integration in order to pre-empt a recurrence of waste pile up.

Sources of Data

The bulk of the data for this study was collected from surveys that were based on nationally representative samples and were carried out by the Ghana Statistical Service between 1991/92 and 2005/06. These are the Ghana Living Standard Surveys (GLSS) and the 2003 Ghana Demographic and Health Survey (GDHS). The GDHS of 2003 was the fourth in a series of surveys and collected data from a total of 5,691 females and 5,015 males who were within an age group of 15 to 59. The survey was undertaken between July and October 2003 (Ghana Statistical Service et al., 2004). On the other hand, each of the three Ghana Living Standards Surveys (GLSS) that provided data for this study was undertaken over full, 12-month periods. The GLSS of 1991/92 covered September 1991 to September 1992 and entailed collection of data from a sample of 4,565 households with more than 20,000 persons (Ghana Statistical Service, 1995). The 1998/99 survey was conducted from April 1989 to April 1999 and involved a sample size of 5,998 households with more than 25,000 members (Ghana Statistical Service, 2000) while the most recent GLSS, the fifth in a series of national surveys, was undertaken between September 2005 and September 2006 (Ghana Statistical Service, 2007). Data from these sources are supplemented with information from other publications on the sanitation situation in Ghana.

Background Information on Ghana

Ghana is a former British colony which gained its political independence on March 6, 1957. It has a total land area of 238,573 square kilometers and its geographical location is defined by Latitudes 5° 36" N and 11° 05" N, and Longitudes 0° 10" E and 0° 05" W. It is bordered to the north by Burkina Faso and the south by the sea known as the Gulf of Guinea. The country is bordered on its western side by La Cote D'Voire and on the east by the Republic of Togo. It has been divided into ten administrative regions (provinces) and one hundred and thirty-eight districts (Ghana Statistical Service et al., 2004). The Ghana 2000 Population and Housing Census count revealed that the country's population rose from 12,296,081 persons in 1984 to 18,912,079 in 2000 (Ghana Statistical Service, 2002). Urban residents comprise 44% of the national population while the rural population makes up 56%. The country has a gender ratio of 98 males per 100 females. Life expectancy for males is 55 years and 60 years for females. The average population growth rate is 2.7% per annum which is lower than the 2.9% average for West Africa (Ghana Statistical Service et al., 2004).

Ghana is largely an agricultural country with up to 59.8% of males and 35.8% of the female population being engaged in the agricultural sector (Ghana Statistical Service et al., 2004). However, there are a number of industrial and mining ventures operating in the country the majority of which are located in the national capital (32%) and in other urban centers (Hens & Boon, 1999). During the last few decades, substantial foreign donor assistance has been provided to support implementation of various water and sanitation projects geared toward improving the proportion of the population with access to potable water and appropriate sanitation facilities (Fuest, 2005; Ghana Statistical Service, 2007; Lane, 2006). Although implementation of these projects made some impact on the sanitation sector, substantial gaps in sanitation coverage remain and the country is still confronted with teething waste management problems. The task of addressing the sanitation problems that plague the society constitutes a huge financial burden on the country. As at 1999, the cost of providing an adequate sewage system for the capital city alone was estimated at US$110 million (Hens & Boon, 1999).

THE SANITATION SITUATION IN GHANA

Sanitation conditions in Ghana are inadequate and contribute to poor health conditions in the society. It has been estimated that environmentally related diseases constitute between 60% and 80% of cases of ill-health that are reported at the national capital's hospitals and health centers alone (Hens & Boon, 1999). Freeman (2010:48) provides a vivid account of the impact of poor sanitation conditions on health and other socio-economic conditions in Ghana:

> Diseases related to poor sanitation take eight lives an hour and account for 80 percent of all ailments reported to clinics and hospitals in Ghana. Diarrhea alone was responsible for 452,250 outpatient deaths in Ghana in 2007, fifth-leading killer. Intestinal worms killed 208,429. Upper respiratory tract infections— twice as likely for those regularly breathing the fumes of raw sewage—came second only to malaria as the country's deadliest outpatient ailment, killing 920,806 Ghanaians in 2007. Disease is not the only serious side effect of Ghana's sanitation crisis. The country's economy, education system, tourism industry and social structures all suffer from the impact of poor sanitation. Ghana's beaches, for example, often serve as toilets, which hurts the tourism industry.

In addition to increased incidence of preventable diseases, poor sanitation has contributed to environmental degradation. One of these environmental problems, is contamination of water bodies including rivers, lakes and lagoons from waste discharges generated as a result of industrial production and mining activities, pollution by agro-chemicals utilized in the agricultural sector and increasing sedimentation in water bodies (Akuffo, 1998; Boadi & Kuitunen, 2002; Hens & Boon, 1999). Pollution of the Volta Lake, the country's main water body, has been on the increase due to waste and effluent discharges from two major textile factories, agricultural activities and settlements located in its basin. Gold, diamond and manganese mining activities have contributed to contamination of major rivers such as Pra and Ankobra (Hens & Boon, 1999). While waste discharges from industrial and mining activities have resulted in poor sanitation, the greater proportion of Ghana's sanitation difficulties emanates from the paucity of disposal facilities, restricted household access to these facilities and inappropriate household waste disposal practices. An assessment of two key components of sanitation namely solid waste disposal and human excreta disposal provides indications of the extent, trend and characteristics of the sanitation situation in Ghana.

Solid Waste Disposal

Disposal of solid waste materials is a major sanitation dilemma that confronts Ghana. Some waste materials are disposed of in garbage bins but significant proportions are deposited at open, unprotected sites, left unattended to or litter various places (Oteng-Ababio, 2010). These constitute blots on the landscape, lead to odor emission and dent the aesthetic scenery. Other waste materials "are dumped into depressions, at roadsides or in gutters from where they are partly washed down the drainage system with stormwater contributing to the anthropogenic pollution load" (Ernie et al., 2010:119). Inappropriate waste disposal behaviors exhibited by both urban and rural households have contributed to the recurring waste disposal difficulties confronting the Ghanaian society. The GDHS revealed varying methods employed by households to dispose of their waste materials with differential impacts on the environment and health conditions in the country. They include waste collection by government agencies, private entities or community associations, burning, burial, composting and dumping in the open.

Table 1. Methods of household waste disposal in Ghana

Methods of waste disposal	Proportion of households using method		
	Urban	Rural	All
Collection by government	30.6	0.4	14.2
Collection by community association	4.2	3.9	4.0
Collection by private company	4.8	0.3	2.3
Dumped in compound	2.5	5.9	4.3
Dumped in street/empty plot	47.3	74.0	61.7
Burned	7.7	6.4	7.0
Buried	2.5	3.5	3.0
Composted	0.2	5.3	3.0
Other	0.2	0.1	0.1

Source: Ghana Statistical Service et al., 2004

As indicated in Table 1, the unhealthy practice of dumping waste in the streets, compounds or other open places was the most preferred method of waste disposal among households in Ghana. Widespread dumping of waste by both urban and rural households was partly a reflection of the paucity of waste collection facilities such as garbage bins into which waste could be deposited and kept until its final collection. The proportion of households engaged in the use of this method was higher among rural than urban households. This could to a certain extent be attributed to availability of alternative methods such as collection by various agencies in urban centers which made dumping a less attractive waste disposal option to them compared to their rural counterparts. The use of open, unprotected and unfenced refuse dumping sites contributed to a worsening of the already precarious health conditions in the country as these areas became breeding grounds for mosquitoes, flies, pests and other disease causing agents and vectors (Boadi & Kuitunen, 2002). Reliance on open dumping by a rather large proportion of households in Ghana therefore impedes government efforts at minimizing public health problems. This practice particularly poses danger to children who turn these sites into playgrounds where they pick up discarded items and use them for various recreational activities, acts which heighten their exposure to various diseases (Akuffo, 1998).

An eventual drainage of waste materials from dumping sites into street drains in urban centers during rainfall results in their getting choked. This, and increased sedimentation of rivers, lagoons and other water bodies that occurs when runoff water deposits waste into them, contribute to widespread flooding during heavy rainfall causing havoc, destruction of property and occasional loss of lives. Apart from these, drainage of refuse from open sites and streets into surface water bodies leads to a deterioration in their pollution levels and contributes to the prevalence of water borne diseases (Akuffo, 1998). These practices as Freeman (2010) notes also constitute eyesores that mar the aesthetic scenery of beaches and other areas that are potential tourist sites. Since tourism is the third most important foreign exchange earning sector in Ghana after cocoa and gold (Government of Ghana, 2002), degradation of potential tourist sites that limits their attraction to foreign visitors obstructs their ability to bring in the much needed foreign exchange for economic growth. This probably provides a partial explanation for the lower than expected proportion of foreigners that travel to Ghana as traditional tourists visiting places of geographical interest (Government of Ghana, 2002).

Waste collection by government was a lot more widespread in urban communities than in rural communities. The most likely explanation for this variation is the tendency for the Metropolitan, Municipal and District Assemblies to establish and operate waste collection systems in urban areas instead of in rural communities. These Assemblies are local government authorities established under the country's decentralization program (Fuest, 2005; Yankson, 2008) and by legislation had ultimate responsibility for provision of basic infrastructure in their areas of jurisdiction (Fuest, 2005). Obviously, the sheer number of rural communities in the country and the distant location of some of them preclude the introduction and operation of financially viable government or private operated waste collection systems. Fuest (2005) states for example, that the remote location of some rural communities and the absence of motorable roads led to an inability of staff of the Assemblies to make contact with them which affected their selection for support under water and sanitation schemes. These factors also explain why though the level of private agency involvement in waste collection was generally low (4.3%) it was virtually non-existent in rural areas. Private agencies will, by their orientation, be a lot more mindful of the profitability of any undertaking they engage in. They will therefore exhibit an understandable reluctance to take on waste collection assignments in rural areas that present little or no financial gain.

Generally, provision of waste collection services by an external agency entailed some expenditure to beneficiary households. It probably explains why only a quarter of the urban population utilized these services. It is most likely these were patronized by the relatively better off urban households that could afford payment of fees associated with such services while the poorer urban households depended upon other methods such as burying, burning and open dumping which entailed no or at most, very minimal expenditure. Since rural areas in Ghana are characterized by higher poverty levels compared to urban centers, with 86% of the poor being rural dwellers as at 2005/06 (Ghana Statistical Service, 2007), the majority of rural households could ill-afford the luxury of incurring any expenses on waste disposal. This also constitutes an additional explanatory factor for the virtual non-existence of external agency led waste collection services in rural areas and the tendency to depend on their own associations for some limited waste collection services.

The involvement of private waste collection agencies in waste collection depicts an increasing shift toward privatization of critically useful utility services previously considered the domain of local government agencies (Agyeman, 2007). Private agency involvement in development service delivery has been on the ascendancy in recent times in Ghana and other countries (Osumanu, 2007) based on the view, largely propounded by international financial institutions, that agencies with private orientation exhibit efficiency, reliability and market focused characteristics that are crucial for effective service delivery but are woefully lacking in public agencies. The lack of these is believed to condition the latter's inability to meet required service delivery standards (Appiah-Kubi, 2001, Cook, 2009). Privatization of such services is also largely as a result of prevailing neoliberal thinking that is "premised on the notion that government functions to facilitate for-profit economic investment in the urban environment". (Mountz & Curran, 2009:1033).

Other waste disposal methods were utilized on a very limited scale by households. These, however, had varying impacts on environmental conditions in the country. Burning which obviously is an effective means of eliminating waste, on the other hand, contributes to air pollution through the release of CO^2 and other gases into the atmosphere. It was practised by a low 7% of households in Ghana, with urban households being slightly in the majority. It has cost implications since it involves access to, and use of, an energy source and might explain why only a few households in both urban and rural areas relied on it. The slightly higher proportion of urban households engaged in this practice might also be attributable to their lack of sufficient space for depositing waste materials hence the need for a more speedy waste disposal. Despite its effectiveness as a waste disposal method, its widespread adoption has the potential danger of generating fire outbreaks whose control in heavily built and populated neighborhoods might be problematic.

Composting was practiced by a relatively small proportion of households in Ghana. Its extensive use is contingent on availability of sufficient land space which is a limiting factor in urban centers. This might be one of the reasons for its relatively restricted use among urban households compared to rural households. Composting is useful since it apart from being an environmentally friendly approach to waste recycling, it yields organic fertilizer that can be applied to enhance farming due to the advantages of enhanced soil fertility, better soil aeration and moisture holding capacity. Given the high cost of imported chemical fertilizers which has been made more expensive as a result of introduction of subsidy withdrawal policies in Ghana under the structural adjustment program (SAP), composting represents a critically beneficial, inexpensive and simple method of enhancing soil nutrient levels for productive farming. Though urban agriculture is practised in urban centers in Ghana such as Kumasi, the second largest city in Ghana (Erni et al., 2010), farming is predominantly a rural activity, hence it is understandable that the majority of households that adopted the method were resident in rural communities.

Human Excreta Disposal

The poor sanitation situation in the country is also illustrated by the limited access to appropriate excreta disposal facilities. A major reason for the low access is the sluggish rate of construction of modern toilet facilities which appears unable to keep pace with population growth. In addition, accessibility remains a problem in several parts of the country partly due to the poor maintenance culture (Akuffo, 1998) that contributes to regular facility breakdown. As a result, whilst some households may in theory have access to excreta disposal systems, in reality their access may be constrained by associated delays, inconveniences, lack of privacy and other limitations especially when these are public toilets or private ones but are shared by a number of households. These continuing accessibility difficulties were manifested in the observed long queues of persons outside public toilets (Freeman, 2010), often by 6.00 am (Geest, 2002). Apart from these, elders encounter difficulties in accessing public toilets since they have to walk for about fifteen minutes and then queue for a period of time before being able to use them (Geest, 2002).

In addition to inadequacy of the current stock of toilet facilities, there is the issue of their quality. Only a relatively small proportion of the population has access to appropriate toilet facilities. As at 2005/06, only 10.2% of households used flush toilets while a lower proportion of 1.3% of households relied on the VIP or KVIP toilets (Ghana Statistical Service, 2007), which although are below the level of quality associated with the flush system, are generally considered appropriate toilet systems. The KVIP (an acronym for Kumasi Improved Ventilated Pit) toilets are odorless, hygienic excreta disposal pit systems that do not rely on water for flushing excreta. Instead each compartment is provided with two alternating pits and when one pit is full, the other is opened for use pending the gradual decomposition of the content of the previous pit into ordinary waste matter over two years which then becomes safe to handle (Whittington, et al., 1993). The VIPs (Ventilated Improved Pits) are similar to the KVIP toilets but differ only in their lack of alternating pits. A little more than one third of the population (37.4%) utilized the pan or bucket system for disposal of human excreta while a further 31.5% relied on the traditional pit system which is of a slightly lower level of quality than the pan latrine system (Ghana Statistical Service, 2007). Both the pit and the bucket systems are poor excreta disposal systems that are affected by significant presence of flies and other insects, odor emission, absence of user privacy and other inconveniences (Geest, 2002). Almost 20% of the population depended upon other inappropriate, environmentally degrading methods that included open defecation for excreta disposal.

Characteristics of Sanitation Conditions in Ghana

Sanitation conditions in Ghana exhibit a number of characteristics. They include regional as well as urban – rural variations, gender differences in responsibility for waste disposal and the influence of wealth on acquisition of various excreta disposal systems. These characteristics have varied influences on efforts to register both qualitative and quantitative improvements in sanitation coverage in Ghana. A detailed analysis of the extent and depth of their impact on waste disposal is therefore essential to provide a better understanding of various aspects of the sanitation situation in the country. Their incorporation into policies and programs on sanitation in Ghana is essential if the country is to make any headway in its efforts to extend access to improved waste disposal.

Spatial Variations

Significant geographical variations exist with regard to access to, and use of, sanitation infrastructure and services in Ghana. Accra is naturally better served in terms of waste collection than other parts of the country whilst Kumasi is also better served than Bolgatanga, Tamale, and Wa which are major urban centers located in the deprived northern parts of Ghana (Amuzu, 1997, cited in Hens & Boon, 1999). Regional variations in sanitation coverage were further noted in a UNICEF (United Nations Children's Fund) study which revealed that "although sanitation coverage rates have improved in the southern and central regions to over 60%, the coverage in the three northern regions is very low, with less than 6 per cent of the population having access to proper sanitation." (Government of Ghana/UNICEF, 2001:15).

Table 2. Rural-urban differences in access to toilet systems (1991/92 – 2005/06)

Toilet types	Urban households			Rural households		
	1991/92	1998/99	2005/06	1991/92	1998/99	2005/06
Flush toilets	17.6	15.5	22.3	1.4	1.3	1.1
Pit toilets	29.6	18.4	15.7	61.2	45.4	43.4
Pan toilets	24.9	12.5	2.6	4.0	3.4	0.3
VIP/KVIP	12.6	44.9	54.1	3.7	19.9	24.8
Other	15.3	8.8	5.3	29.7	30.0	30.5

Source: Compiled from Ghana Statistical Service, 2007

There were significant urban-rural variations in sanitation conditions. In Ghana, urban and rural communities are defined by their population sizes. Urban centers are settlements with at least 5,000 residents while those with populations less than that are classified as rural (Yankson, 2008). As indicated in Table 2, residents in rural areas were characterized by lower access to toilet facilities compared to those in urban areas. For example, in 2005/06, up to 30.5% of rural households had no access to toilet facilities compared with only 5.3% of urban households. It must also be noted that the poor rural sanitation coverage relative to that of urban centers mirrors the rural-urban disparity in accessibility to basic infrastructure and social services in Ghana (Arku & Arku, 2010). In addition to their limited access to toilet facilities, rural households depended on facilities that were of inferior quality compared with what was available to their urban counterparts. During 2005/06, up to 22.3% of urban households used flush toilets as against only 1.1% of rural households while the proportion of urban households that used KVIP toilets was almost twice that of rural households. Reliance on inferior quality toilets such as pit and pan toilet systems was a lot more widespread among rural households, with almost half of them utilizing such toilets compared with 15.7% of their urban counterparts. A similar dependence on inappropriate sanitation facilities in rural areas was observed in Swaziland where as a result of poor access to improved toilets, traditional pits were the most prevalent toilet systems (Mwendera, 2006).

Trends in Acquisition of Toilet Systems

Although Ghana recorded some improvements in the use of better quality excreta disposal facilities over the period of 1991/92 to 2005/06, the increase in access to flush toilets was marginal, from 7.1% to only 10.2%. At the same time the proportion of households that utilized pit toilets and open defécation declined, indicating a natural inclination to shift from poor to improved excreta disposal facilities over time (Ghana Statistical Service, 2007). Analysis of trends in access to toilet facilities based on a rural - urban split indicates some similarities as well as differences. As Table 2 shows, even though access of rural households to flush toilets showed a steady decline, it contrasted with an almost seven fold increase in their use of VIP and KVIP toilets over the period 1991/92 to 2005/06. At the same time, their dependence on the inferior pan and pit toilets waned. This is suggestive of a tendency on the part of rural households that abandon relatively inferior toilets to embrace VIP and KVIP toilets even though they could not acquire flush toilets due to other inhibiting factors. Thus the trend over the years portrays a declining reliance on inferior toilets and a simultaneous increase in the use of better toilet systems by rural households in line with the national level trend. In the case of urban households there is a similar pattern of an increased use of KVIP as well as flush systems being associated with a reduced reliance on pan and pit toilets, implying, just as in the case of their rural counterparts, that greater access to improved toilets translates into a lessened dependence on relatively inferior excreta disposal systems. Their significant disengagement from use of open defecation which is the most inferior excreta disposal method, however, contrasts with the trend in rural communities, where the proportion of rural households that engaged in it rather registered an increase, albeit a minor, insignificant one.

The Factor of Well-Being

Other variables apart from urban – rural variations explain the differential access to sanitation facilities in Ghana. One such factor is levels of well-being ((Boadi & Kuitunen, 2002; Ghana Statistical Service, 2007). As indicated in Table 3, there was a significant gap in levels of access to flush toilets between wealthy and poor households in urban centers which could be attributed to affordability. Moreover, the proportion of well-to-do urban households engaged in open defecation was less than half that of poor households during the 2005/06 period. The gap between the former and very poor households in terms of patronage of open defecation was even more extensive. These are indicative of a tendency to shun inappropriate excreta disposal systems as income levels improve. In rural areas, the proportion of the well-to-do that depended on KVIP toilets was also more than double that of the very poor. The proportion of the latter that engaged in open defecation was almost three times higher than well-to-do rural households, which is a further confirmation of the tendency of higher income households to avoid inferior excreta disposal systems. Among rural households, none of the very poor could not afford any flush toilets while some, although a small proportion, of the well-to-do ones were able to do so. These are indicative of the critical influence of wealth on the quality of toilet facility used by households in Ghana.

Table 3. Toilet facilities used by households in Ghana 2005/06

Type of toilet facility	Poverty status (urban)			Poverty status (rural)		
	Very poor	Poor	Non poor	Very poor	Poor	Non poor
Flush toilet	3.5	2.9	23.8	0.0	1.4	1.4
Pit latrine	17.1	19.2	15.5	26.8	47.4	47.4
Pan latrine	0.0	3.0	2.7	0.0	0.3	0.3
KVIP	55.0	64.1	53.6	11.0	28.9	28.9
Other	24.4	10.8	4.4	62.2	22.0	22.0

Source: Ghana Statistical Service, 2007

Osumanu's research (2007) in Tamale metropolitan area similarly reveals the influence of differential levels of well-being in observed variations in sanitation coverage. A recent United Nations assessment of global progress toward attainment of the MDGs mentions income inequalities as a dominant factor in the acquisition and use of sanitation facilities and reveals that as much as 77% of the population in the richest 20% of households had access to improved sanitation facilities compared with only 16% of those in the poorest households (United Nations, 2010). The implication is that acquisition of flush and KVIP toilets that would ensure improved human excreta disposal is contingent on Ghana making considerable progress toward poverty reduction.

Reliance on Shared Toilet Facilities

The majority of households in Ghana exhibited a tendency to rely a lot more on public and shared toilet facilities than their own privately owned toilets. The observation by Gordon (2008:244-5), in his assessment of the sanitation situation in Bofano (pseudonym) district in Brong Ahafo Region, that "only 0.7% of households have their own or shared toilet facilities, with 90% of households reliant on public latrines" illustrates this situation. As at 2003, a disturbing 69% of households in Ghana depended on shared toilet facilities, with a greater proportion of urban residents (79%) utilizing shared toilet facilities compared with a relatively smaller proportion of rural dwellers (61%) (Ghana Statistical Service et al., 2004). This implied that the relatively greater urban access to superior sanitation facilities notwithstanding, a higher proportion of rural dwellers utilized their own toilet facilities. Since privately owned toilet systems tend to be better managed than shared or publicly owned ones, there is the likelihood that the anticipated benefits of privacy, convenience, odorless conditions and prompt access that are expected to accrue to households using improved toilet facilities may not be available due to their dependence on shared facilities. Apart from a likely loss of these benefits, reliance on shared facilities entails potentially hazardous health risks since it means a heightened exposure of user households to disease transmission from infected users who patronize the facilities.

Gender Influences

Gender, which refers to socially determined differential roles and responsibilities that are separately assigned to males and females, has a major influence on environmental sanitation in Ghana. Women are saddled with a lot more sanitation improvement responsibilities than men due to socio-cultural perceptions on their gender (Mensah & Whitney, 1992). This is evidenced by the fact that as at 1998/99, up to 71% of females were responsible for sweeping compared to 25% of males while 48% of females were in charge of household waste disposal as against only 18% of males (Ghana Statistical Service, 2000). These gender differences in performance of sanitation related tasks are based on prevailing socio-culturally perceptions. Mensah and Whitney (1992:143) argue "it is culturally unacceptable for a man to sweep around the house or carry garbage to the dump site." Sarris and Sham (1991) also note that it is considered demeaning for a male in the Ghanaian society to be seen carrying materials, resulting in women being saddled with such tasks. These views are reinforced by Osumanu's study (2007) which reveals that men are less exposed to sanitation situations due to a habitual tendency desist from participation in sanitation related tasks and this explains an inclination on their part to place less emphasis on sanitation problems such as lack of waste disposal facilities compared with women. This gender-based division of responsibilities for sanitation improvement affects the extent of waste disposed as by placing the burden principally on females, the Ghanaian society effectively curtails the proportion of household members who could contribute to waste disposal.

Joint Implementation of Sanitation and Water Supply Schemes

Sanitation and water supply schemes have been implemented jointly over the years due to their obvious linkages and the possibility of their combined implementation impacting better on health, environment and socio-economic development generally. A number of foreign donor funded water and sanitation programs have therefore been implemented in Ghana to ensure an eventual realization of the linked benefits that accrue from their coordinated and simultaneous implementation. They include the DANIDA funded Volta Region Community Water and Sanitation Program (Kwashie, 2006), the UNDP (United Nations Development Programme) assisted Rural Water and Sanitation Project (Dotse & Laryea, 2006) and various projects funded by UNICEF, the World Bank's IDA (International Development Association), the EU and KfW (Kreditanstalt für Wiederaufbau, meaning Reconstruction Credit Institute) of Germany. Others are France's AFD (Agence Francaise de Development or French Agency for Development), UK's Department for International Development (DFID) and Canada's Canadian International Development Agency (CIDA) (Fuest, 2005).

Arguably, implementation of these projects made substantial contributions to water supply in Ghana which resulted in an extended coverage from half to two thirds of rural households between 1991/92 and 2005/06 (Ghana Statistical Service, 2007). However, a similar improvement in sanitation coverage did not materialize (Ghana Statistical Service, 2007; Lane, 2006). Whittington and others (1993:733) for example lament that though the International Decade of Water Supply and Sanitation's water supply and sanitation goals were generally not met, it was "particularly troublesome that improvements in sanitation lagged far behind those in water supply". The 2006 Water and Sanitation Plan for Akwapim South District of Ghana's Eastern Region reveals that whilst 108 out of the district's 234 communities had access to potable water, only 48 communities (representing less than half of those with access to water) had access to KVIP toilets which were the improved toilet facilities that were dominantly utilized in the district. The report captures the prevailing situation with this statement

> In the area of sanitation, it should be noted that the situation is far from adequate. Apart from Nsawam, Aburi and other few urban centres, which can boast of some modern decent toilet facilities, majority of the rural areas do not have modern/decent toilet facilities. Most of these rural communities adopt what is known in the local parlance as "free-range" or going to toilet any where in the bush, with its attendant health and development problems." (Akwapim South District Assembly, 2006:1).

There is a likelihood that water supply schemes recorded far greater accomplishments than sanitation despite the institution of combined schemes and their joint implementation either due to a higher community preference for them or a systemic failure to ensure a similar focus on sanitation schemes during program formulation stages.

CONSTRAINTS TO SUSTAINABLE WASTE MANAGEMENT

Given the above scenario, there is the need to initiate various measures to secure long-term improvements in sanitation conditions in Ghana. These measures can however be derailed by various constraining variables that constitute bottlenecks to planned efforts toward ensuring an expanded household access to waste disposal facilities. These constraints should also be considered challenges that would test the determination and creativity of various actors engaged in waste management in Ghana. The fact that a number of them exhibit a high degree of inter-relatedness implies that frontal attacks on them should be joined in order to be effective.

Space Constraints

A key factor that affects sustainable waste management especially human excreta disposal is space limitations. Acquisition of improved toilet systems particularly flush toilet systems was adversely affected by lack of adequate space around houses in urban centers. Akuffo (1998) finds lack of adequate space as a major explanation for the failure of households to comply with legal provisions requiring mandatory installation of flush toilets in all houses in Accra. This is due to the congested nature of various parts of the city which have been described as fully built up with most available places being suitable for only pedestrian use. This constitutes a major constraint to efforts to install flush systems since it precludes the laying of pipes and construction of the needed sewage systems that link the toilets to the main sewers (Akuffo, 1998). However, it must be noted that additional factors other than spatial restrictions account for the repeated failure to provide better toilet facilities in Ghana since rural areas and urban fringes with adequate space similarly lack toilet facilities.

Low Interest in Waste Management

Limited government focus and low community interest in waste management slow down sanitation improvement in Ghana (Freeman, 2010). It has been argued that waste disposal is a lot more problematic in developing countries due primarily to less premium placed on it by their governments (Afon & Okewole, 2007), compared with developed countries with greater waste generating capacities but which have been able to reduce waste to the barest minimum due to concerted efforts toward waste disposal. Community interest in waste management has also been found to be on the low side compared with water, and is regarded as one of the limiting factors that account for the persistence of sanitation problems in Ghana (Lane, 2006). Preferences for water supply and low interest in sanitation also appear to be quite widespread. In Cameroon, communities were said to have persistently exhibited priority interest in water schemes ahead of other development projects (Fonjong et al., 2004). The European Union similarly demonstrated greater interest in water supply schemes than sanitation services delivery. Hemson (2008:42) notes with regard to the relatively high EU interest in water supply that "this emphasis is confirmed by the announcement by the European Union on World Water Day 2004 that water was firmly placed at the top of its political agenda. There is, unfortunately not an equal apportionment of attention to sanitation…". This factor provides a plausible explanation for the persistent low sanitation coverage relative to water supply, which calls for periodic public education on the adverse impacts of poor waste disposal on health and other socio-economic conditions.

Poor Management

Allied to the problem of low community and government interest in waste disposal is the adoption of inappropriate management measures. Gordon (2010:99) for example attributes the acute sanitation difficulties the Ghanaian society is experiencing to poor waste management, explaining further that "the public toilets that exist are said to have become 'health hazards' due to a complete lack of waste management by the District Assembly, and while the construction of private household latrines is the current sanitation strategy, they remain very few in number". Laube (2007) lists a litany of management shortcomings such as endemic corruption, an inability to ensure enforcement of existing legislations, lack of transparency, nepotism and political clientelism as constraining factors that place limitations on local resource management efforts in Ghana. Since local governments such as municipal authorities have traditionally been the lead responsible organizations for waste disposal (Oteng-Ababio, 2010; Satterthwaite, 2002), the onus of instituting better waste management measures rests upon them and accordingly the blame for the poor state of sanitation in Ghana should be laid at their doorsteps.

Inadequate Water Supply

Inadequate water supply in Ghana affects the use of flush toilets. Despite the recorded expansion in water supply in Ghana in the last few decades (Ghana Statistical Service, 2007), some areas are still not covered while in areas that have access to water, supply is constrained by interrupted flows due to frequent facility breakdowns and other technical bottlenecks (Erni et al., 2010; Osumanu, 2007). Household connections that can ease inconveniences associated with water supply are also embarrassingly low (Osumanu, 2007). These affect acquisition of flush toilets in particular since their continuous use is contingent on uninterrupted and adequate supply of water. Thus well-to-do residents of urban neighborhoods whose water supply is characterized by erratic and therefore unreliable flow may be unwilling to acquire flush toilets in their homes even though they could afford their installation. As a result of the uneven flow, estimated urban water use in Ghana is only about 50 l/cap/d (Boadi et al., 2005; Erni et al., 2010). This compares unfavorably with the average level of domestic water consumption in United States which is estimated at 400 l/cap/d (Postel, 1996, cited in Pimentel, 1999) and those of Western Europe which range between Belgium's 115 l/cap/d and Spain's 260 l/cap/d (EEA 2003, cited in Erni et al., 2010). Although low levels of rural income restrict the ability of rural households to acquire flush toilets, insufficient water supply constitutes an additional constraining factor. This is because notwithstanding the fact that a significant proportion of them have water sources such as boreholes in their communities, they lack ready access to water supply in their own homes due to limited household connections.

Lack of Waste Disposal Facilities

Inadequate supply of waste disposal facilities such as garbage bins into which waste materials can be deposited contributes to waste management problems in Ghana (Akuffo, 1998; Mensah & Whitney, 1992). As at 1998, garbage bins provided in Accra served only 4% of the population (Akuffo, 1998). The only waste processing facility in the city was established as far back as 1979 and currently operates at 10% capacity due to continued reliance on a set of dilapidated, malfunctioning equipment (Oteng-Ababio, 2010). As a result of these, some residents are compelled to litter the streets with garbage which after rainfall gradually drains into streams and other water bodies as well as into gutters (Boadi & Kuitunen, 2002).

Inadequate Institutional Arrangements

Fragmented institutional responsibilities for the sanitation sector and lack of institutional coordination mechanisms have been the bane of sanitation improvement in Ghana (Akuffo, 1998). Institutional responsibility for sanitation improvement remains fractured between the Community Water and Sanitation Agency (CWSA) established in 1994, the Environmental Protection Council set up in 1974 (Hens & Boon, 1999) and later restructured into an agency, the Ministry of Local Government, Rural Development and Environment (MLGRDE) and finally the Environmental Health Divisions of the various Assemblies which as local government agencies are considered as having ultimate responsibility for sanitation improvement in communities under their jurisdiction. Currently, CWSA is engaged in implementation of the National Community Water and Sanitation Programme formulated in 1993 with technical assistance from the donor community particularly the World Bank (Fuest, 2005) whilst implementation of the National Environmental Sanitation Policy designed in 1999 by MLGRDE is also on-going (Oteng-Ababio, 2010). In addition to these agencies that have direct responsibility for sanitation improvement, other public agencies such as the Ministry of Lands, Forestry and Mines, the Ministry of Food and Agriculture, the Ministry of Trade and Industries and the Ministry of Tourism have additional responsibilities in the sanitation sector and were mentioned in a major government policy document as having critical roles to play in ensuring improved environmental conditions (Government of Ghana, 2002). These lead to divided attention, poor resource allocation, duplication of efforts and inter-agency squabbles for dominance which affect sustainability of waste management. As stated in a UNICEF publication on Ghana (2001:111), "efforts to increase water, sanitation, hygiene education are hampered by institutional difficulties. Expenditure and responsibility for environmental services is fragmented among several agencies .....This leads to a lack of coordination and effectiveness in the formulation of strategies and delivery of services."

Urbanization

Due to declining socio-economic opportunities and a desire on the part of rural dwellers to escape depressing conditions in their communities, rural-urban migration has been on the ascendancy in Ghana (Boadi et al., 2005; Eyong & Mbuagbo, 2002; Jorgensen, 2008). Persistent poverty conditions contributed to the significant net migration of more than 310,000 per 1,000 persons to the urbanized Greater Accra Region during the period of 2005/06 (Ghana Statistical Service, 2007). The ever-increasing migration of people to urban centers constitutes one of the main causes of the mounting waste management troubles in Ghana as local government agencies struggle unsuccessfully to find a panacea to the accumulated waste (Freeman, 2010; Mensah & Whitney, 1992). Ghana has witnessed persistent increases in its urban population since 1960s and these increases have been concentrated in the principal cities of Accra, Kumasi, Sekondi-Takoradi, Tema and Tamale (Laryea-Adjei, 2000). Increased population concentration in urban centres contributes to a rise in waste generation (Kgathi & Bolaane, 2001) as urban populations have greater wealth ((Laryea-Adjei, 2000) that provides them with a relatively higher purchasing power than rural populations. They therefore engage in greater consumption of a wide variety of goods, which generates huge waste materials (Boadi et al., 2005).

Industrial Activities

Apart from household waste, industrial waste contributes to sanitation problems. Despite being a predominantly agricultural country, a number of industries as well as mining companies have been established in Ghana and their activities generate waste materials which pose various disposal difficulties. Since the majority of industrial ventures are located in Accra and other urban centers (Hens & Boon, 1999; Laing, 1994, cited in Boadi et al, 2005), the industrial waste materials they generate contribute to a worsening of urban sanitation conditions. Some of their waste materials are deposited into water bodies in urban centers whilst the rest are deposited at nearby dumping sites (Adomako et al., 2008; Akuffo, 1998; Hens & Boon, 1999), activities that are environmentally degrading. Food processing industries and other industrial ventures engaged in the manufacture of various chemicals and asbestos products empty their waste products into water bodies in Accra particularly the Odaw River and the Korle Lagoon (Boadi & Kuitunen, 2002). Studies in Kumasi reveal that largely as a result of concentrated discharge of arsenic, chromium, cadmium and aluminium compounds as well as other industrial effluents from lumber processing and preservation firms, the River Subin which flows through the city is heavily contaminated and its water is toxic to both animals and humans (Adomako et al., 2008). Excessive industrial pollution has damaged the natural ecology of these water bodies and destroyed all fish life. Continuous deposition of industrial waste has also made them shallow. This contributed to the frequent floods encountered in Accra as the water bodies tended to overflow their banks after excessive rainfall (Boadi & Kuitunen, 2002).

The Way Forward

Sustainable management of Ghana's waste materials is possible. However, it will require increased attention from government agencies, streamlining of existing institutional mechanisms to ensure better focus on waste management, increased resource support to poor households toward accelerated acquisition of sanitation infrastructure and a changed attitude to waste disposal by households. Implementation of the suggested strategies below could make a considerable contribution to sustainable waste management.

Composting

The four critical components of resource conservation – reduce, re-use, recover and recycle waste should be the guiding principles of sanitation improvement schemes in Ghana. Waste materials could be recycled and profitably used after composting. Huge proportions of waste that are generated in both rural and urban areas are suitable raw materials for the production of manure through composting. Waste would then be converted into usable forms for soil fertility enhancement that could promote organic farming and minimize expenditure on imported, chemical fertilizers. Currently undertaken on a very minimal scale, composting can be expanded through public education campaigns on its usefulness and provision of extension service support by technical staff of the Ministry of Food and Agriculture to ensure proper composting. It is a task that entails minimal cost requirements. It will be beneficial to the rural economy which has farming as the predominant economic activity. Urban communities also stand to gain since urban agriculture is on the increase in Ghana (Erni et al., 2010) and in other Sub-Saharan Africa countries due to the numerous benefits such as supply of vegetables, employment generation and opportunities for additional incomes (Keraita et al., 2008). Reducing and re-using waste could minimize their dumping in the open, reduce expenditure on waste collection by external agencies and discourage a resort to other unhealthy disposal practices. In addition, government could also partner the private sector to establish and operate waste recycling ventures to recycle waste materials into usable forms.

Public Education on Sanitation

Attitudinal change is possible through public education (Oteng-Ababio, 2010). Foremost among the series of attitudinal changes that are required to promote appropriate sanitation conditions is that of open defecation. This is critical due to the threats it poses to health, environmental, child survival, tourism and other essential aspects of societal development. Described by the UN (2010:61) as "an affront to human dignity", it unfortunately appears to be a widespread practice not only in Ghana but in other developing countries similarly plagued with the problem of limited sanitation facilities and services, with an estimated 1.1 billion currently engaged in it (UN, 2010). In similar vein, dumping of waste in open, unprotected sites is environmentally harmful and should be discouraged. In the interim, these sites should be fenced off to limit animal entry, minimize drainage of waste into gutters and streams after rainfall and prevent children from gaining access to them and turning them into playgrounds. Periodic fumigation should be undertaken. Vigorous educational campaigns on the dangers of open defecation should be pursued alongside support in construction of toilet facilities. Community level public education campaigns such as community and focus group meetings should be the dominant approaches in distant rural communities where access to mass media sources is constrained by lack of electricity supply and other infrastructural bottlenecks.

Sanitation Tribunals

Public educational campaigns on adverse consequences of inappropriate waste disposal behaviors should also be backed by regulatory measures whose enforcement could elicit the required attitudinal change. Without an effective enforcement system, conformity with standard environmental practices is unlikely to occur. Satterthwaite (in Desai & Potter, 2002) for example, recommends subjecting both households and enterprises to regulatory mechanisms to ensure sound environmental management. With regard to Ghana's sanitation sector, the relevant by-laws are in place, some dating as far back as 1975 (Akuffo, 1998). A number of these make it an offence to throw waste materials into the open and engage in other inappropriate waste disposal acts. One essential by-law, for example, requests all landlords in the capital city to install flush toilet systems in their houses (Akuffo, 1998). These by-laws specify penalties whose imposition could constitute deterrents to further harmful household waste disposal behaviors. The critical issue is limited enforcement capacities as the application of these by-laws is contingent on the arrest, trial and sentencing of offenders. As Akuffo (1998:16) argues, "the debility of law enforcement …made sanitation an intractable problem".

The enforcement system in Ghana is unfortunately weak and inadequate (Government of Ghana, 2002) hence an accelerated establishment of sanitation tribunals to handle such cases will be helpful since it will also free the normal courts to focus on other criminal offences they are already saddled with. The Ghana Government has indicated a commitment to the establishment of such tribunals in every district (Ghana Government, 2002). The process, however, has been unhurried and needs to be accelerated. Without an effective enforcement of sanitation legislations any attitudinal changes elicited through education cannot be sustained, as they would be temporary and ineffective.

Contributions to sanitation costs

Given the fact that industrial ventures have contributed to a worsening of sanitation conditions, another approach to reducing the huge piles of solid waste and pollution of water bodies is to request their mandatory involvement in cost sharing waste disposal expenses (Akuffo, 1998). The Assemblies should liaise with industries and mining ventures located in their areas of jurisdiction to make appropriate financial contributions toward management of waste that they contributed to generating. This is in line with the 'polluter pays principle' (PPP) approach enshrined in Principle 16 of the Rio Declaration, an output of the 1992 Earth Summit that was convened under the auspices of the UN. Whilst financial contributions will be helpful in waste management efforts, a more sustainable approach should be for industrial establishments to reduce and recycle waste. The idea of 'corporate responsibility' could be invoked to compel companies to focus on minimizing their environmental footprint. This could lead to a "more socially-responsible corporate behaviour" as advocated by Sadler and Lloyd (2009:613). The Department of Factory Inspections should pick up the responsibility of liaising with industrial ventures to strategize on effective management of industrial waste through waste minimization and recycling.

Other Strategies

In view of the inter-relatedness of water and sanitation, implementation should stay joined to maximize benefits. However, the sanitation component should incorporate additional measures that would ensure expanded coverage. Under the UNDP assisted Rural Water and Sanitation Program, the key sanitation improvement strategy was the construction of 750 VIP toilets as demonstration units (Dotse & Laryea, 2006), based on a conviction that a visible observation of their merits over the traditional pit and pan toilets would generate sufficient interest in their acquisition and consequently, lead to their widespread adoption. While additional 250 toilets that were constructed by households without direct project input support indicated that the promotional strategy worked to some extent (Dotse & Laryea, 2006), it is obvious that a lot more of such improved toilets should have been constructed by households to make any real impact on sanitation conditions. This implies that additional supportive measures, including financial incentives and technical assistance to households are needed to facilitate the process.

Households resident in areas experiencing water supply constraints such as infrequent flow and limited household connections should be assisted to acquire KVIP toilets that do not require water to operate while flush toilets are promoted in neighborhoods households that do not experience these difficulties. Households outside the high income bracket would be more amenable to acquisition of KVIP toilets since their construction entails less expenses than flush toilets. Unlike flush toilets, construction of KVIP toilets is undertaken with less costly, locally available materials. Poorer households might require with financial incentives such a subsidy of 40-50% of total cost or a micro credit facility with flexible repayment terms to enable them to acquire improved toilet facilities. The state could train more artisans in the construction of KVIP and flush toilets to enable them construct an increased number of these facilities. Acquisition of privately owned toilets will minimize shared use of toilets and also ensure decreased reliance on open defecation. Financial support should initially target regions in the northern part of Ghana that lag behind in sanitation coverage to minimize regional inequalities.

A way out of the problem of spatial constraints in urban centers is to convert an existing room into a toilet facility. Another would be explore to the possibility of a pooled scheme involving two or three households that are resident in the same vicinity who can unite efforts, resources and the limited spaces between their houses to construct a toilet and institute workable arrangements for their collective use and management. Whilst this implies sharing facility use, it involves only few households and therefore entails fewer health hazards to them.

CONCLUSION

Improved sanitation is critical to Ghana's development in view of its significant linkages with health and other aspects of socio-economic development. However, sanitation infrastructure and services in the country are not in an optimal state. Proliferation of both open defecation and open dumping of wastes, reliance on shared toilet facilities and other constraints are disturbing revelations that need to be addressed. Public education on the dangers of inappropriate disposal behavior, institution of punitive sanctions backed by an enforcement capacity and significant government intervention in provision of facilities such as garbage bins, and extension of financial and technical support to poor households toward acquisition of better excreta disposal facilities will be critical. The benefits of composting should feature prominently in public education campaigns. Its adoption will also require technical assistance. Among the various sanitation problems, those that are of critical concern and need to be addressed urgently are the twin practices of depositing waste in the open and open defecation. A coordinated, simultaneous implementation of strategies outlined above would minimize these and other poor waste situations in various parts of Ghana on a more sustainable basis. Their implementation requires the active involvement of government and private agencies, non-governmental organizations, community associations and households. External support from the donor community and financial contributions from industrial ventures would be required.

The strength of the relationship between sanitation and water supply is demonstrated by the limiting impact of irregular water supply on acquisition of flush toilets. The influence of well-being on the rate and extent of acquisition of appropriate toilet facilities, similarly indicates the linkages between sanitation and poverty reduction schemes. These and other intersectoral linkages must be recognized and factored into the design and implementation of sanitation improvement schemes to ensure their sustainability. Whilst joint implementation of water and sanitation schemes remains the preferred option, there is the need for an increased focus on sanitation to ensure the sector receives the quantum of investible resources, policy guidelines and a streamlined institutional arrangement it needs to leap forward. The critical issue is implementation of measures which ensure that the gains are not only forthcoming but are long lasting and comprehensive enough to impact not just on health but on other essential socio-economic spheres of life. Strategies outlined above will make a productive contribution to, and sustain these objectives.

REFERENCES

Adomako, D., Nyarko, B. J. B., Dampare, S. B., Serfor-Armah, Y., Osae, S., Fianko. J. R., & Akaho, E. H. K. (2008). Determination of toxic elements in waters and sediments from River Subin in the Ashanti Region of Ghana. *Environmental Monitoring and Assessment,* Vol. 141, Issue 1-3, 165-175

Afon, A. O. & Okewole, A. (2007). Estimating the quantity of solid waste generation in Oyo, Nigeria. *Waste Management & Research*, Vol. *25*, No. 4, 371-379

Agyeman, K. (2007). Privatization of water in Ghana: Stopped in its tracks or a strategic pause?. *International Journal of Environmental Studies*, Vol. 64, No. 5, 525 — 536

Akuffo, S. B. (1998). *Pollution Control in A Developing Economy: A Study of the Situation in Ghana.* 2nd Edition, Accra: Ghana Universities Press

Akwapim South District Assembly, (2006). *Akwapim South District Assembly District Water and Sanitation Plan 2006 Review.* Report submitted to Community Water and Sanitation Agency, Eastern Region, Koforidua-Ghana

Amuzu, A. T. (1997). *Environmental Management: A case study on water resources in Ghana.* In E. K. Boon & L. Hens, (Eds.). Environmental management in West Africa, (Pp. 242–252). Human Ecology Department, VUB, Brussels. Belgium.

Anand, P. B. (2007). Semantics of Success or Pragmatics of Progress? An Assessment of India's Progress With Drinking Water Supply. *The Journal of Environment & Development,* Vol. 16, No. 1, 32-57

Appiah-Kubi, K. (2001). State-owned enterprises and privatization in Ghana. *The Journal of Modern African Studies*, Vol. 39, No. 2, 197-229

Arku, F. S., & Arku, C. (2010). I cannot drink water on an empty stomach: a gender perspective on living with drought. *Gender & Development*, Vol. 18, No 1, 115 — 124

Boadi, K., Kuitunen, M., Raheem, K. & Hanninen, K. (2005). Urbanisation without development: Environmental health implications in African cities. *Environment, Development and Sustainability*, Vol. 7, No. 4, 465–500

Boadi, K., & Kuitunen, M. (2002). Urban waste pollution in the Korle Lagoon, Accra, Ghana. *The Environment,* Vol. 22, No. 4, 301–309

Busari, O. (2009). Water, sanitation and sustainability: lessons from a community project. *Environment, Development and Sustainability*, Vol. 11, No.1, 7[illegible]–83

Cook, I. R. (2009). Private sector involvement in urban governance: The case of Business Improvement Districts and Town Centre Management partnerships in England. *Geoforum,* Vol. 40, No. 5, 930–940

Dotse, F. M. & Laryea, N. O. (2006*). Multi-faceted participatory approaches'* Paper presented at 30th WEDC International Conference – People-Centred Approaches To Water And Environmental Sanitation, Vientiane, Lao PDR

DWAF, (2008). *Water and Sanitation coverage in South Africa.* Department of Water Affairs and Forestry.

EEA, (2003). *Europe's environment: the third assessment. Environmental assessment report No 10.* European Environment Agency, Copenhagen, Denmark. (http://reports.eea.eu.int/environmental assessment report 2003 10/Chapter8; accessed on 19 July 2007)

Erni, M., Drechsel, P., Bader, H., Scheidegger, R., Zurbruegg, C. & Kipfer, R. (2010). Bad for the environment, good for the farmer? Urban sanitation and nutrient flows. *Irrigation Drainage System*, Vol. 24, No.1-2, 113–125

Eyles, J. & Consitt, N. (2004). What's at Risk: Environmental influences on Human Health. *Environment*, Vol. 46, No. 8, 25-39

Eyong, C. T. & Mbuagbo, O. T. (2002). Assessment Of The Hygiene And Sanitation Campaign in Cameroon. *Environment, Development and Sustainability,* Vol. 4, No.4, 391–402

Fonjong, L. N., Emmanuel, N. N. & Fonchingong, C. C. (2004). Rethinking the Contribution of Indigenous Management in Small Scale Water Provision among selected Rural Communities in Cameroon. *Environment, Development and Sustainability,* Vol. 6, Issue 4, 429-451

Freeman, F. (2010). Ghana: The Waste Land' *World Policy Journal.* Vol. 27, No. 2, 47-53

Fuest, V. (2005). *Policies, Practices and Outcomes of Demand-oriented Community Water Supply in Ghana: The National Community Water and Sanitation Programme 1994 – 2004.* ZEF Working Paper Series No. 5, Center for Development Research, University of Bonn

Garande T, & Dagg, S. (2005). Public participation and effective water governance at the local level: a case study from a small under-developed area in Chile. *Environment, Development and Sustainability,* Vol.7, No.4, 417–431

Garcıa, R. C. E. & Brown, S. (2009). Assessing water use and quality through youth participatory research in a rural Andean watershed. *Journal of Environmental Management* xxx, 1–8

Geest, S. V. D. (2002). The night-soil collector: bucket latrines in Ghana *Postcolonial Studies,* Vol. 5, No. 2, 197–206,

Ghana Statistical Service, (2007). *Pattern and Trends of Poverty in Ghana 1991-2006.* Accra: Ghana Statistical Service

Ghana Statistical Service, Noguchi Memorial Institute for Medical Research, ORC Macro. (2004). *Demographic and Health Survey 2003.* Accra: Ghana Statistical Service

Ghana Statistical Service, (2002). *2000 Population and Housing Census – Summary Report of Final Results.* Accra: Ghana Statistical Service

Ghana Statistical Service, (2000). *Ghana Living Standards Survey: Report of the Fourth Round (GLSS 4).* Accra: Ghana Statistical Service

Ghana Statistical Service, (1995). *Ghana Living Standards Survey: Report on the Third Round (GLSS 3) September 1991- September 1992.* Accra: Ghana Statistical Service

Gordon, C. (2010). Decentralisation and struggles for basic rights in Ghana: opportunities and constraints. *The International Journal of Human Rights*, Vol. 14. No. 1, 92 - 125

Gordon, C. (2008). Decentralization and the Limits to Poverty Reduction: Findings from Ghana. *Oxford Development Studies*, Vol. 36, No. 2, 235 — 258

Government of Ghana, (2002). *Ghana Poverty Reduction Strategy 2002-2004.* Accra: Government of Ghana

Government of Ghana/UNICEF, (2001). *Master Plan of Operation and Programme Plans of Operation 2001-2005* Accra: Government of Ghana/UNICEF

Hens, L. & Boon E. K. (1999). Institutional, Legal, and Economic Instruments in Ghana's Environmental Policy. *Environmental Management,* Vol. 24, No. 3, 337–351

Hemson, D. (2008). Water for all: from firm promises to 'new realism'?. In D. Hemson, K. Kulindwa, H. Lein, & A. Mascarenhas. (Eds.), *Poverty and Water: Explorations of the Reciprocal Relationship* (Pp. 13-46). London & New York: Zed Books

Hoko, Z. & Hertle, J. (2006). An evaluation of the sustainability of a rural water rehabilitation project in Zimbabwe. *Physics and Chemistry of the Earth,* Vol. 31, 699–706

Jørgensen, S. H. (2008). Some perspectives on the geographies of poverty and health: A Ghanaian context. *Norsk Geografisk Tidsskrift - Norwegian Journal of Geography*, Vol. 62, No. 3, 241 — 250

Kazelli, H., Chrisofides, T. & Keravnou E. (2003). Water Management in Cyprus Through a Decision – Support System. In P. Koundouri, P. Pashardes, T. M. Swanson and A. Xepapadeas (Eds.), *The Economics of Water Management in Developing Countries: Problems, Principles and Policies* (Pp. 215-236). Chelteham and Northampton: Edward Elgar Publishing Ltd.

Keraita, B., Drechsel, P., Konradsen, F. & Vreugdenhil, R. C. (2008). Potential of simple filters to improve microbial quality of irrigation water used in urban vegetable farming in Ghana. *Journal of Environmental Science and Health Part A,* Vol. 43, No. 7, 749–755

Kgathi, D. L. & Bolaane, B. (2001). Instruments for sustainable solid waste management in Botswana. *Waste Management & Research,* Vol. 19, No. 4, 342–353

Kwashie, H. B. (2006). The Motivation and Orientation of Persons Managing Community Water Supply and Sanitation Programmes in the Volta Region of Ghana: An Empirical Study. *Ghana Journal of Literacy and Adult Education*, Vol. 3, No.1, 66-82

Laing, E. (Ed.), (1994). *Ghana Environmental Action Plan, 2.* Accra: Environmental Protection Agency.

Lane, J. (2006). Improving Water and Sanitation Services in Rural Areas: Lessons Learned from Ghana, Lesotho and South Africa. In L. Fox & R. Liebenthal (Eds.), *Attacking Africa's Poverty: Experience from the Ground* (Pp. 257-284). Washington: The World Bank

Laryea-Adjei, G. (2000). Building capacity for urban management in Ghana: some critical considerations. *Habitat International,* Vol. 24, Issue 4, 391-401

Laube, W. (2007) The Promise and Perils of Water Reforms: Perspectives from Northern Ghana. *Africa Spectrum,* Vol., 42, Issue 3, 419-437

Malley, Z. F. U., Taeb, M., Matsumoto, T. & Takeya H. (2008). Environmental sustainability and water availability: Analysis of the scarcity and improvement opportunities in the Usangu plain, Tanzania. *Physics and Chemistry of the Earth,* Vol. 33, Issue 5, 330-339

Manase, G., Nkuna, Z.& Ngorima, E. (2009). Using water and sanitation as an entry point to fight poverty and respond to HIV/AIDS: The case of Isulabasha Small Medium Enterprise. *Physics and Chemistry of the Earth,* Vol. 34, Issues 13-16, 866–873

Mausch, K., Mithöfer, D., Asfaw, S., & Waibel, H. (2006). *Impact of EurepGAP Standard in Kenya: Comparing Smallholders Large-Scale Vegetable Producers.*Conference on International Agricultural Research for Development, University of Bonn, October 11-13, 2006

Mensah, J., & Whitney, H. A.(1992). Urban environmental perception and behaviour in Africa: what gender engenders. *International Journal of Environmental Studies*, Vol. 40: No. 2, 141 — 149

Mountz, A., Curran, W. (2009). Policing in drag: Giuliani goes global with the illusion of control. *Geoforum,* Vol. 40, Issue 6, 1033–1040

Mwendera ,E. J. (2006). Rural water supply and sanitation (RWSS) coverage in Swaziland: Toward achieving millennium development goals. *Physics and Chemistry of the Earth,* Vol. 31, Issue 15-16, 681–689

Newell, P. (2002). Environment. In P. Burnell, & V. Randall, (Ed.), *Politics in the Developing World.* (Pp. 221-236), Oxford: Oxford University Press

Ngwenya, B., Kgathi, D. (2006). HIV/AIDS and access to water: a case study of home-based care in Ngamiland, Botswana. *Physics and Chemistry of the Earth*, Vol. 31, Issues 15-16, 669-680

Oteng-Ababio, M. (2010). Missing links in solid waste management in the Greater Accra Metropolitan Area in Ghana. *GeoJournal*, (Online first) DOI 10.1007/s10708-010-9363-9

Osumanu, I. K., (2007). Perceptions of Environmental Health Risks and Care Seeking Behaviours of Vulnerable Urban Households in the Tamale Metropolitan Area. *Ghana Journal of Development Studies,* Vol. 4, No. 2, 60-75

Pimentel, D., Bailey, O., Kim, P., Mullaney, E., Calabrese, J., Walman, L., Nelson F. & Yao X. (1999). Will Limits Of The Earth's Resources Control Human Numbers?. *Environment, Development and Sustainability,* Vol. 1, No. 1, 19–39

Postel, S. (1996). *Dividing the Waters: Food Security, Ecosystem Health, and the New Politics of Scarcity, Vol. 132*, Washington, DC, Worldwatch Institute.

New Partnership for African Development (2002) *Comprehensive African Agriculture Development Programme* http://ww.fao.org/docrep/005/Y683e-02.htm (19.10.04)

PAHO, (2001). *Regional Report on the Evaluation 2000 in the Region of the Americas: Water Supply and Sanitation.* PAHO Report, Washington, DC

Sadler, D. & Lloyd, S. (2009). Neo-liberalising corporate social responsibility: A political economy of corporate citizenship *Geoforum,* 40, Issue 4, 613–622

Sarris, A. & Sham, H. (1991). Ghana Under Structural Adjustment – The Impact on Agriculture and the Rural Poor. New York: NYUP/IFAD

Satterthwaite, D. (2002). Urbanization in developing countries. In V. Desai & R. B. Potter, (Eds.), *The Companion to Development Studies* (Pp.243-247). London: Arnold

Shiva, V. (2004). The Impoverishment of the Environment: Women and Children Last. In M. Zimmerman, M., Callicott, J. B. Warren, K. J. Klaver, I., Clark, J., (Eds.), *Environmental Philosophy: From Animal Rights to Radical Ecology, Fourth Edition* (Pp. 178-195). Upper Saddle River: Pearson/Prentice Hall

Tornqvist, R., Norstrom, A., Karrman, E. & Malmqvist, P. A. (2008). A framework for planning of sustainable water and sanitation systems in Peri-urban areas. *Water, Technology, Science and Technology,* Vol. 58, No. 3

UNICEF/WHO, (2006) *Global water supply and sanitation assessment 2006 report.* New York: UNICEF/WHO

UNICEF/WHO, (2002). *Global Water Supply and Sanitation Assessment 2002 Report.* New York: UNICEF/WHO

United Nations, (2010). *The Millennium Development Goals Report 2010.* New York: UN

Whittington, D., Lauria, D. T., Choe, K., Hughes, J. A. & Swarna, V. (1993). Household Sanitation Practices in Kumasi, Ghana: A Description of Current practices, Attitudes, and Perceptions. *World Development,* Vol. 21, No. 5, 733-748

WHO-UNICEF, (2004). *Meeting the Millennium Development Goals Drinking Water and Sanitation Target: A mid-term assessment.* Geneva: WHO-UNICEF Joint Monitoring Program.

WHO, (2002). *WHO Statistical Information System (WHOSIS).* Geneva, Switzerland.

WHO, (2000). *Global Water Supply and Sanitation Assessment 2000 Report.* World Health Organization and United Nations Children's Fund

World Bank, (2005). *The Little Green Data Book.* Washington: The World Bank.

Yankson, P.W.K. (2008). Decentralisation and poverty reduction in the Gomoa District of Ghana. *Norsk Geografisk Tidsskrift - Norwegian Journal of Geography,* Vol. 62, No. 3, 230 — 240

In: Potable Water and Sanitation
Editor: Joel M. McMann

ISBN 978-1-61122-319-4

Chapter 6

SANITATION DEVELOPMENT AND ROLES OF JAPAN

Yoshiaki Tsuzuki*
Cooperative Researcher, Research Center for Coastal Lagoon Environments, Shimane University, Matsue, Shimane 690-8504, Japan

ABSTRACT

Water and sanitation sector development is demanded in face of the Millennium Development Goals (MDGs) target year of 2015. In Japanese sanitation sector, a zoning planning or geographical division scheme of centralized and on-site municipal wastewater treatment systems have been developed in several decades. On the contrary, there are countries with mandatory on-site treatment systems even in the areas served with centralized wastewater treatment systems to prevent infectious diseases. In the Introduction section, recent development especially in sanitation sector after 1990, the completion year of the Water Supply and Sanitation Decade, was chronologically and briefly summarized focusing on resources and benefits of sanitation in the world. In the second section, resources or investments and profits of sanitation were discussed. In the third section, Japanese sanitation scheme was described including typical on-site treatment systems, *johkasou*, which treats both black and gray water. Not only technological aspects but also regulation and institutional aspects are important for on-site treatment system development because of necessity of operation and maintenance. In the fourth section, differences between zoning scheme and mixture situations of centralized and on-site treatment systems were discussed. In the fifth section, technological alternatives in consideration with pollutant discharge reduction function of wastewater treatment systems and investments were discussed. Japanese experiences in sanitation sector will contribute to sanitation sector development in the world together with a huge stock of world experiences. For sanitation perspective, the author summarized his personnal ones on the basis of the government documents and academic literature for the sector development and the roles of Japan which are conducting huge investments in the sector.

* Corresponding author: E-mail: tsuzuki.yoshiaki@gmail.com, URL: http://sites.google.com/site/yoshiakitsuzuki/

1. INTRODUCTION

For sanitation, there are many aspects including hygiene, infectious diseases and mortality (Bartram et al., 2005; Fewtrell et al., 2005), children growth (Checkley et al., 2004) and food processing hygiene (Norman and Gravani, 2006). Cost of food sanitation was estimated as US$ 229-610 million for E. coli O157:H7 outbreak for beef parties industry of the fast food restaurant, US$ 263 million for benzene contamination in a popular brand imported bottled water, and US$ 10-50 million for inadvertent defective glass of export bottles by a European beer maker.

Every year, 1.8 million people die from diarrhoeal[1] diseases including cholera, among which 90% are children under 5, mostly in developing countries (World Health Organization, 2004). Unsafe drinking water and inadequate sanitation and hygiene were attributed to 88% of diarrhoal disease. Approximately 3.1% of 1.7 million deaths and 3.7% of disability-adjusted life years (DALYs) of 54.2 million in the world were estimated to be attributable to unsafe drinking water and inappropriate sanitation and hygiene (Water Supply and Sanitation Collaboration Council and World Health Organization, 2005). The daily mortality number, about five thousand mortality per day, corresponds to passengers of 10 large-size airplanes (Harada, 2010; Original source is WHO).

This chapter manuscript focuses more on water and sanitation sectors. Anticipated readers are from university undergraduate and graduate students in the fields of environment engineering, environment economics, development economics, environment policy and related fields, and professionals and specialist in these fields. Some statistical data and information were collected using Internet. This will help readers to easily find further references in English and in Japanese. Some URL of the references may be different at the time of publication. Then, the readers can use authors' name and titles of references to search literatures and documents. For Japanese references, the readers should know authors' name and titles in Japanese when try to find by use of search engines. The readers can freely contact the author of this chapter manuscript by e-mail in such cases. Moreover, a reference list in Japanese will be uploaded on the author URL after publication of the book. For sanitation perspectives, the author summarized his personnal ones on the basis of the government documents and academic literature for the sector development and the roles of Japan which are conducting huge investments in the sector.

After the Water Supply and Sanitation Decade in 1981-90, increasing percentages of people with safe drinking water supply and appropriate sanitation have been targeted in the water and sanitation sectors in the world. The percentage without access to improved water sources was 23% in 1990, decreased linearly to 13% in 2008, and estimated to reach to 9% in 2015, the targeted year of the Millennium Development Goals (MDGs) (WHO/UNICEF Joint Monitoring Programme on Water Supply and Sanitation, 2010). The water supply MDGs target of 12% is estimated to be achieved with continuing efforts in water sector. On the contrary, the MDGs sanitation indicator was still 39% in 2008 and projected sanitation indicator is 36% in 2015 whereas the MDGs sanitation target aims for 23%. The MDGs sanitation indicator is considered to be hard to be achieved. There should be more efforts in sanitation sector to prevent infectious diseases and to improve people's lifestyles and ambient water quality.

[1] Both diarrhea and diarrhoea are used in this manuscript mostly based on the source references.

After the Water Supply and Sanitation Decade, conceptual development has been achieved in the sanitation sectors (Table 1).

Franceys et al. (1992) comprehensively summarized technical options for on-site sanitation and financial and institutional requirements for sanitation development especially for developing countries. The technical options included open defecation, shallow pit, simple pit latrine, borehole latrine, ventilated pit latrine, pour-flush latrine, single or double pit, composting latrine, septic tank, and aqua-privy. Removal systems for excrete included in their summary were overhung latrine, bucket latrine, vaults and cesspits and sewerage. Planning and designing of on-site sanitation were also discussed including designs of pit latrine, septic tank, aqua-privy and composting toilets, and planning schemes from project definition to institutional, economic and financial factors.

Serageldin (1994) compared financial sources of the World Bank-financed water and sanitation projects. Whereas there were regional differences of the financial sources among the regions, overall percentage of projects financed by the internal cash generation decreased by time: 34% in 1988, 22% in 1989, 18% in 1990 and 10% in 1991. In the "new agenda", it was important to consider not only total budget but also who benefits from subsidized from water and sanitation services. Analysis results of water sector projects in the Latin American countries revealed that water sector projects were pro-poor, i.e. poor people were beneficial, in countries with large per capita Gross National Product (GNP) and pro-rich, i.e. rich people were beneficial, in low-income countries. On the contrary, sanitation sector projects were pro-rich in all the analyzed countries. Serageldin (1994) also discussed on community based organizational structures for water and sanitation sector development, i.e. Ruhverband in the Ruhr area in Germany, nation-wide River Basin Financing Agencies in France, and Condominal Sewerage System in Brazil.

Mara prepared two books focusing on affordability of sanitation (Mara, 1996a, 1996b). The textbooks are useful and benefitial to improve understanding of sanitation development which requires a mixture of resources including technological, financial and institutional resources after experiences of the Water Supply and Sanitation Decade in 1981-1990. McGranahan et al. (2001) discussed on more broad socioeconomic aspects of urban sanitation development based on the study results of the Stockholm Environment Institute conducted on environment burdens in Accra, Ghana, Jakarta, Indonesia, and Sao Paulo, Brazil, concerning the relationship between urban environmental health and sustainability. In their study, narrow meaning of “environment”, “sustainability” and “health” were defined in largely physical terms. In the context of sustainable development, these words sometimes include wider meaning. Therefore, definitions of the technical terms are important. They expanded their analysis areas to include “green” and “blown” concepts and argued necessity of a more politically and ethically explicit approach.

The Government of India established ambitious goals to provide potable drinking water and to reduce infant mortality rate across the country under the conditions with lacking or completely missing infrastructure for providing safe drinking water and effective management of human wastes (Pattanayak et al., 2005). The World Bank assisted Jalswarajya Project in Maharashtra State to meet these goals. The Project is supported by voluntary participation of communities. Water supply and sanitation services were provided by project administrators.

A demand-driven institution, the Bolivian Social Investment Fund, was established in 1991 after the Emergency Social Fund established in 1986 to improve coverage and quality of basic services in education, health, water and sanitation (Newman et al., 2002). A comparative evaluation of water and sanitation interventions were conducted between two intervention areas, five provinces in Chaco and 17 provinces in Resto Rural, and control areas. Comparison was conducted using a baseline dataset in 1993 and a follow-up data in 1997-98. Distance to water sources and under-age-five mortality were reduced significantly in the two intervention areas, and sanitation facilities improvement was found in Resto Rural. Drinking water quality did not improve significantly.

Table 1. Conceptual development in sanitation sector after 1990

Parameters Resources for sanitation	Benefit of sanitation	Explanation	Reference
Labour, Building materials, Water, Land use, Timing of costs, Design life of the system, Emptying and disposal, Groundwater pollution, Sullage disposal, Water reuse, and Administrative management.	Enhanced privacy, Convenience for the users, Environmental protection, Reduction and anticipated eventual elimination of excrete-related disease, Increase in productive life expectancy, Increase in work capacity, Reduction of demand for medical facilities and drugs, Reuse of composted or digested excrete for agricultural purposes, and Production of biogas for energy needs	Standard duration of cost analyses were recommended as 10-15 years. TACH[a] should be estimated for the standard duration time including all the necessary financial factors. The least-cost analysis approach was appreciated.	Franceys et al. (1992) WHO.
Financing Who pays for water and sanitation projects; Pro-rich or pro-poor; Community-based structures for water and sanitation sector investments; Tariff paid by the customers and connection rate; and Reciepts and expenditures for industry and public.		Water sector projects were pro-poor, i.e. poor people were beneficial, in countries with large per capita Gross National Product (GNP) and pro-rich, i.e. rich people were beneficial, in low-income countries. On the contrary, sanitation sector projects were pro-rich in all the analyzed countries.	Serageldin (1994) World Bank.
	RR[b] and DALY[c] of diarrhea and other diseases	DALY[c] was estimated based on the regional RR[a] data using exposure scenarios. Disease burden from water, sanitation and hygiene to be 4.0% of all deaths and 5.7% of the total diseases burden (in DALYs) in the world in consideration with diarrheal disease, schistosomiasis, trachoma, ascariasis, trichuriasis and hookworm diseases.	Prüss et al (2002) Environ Health Perspect.

Table 1. (Continued)

Parameters Resources for sanitation	Benefit of sanitation	Explanation	Reference
	Access to water and sanitation infrastructure, Availability of water, Improved health behaviors, Diarrhea rates in children under five, and Under-age-five mortality rate.	Distance to water sources and under-age-five mortality were reduced significantly in the two areas, and sanitation facilities improvement was found in one of the two treatment area. Drinking water quality did not improve significantly.	Newman et al. (2002) World Bank Economic Review.
	Access to water and sanitation infrastructure, Improved health behaviors, Diarrheal rates in children under five, Athropomorphic measures in children under five, and Expenditure levels on water, and income.	Water supply and sanitation intervention benefits on child health and income will be calculated using a difference-in-difference estimator to measure differences between the villages with the Jalswarajya Project (treatments) and non-project villages (controls).	Poulos (2006) (Original source: Pattanayak et al., 2005) World Bank.
Coherence, Predictability and credibility, and Legitimacy and accountability.		Good design will incorporate lessons from international experience including working with the existing organizational framework, creating an appropriate role for politics, limiting the discretion given to regulatory decision makers, and trading off sophistication in favor of simplicity. Regulations should be prepared to be appropriate for unique social and economic conditions by country.	Ehrhardt et al. (2007) World Bank.

[a] Total annual cost per household.

[b] Relative risk.

[c] Disability-adjusted life year.

Ehrhardt et al. (2007) conducted qualitative analyses on financial, institutional and managerial aspects of water and sanitation sectors in the world. Organizational frameworks and regulations of water and sanitation sectors should be prepared to be appropriate for unique social and economic conditions by country. There were examples of successes and failures among water and sanitation sector development in some countries and areas including Botswana, Burkina Faso, Colombia, the United Kingdom, Guyana, New Zealand, Australia, the United States, Senegal, Vanuatu, Bolivia, Phnom Penh, Manila, Florida, Buenos Aires, and municipalities in the Netherlands, New Zealand and Scandinavian countries.

2. RESOURCES AND PROFITS OF SANITATION DEVELOPMENT

Recent studies have revealed that necessary resources for sanitation development are not only technology and investment but also institutional framework and appropriate regulations (Figure 1).

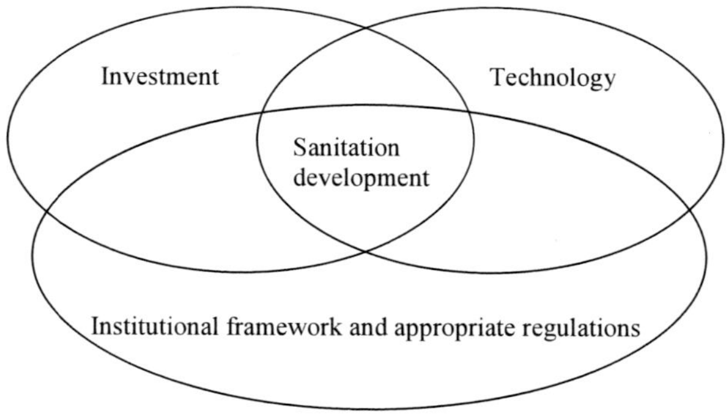

Figure 1. Resources necessary for sanitation development.

World annual investments in sanitation sector for the "off-track countries" were estimated as US$ 3.8 billion. Total of official development assistance (ODA) to the water and sanitation sectors by Development Assistance Committee countries and multilateral agencies increased to more than US$ 6 billion in 2005 after declines in 1990's (World Water Assessment Programme, 2009). Annual commitments of official development assistance from bilateral and multilateral agencies in water and sanitation sectors were from US$ 3.1 to 4.4 billion in 2004-06. The amounts consisted of 5.4-6.2% of total ODA investments. In Uganda, water supply service coverage increased from 47% in 1998 to 71% in 2007 using combination of measurements including timely and rational expansion of water production and network facilities, an optimal mixture of technology alternatives, introducing and subsidizing public access to water connections, innovative capital financing mechanisms, output-based investment approaches to strengthen service targeting, community-based approaches and mainstreaming consumer preferences, stakeholder coordination forums at different levels, and equitable providing water and sanitation infrastructures with an emphasis on sanitation (World Water Assessment Programme, 2009: Box14.23). Income and profit of water delivery service after depreciation increased from 21.9 to 70.4 billion Uganda Shillings and from -2.0 to 6.5 billion Uganda Shillings, respectively, in 1998-2007. Income and profit per employee also increased during the period.

Sanitation benefit in the world was estimated as US$ 35 billion, about nine times of investment, i.e. US$ 9 from US$ 1 (UN Water, 2008). The sanitation benefit comprised of 90.0% time savings due to access to improve sanitation, 5.0% deaths avoided, 3.1% productive healthy days gained by people with avoided illness, 1.6% health sector benefits due to avoided illness, and 0.2% patients' expenses avoided.

In Japan, costs and benefits of some agriculture community wastewater treatment plants (WWTPs)[2] are publicized by the government (Ministry of Agriculture, Forestry and Fisheries,

[2] There are several kinds of centralized WWTPs in Japan depend on their population equivalent of influent, area served with WWTPs and ministries in charge of WWTPs as described in the next section.

2009b). These kinds of analyses are required to explain fairness for public expenditures of publicly funded projects including agriculture community WWTPs. Table 2-1 and 2-2 summarize recent cost-benefit analysis results of agriculture community WWTPs. Average and standard deviation of estimated total cost among the twelve agriculture community WWTPs projects during the evaluation periods of 31-50 years was Japanese Yen (JY) 2,741±1,147 million, total benefit was JY 4,481±3,214 million and benefit-cost ratio was 1.56±0.52 (Table 2-1). Project benefits included in the estimation were agricultural product quantity and quality increases, agriculture cost decreases, change in maintenance costs of municipal wastewater treatment and irrigation waterway, reuse benefit of treated wastewater and sludge, lifestyle improvement in households by bathing and kitchen environment improvement and by using flash latrine, comfortableness increase of lifestyles in households and in a village, and public ambient water quality improvement (Table 2-2). Benefit estimations were mostly based on contingent valuation method (CVM) and replacement cost method (RCM) (Ministry of Agriculture, Forestry and Fisheries, 2008a; Ohno, 2005). Largest benefit was improving lifestyle by using flash latrine instead of septic tank style latrine, JY 135,122±123,552 million (49%) among annual total benefit of JY 275,670±191,279 of agriculture community WWTPs. Lifestyle improvement in rural areas is one of major challenges in domestic socioeconomics from the perspective of equity in the country.

Table 2.1. A summary of cost-benefit analyses during the evaluation periods of 31-50 years among twelve agriculture community WWTPs in Japan (Unit: thousand Japanese Yen) (Prepared by the author based on Ministry of Agriculture, Forestry and Fisheries, 2009b)

Cost and benefit	Number of cases	Minimum	Maximum	Average	Standard deviation
Total cost (current value)	12	1,514,600	5,226,001	2,741,300	1,146,840
Project cost[a]	8	1,318,900	4,142,859	2,026,409	944,079
Other cost[a]	8	195,700	1,083,142	384,159	289,204
Total benefit (current value)	12	1,304,643	12,246,676	4,481,299	3,214,136
Benefit cost ratio	12	1.03	2.60	1.56	0.52

[a] For other four cases, benefit and cost estimations were conducted with the other methods than the method applied for the eight cases.

Japan implemented US$ 4.6 billion of bilateral ODA in the fields of water and sanitation in 2000-04, which corresponded to 41% of total bilateral donors' ODA amounts in the fields (Ministry of Foreign Affairs, 2006; Harada, 2010). Besides large amount of Japanese financial contribution in water and sanitation sectors in the world, such contribution does not seem to be regarded as very important or beneficial in the water and sanitation sectors. Harada (2010) pointed out Japanese ODA in sanitation sector have concentrated to activated sludge technology which is popular in centralized WWTPs in Japan whereas "Appropriate Technologies[3]" are needed in developing countries, and emphasized importance of "Self-sustainable[4]" aspect of "Appropriate Technology".

---

[3] The words are described in English in the article of Journal of Japan Society on Water Environment in Japanese.

[4] The words are also described in English in the article.

Follow-up of the intervention projects are also important. Hoque et al. (1996) conducted follow-up survey of an integrated water supply, sanitation and hygiene (WSH) education intervention project which was run by the International Centre for Diarrhoeal Disease Research, Bangladesh, in 1983-87. The pumps in good functional condition declined with time, 94% in 1987 and 82% in 1992. Latrines functional also decreased with time, 93% in 1987 and 64% in 1992. However, adults using these sanitary latrines in the projected areas were 84% in 1992 comparing to 7% in the control area without project. Infectious disease knowledge were poor in both intervention and control areas, whereas diarrhoel disease prevalence in control areas was twice as large as that in intervention areas. Although people's awareness on sanitation continued to be high in intervention area, hardware could not be properly maintained. Self-sustainability or additional supports should be considered after investment projects.

Table 2.2. A summary of annual benefit estimation of twelve agriculture community WWTPs in Japan (Unit: thousand Japanese Yen) (Prepared by the author based on Ministry of Agriculture, Forestry and Fisheries, 2009b)

Benefit	Explanation	Number of cases	Minimum	Maximum	Average	Standard deviation
Agriculture product increase	Agriculture product increase because of water quality improvement	10	388	13,889	3,210	4,101
Agriculture product quality increase	Agriculture product quality increase because of water quality improvement	7	312	13,060	2,492	4,682
Agriculture cost decrease	Avoiding crop falling down by excess growth because of too much nutrients in water	10	312	40,137	6,547	12,282
Decrease of maintenance cost of irrigation waterway	Maintenance cost decrease because of decrease of sludge and wastes accumulation in waterway	12	741	3,177	1,718	825
Reclaimed wastewater reuse	Reclaimed wastewater is discharged to irrigation waterway and reused.	10	3,781	167,117	30,339	49,050
Reuse of wastewater sludge	Decrease of sludge disposal and fertilizer cost because wastewater sludge is reused to fertilizer	9	249	3,557	860	1,027
Psychological improvement of working environment	Working environment improvement because of water quality improvement of irrigation waterway	12	82	6,424	1,103	1,836
Physical improvement of working environment	Reducing cleaning activity of irrigation waterway	12	52	2,820	388	770
Improving life environment of toilet	Using flush latrine instead of septic tank with sludge collection	12	34,581	461,929	135,122	123,552
Improving life environment of bathing and kitchen	Improving of bathing and kitchen environment	12	2,005	53,317	26,714	15,140

Table 2.2.(Continued)

Improving life environment of waterway in the village	Improving waterway environment and reducing appearance of vector insects such as housefly	12	5,634	69,876	22,494	18,318
Reducing maintenance cost other than agriculture products	Maintenance cost decrease because of changing from sludge collection toilet to flash latrine	12	-36,590	50,348	4,880	21,452
Public water quality improvement	Increase of recreational opportunity and reducing treatment cost of drinking water by public water quality increase	12	6,088	58,102	19,920	12,959
Comfortableness increase in the village environment	Comfortableness increase because of water quality improvement and scenery protection	12	11,204	86,520	27,528	19,717
Effect on agriculture village roads	Improvement of safety because of road development accompanied by WWTP development and reducing transportation cost of agriculture products	1	3,496	3,496	3,496	NA[a]
Total		12	136,346	797,496	275,670	191,279

[a] Not available.

3. Japanese Model of Sanitation Development

3.1. Centralized and On-site Municipal Wastewater Treatment in Japan

After the World War II, Japan has experienced rapid economic development. Sanitation system has also developed during the period. Comparing to other developed and developing countries, Japan currently has a kind of a unique and independent sanitation system in the world, which consists of centralized wastewater treatment plants (WWTPs) and onsite treatment systems called combined *johkasou*[5]. Newly installations of municipal wastewater treatment facilities have been recently limited by regulations only to centralized WWTPs and combined *johkasou*, however, we still have large stock of old type sanitation systems, namely simple *johkasou* and night soil treatment systems. In this section, Japanese sanitation system situations including the treatment facility types and methods are summarized.

In Japan, for the river basin wide WWTPs in 1980's, economics feasibility of centralized and on-site municipal wastewater systems were discussed as well as river water deficiency

[5] *Johkasou* is a common name of on-site treatment system in Japan and also used as a trade name of a company product (http://www.gec.jp/WATER/, http://www.wepa-db.net/technologies/individual/list_b_onsite.htm, accessed on 13 April 2010). There are two types of *johkasou*, combined *johkasou* and simple *johkasou*. Combined *johkasou* treats both black and gray water, and simple *johkasou* treats only black water. Only a word, *johkasou*, currently means combined *johkasou*. *Johkaso* and *jokaso* are also sometimes used.

between a water intake point of water supply plants and a water discharge point of WWTPs (Nakanishi and Okinoto, 1982). Zoning or geographical dividing concept of centralized and on-site municipal wastewater treatment systems has been established abreast of the discussions. Wastewater treatment system zoning is mainly based on population density and specific characteristics of subjected areas including agriculture and fishery communities. The zoning is generally conducted at a prefecture[6] scale. Centralized WWTPs with piped-connection systems to collect municipal and some commercial and industrial wastewater are developed in urban areas, and on-site treatment systems called combined *johkasou* are constructed in less-habitat or rural areas (Infrastructure Development Institute, Japan, 2004; Tsuzuki, 2010). The population density based scheme of centralized and on-site treatment zoning is explained in the official guidelines of the Ministry of Land, Infrastructure, Transportation and Tourism (Basin Wide Sewage Comprehensive Planning System Design Council, 2008).

The actual planning of municipal wastewater treatment and sewerage at prefectural scale is consisted of the areas for basin wide WWTPs, local government WWTPs, specific environment protection public WWTPs, fishery community WWTPs, agriculture community WWTPs, community plants and combined *johkasou*. Community plants are something like large scale on-site treatment systems. Regulations of these centralized WWTPs are managed by each corresponding central government ministries. There were 166 basin wide WWTPs, 1200 local government WWTPs and 736 specific environment protection public WWTPs at the end of March, 2009 (Ministry of Land, Infrastructure, Transportation and Tourism, Japan, 2009). These three kinds of WWTPs are generally categorized as public WWTPs. For agriculture community WWTPs, development and construction was completed at 5,060 communities among 5,400 communities with their planning and design in 2007 (Ministry of Agriculture, Forestry and Fisheries, 2008b).

The areas of several kinds of WWTPs are zoned or geographically divided depending on population density and geographical characteristics. For example, population with "appropriate" sanitation was 1,216 thousand among 2,374 thousand total population in fishery villages in 2009 (Ministry of Agriculture, Forestry and Fisheries, 2009a). The definition of "appropriate" sanitation in Japan is not the same as "appropriate" sanitation of the world sanitation perspectives. Therefore, "advanced" sanitation is used to describe "appropriate" sanitation of the current Japanese sanitation perspectives in this chapter manuscript. Simple *johkasou* and night soil treatment systems, which treat only black water[7], are not included in "advanced" sanitation in Japan (Tsuzuki, 2006). Gray water[8] is directly discharged to ambient water without treatment by simple *johkasou* and night soil treatment systems. Among the population with advanced sanitation in fishery villages of 1,215 thousands, 641 thousands were with public WWTPs (53%), 319 thousands with combined *johkasou* (26%), 173 thousands with fishery community WWTPs (14%), and 82 thousands with agriculture community WWTPs (7%) at the end of March, 2009. These figures show that the geographical planning or zoning of municipal wastewater treatment is designed in detail and fishery villages are consisted of miscellaneous areas in regards to population density and geographical characteristics. National average of population percentage with "advanced"

[6] There are 47 prefectures in Japan.

[7] Toilet wastewater.

[8] Wastewater from households other than toilet including kitchen, bathing and washing clothes and so on.

sanitation was 85% and that in fishery villages was 51% in March, 2009. Populations with "advanced" municipal wastewater treatment including public WWTPs, agriculture community WWTPs and combined *johkasou* have been increasing every year (Table 3). From the perspective of the MDGs sanitation indicator, people with appropriate sanitation is already 100% in Japan. The MDGs sanitation indicator figure in Japan also includes both simple *johkasou* and night soil treatment systems.

Table 3. People with advanced municipal wastewater treatment in Japan in 2003-05

"Advanced" wastewater treatment method	Population served with "advanced" municipal wastewater treatment (1,000 persons)		
	March 2003	March 2004	March 2005
Public WWTPs[a]	82,570	84,580	86,360
Agricultural community WWTPs[b]	3,110	3,280	3,440
Combined *johkasou*	9,930	10,300	10,620
Community plants	380	380	370
Total[c]	95,990	98,540	100,790
Proportion of people served with advanced municipal wastewater treatment (%)	75.8%	77.7%	79.4%
Total population[d]	126,690	126,820	126,870

[a] Basin wide WWTPs and local government WWTPs.
[b] Agriculture community WWTPs, fishery community WWTPs, forest community WWTPs, and specific environment protection public WWTPs.
[c] Total population with advanced municipal wastewater treatment; and d: Total population in Japan.
Prepared by the author based on Ministry of the Environments (2005).

Figure 2 shows flow of black and gray water and sludge by use of centralized WWTPs, combined *johkasou*, simple *johkasou*, and night soil treatment systems. Among three on-site wastewater treatment methods, only combined *johkasou* can be newly constructed by the current regulations in Japan. However, simple *johkasou* and night soil treatment systems which treat only black water are still working and re-construction of these old type on-site treatment systems into combined *johkasou* is a challenging topic in rural area (Tsuzuki, 2006). In urban area with existing centralized WWTPs, piped connection development is a challenging topic, which changes on-site treatment systems to centralized WWTPs.

Construction and maintenance costs of centralized municipal wastewater treatment systems including WWTPs and piped collection systems are covered by wastewater treatment tariff which is usually calculated based on water supply usage amounts and government subsidies. For on-site treatment systems, combined *johkasou*, construction costs are covered by house owners and subsidies by governments. Subsidies are provided by governments for on-site treatment systems because of financial efficiency. On-site treatment systems in rural areas with low population density are cost effective comparing to centralized wastewater treatment systems with piped collection systems. Operating and maintenance cost of on-site treatment systems including electricity cost for aeration blower operation and payment for de-sludging are usually covered by house owners. Construction and maintenance costs per wastewater treatment volume of several kinds of municipal wastewater treatment systems in Japanese Fiscal Year 1995-98 were JY 15-1,516 m^{-3}, and JY 43-1,148 m^{-3} in 1998 (Figure 3) (Sewage Works Project Management Research Group, 1997-2000). There were also differences in costs per wastewater treatment volume even in the same treatment methods. Therefore, benefit-cost analyses are conducted for each WWTP project in the case of agricultural community WWTPs (Table 2).

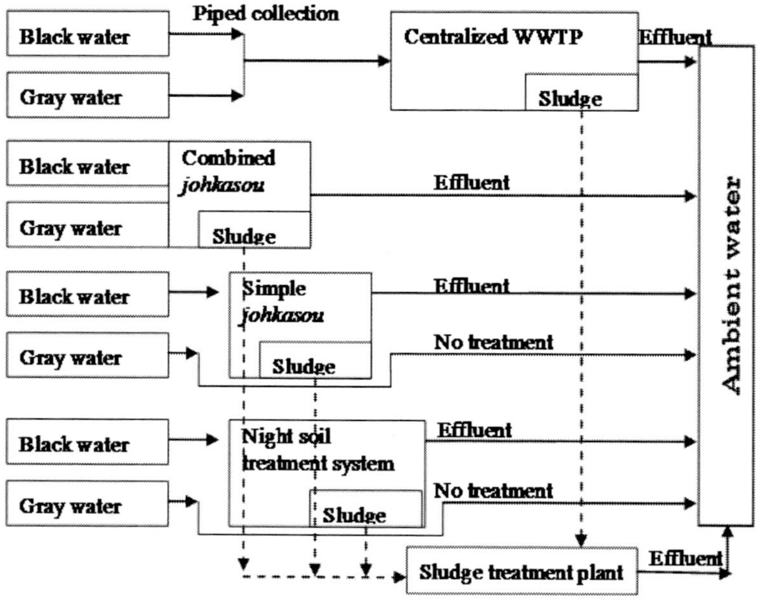

Figure 2. Flow of black and gray water with municipal wastewater treatment in Japan.

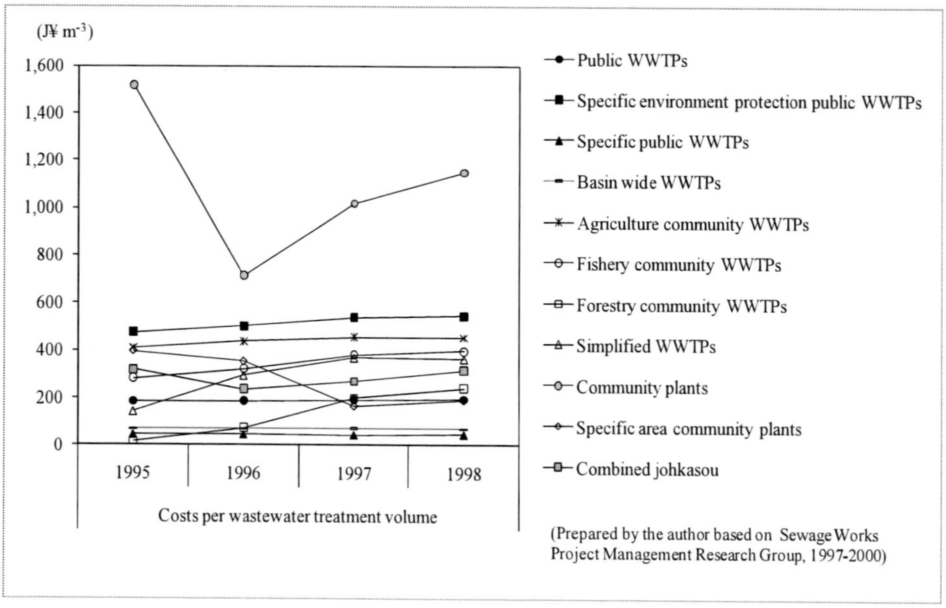

Figure 3. Construction and maintenance costs per wastewater treatment volume in Japanese Fiscal Year 1995-98 (JY m^{-3}).

3.2. Institutional and Regulation Structures of Municipal Wastewater Treatment in Japan

Institutional structures have been established with several organizations in Japanese sanitation sector (Hashimoto, 2009). These organizations correspond to several kinds of centralized and on-site wastewater treatment systems. There are multiple organizations for certain wastewater treatment systems, e.g. for public WWTPs, an organization deals with technology research and development, another organization deals with information dissemination, another organization deals with piped-collection systems and so on. Japan Sewage Works Agency is funded by local governments and keeps advancement of engineering in the fields of sewage works (Japan Sewage Works Agency, 2010). Japan Sewage Works Association works for information dissemination on sewage works (Japan Sewage Works Association, 2009). Japan Environmental Sanitation Center pursues environmental protection and conservation especially in the fields of solid wastes, night soil treatment system and acid rain (Japan Environmental Sanitation Center, 2009). Japan Education Center for Environment Sanitation focuses on *johkasou* technology including simple and combined *johkasou* and conducts education and research for training technicians of installation, maintenance and de-sludging of the on-site treatment systems (Japan Education Center for Environment Sanitation, 2010; Ogawa, 2006). Such kind of complicated structure of organizations in the fields of municipal wastewater treatment has been developed in Japanese sanitation sector for several decades.

For infectious disease prevention, disinfection of effluent is necessary especially for biological wastewater treatment effluent. Chlorination using chlorine tablets or liquids are sometimes applied for biologically treated wastewater of *johkasou*. Relationship between solubility and disinfection efficiency of chlorine tablets were studied by Yahya (1994). Other disinfection methods such as ultraviolet and ozone disinfection are also applied for effluent of centralized and on-site wastewater treatment systems. For basin wide WWTPs, local government WWTPs and specific environment WWTPs, there were 1964 WWTPs with chlorination using chlorine tablets and liquid type chlorine, 110 WWTPs with ultraviolet disinfection, and 28 WWTPs with ozonation at the end of March 2006 in Japan (Japan Sewage Works Association, 2007).

For pollutant removal efficiencies of combined *johkasou*, regulations of small scale *johkasou* including combined *johkasou* and simple *johkasou* were firstly established as double standards: (1) BOD removal rate of more than 65% and effluent BOD is less than 90 g-BOD m^{-3} in the Notification of Ministry of Construction (Currently, Ministry of Land, Infrastructure, Transportation and Tourism)[9], and (2) BOD removal rate of more than 90% and effluent BOD is less than 20 g-BOD m^{-3} in the Notification of Architecture Guidance Section of Housing Bureau in the Ministry of Construction[10] in 1988 (Kitao, 2006). The former Notification subjected both combined and simple *johkasou*.

Pollutant removal ratios of small scale *johkasou* have been increased by research and development and currently as large as those of centralized WWTPs. Building Standards Act was revised to include performance evaluation system into the existed structure based

[9] Notification 1292 of Ministry of Construction on 8 March 1988.

[10] Notification of Architecture Guidance Section of Housing Bureau in the Ministry of Construction on 8 March 1988.

evaluation system in 1998[11]. Percentage of newly installation of performance based combined *johkasou* has been increased from around 20% in 1998 to 93% in 2005. National Combined *Johkasou* Promotion Local Governments Union classified four categories among performance based evaluation standards of combined *johkasou* by effluent concentrations, 1) BOD removal type: 20 g-BOD m^{-3}, 2) nitrogen removal types (three types): 20 g-BOD m^{-3} and 20 g-total nitrogen (TN) m^{-3}; 10 g-BOD m^{-3} and 20 g-TN m^{-3}; and 10 g-BOD m^{-3} and 10 g-TN m^{-3}, 3) nitrogen and phosphorus removal type: 10 g-BOD m^{-3}, 10 g-TN m^{-3} and 1 g-total phosphorus (TP) m^{-3}, 4) BOD advanced treatment type: 5 g-BOD m^{-3} and 10 g-TN m^{-3}. Pollutant removal ratios of combined *johkasou* with these performance based standards especially for nutrients (nitrogen and phosphorus) removal types can compare with those of centralized WWTPs with tertiary or advanced treatment.

Operation of *johkasou* is conducted by registered vendors and operators. The number of the vendors, operators and technicians were 38 *johkasou* manufacturers, 65 specified examination agencies, 77,728 *johkasou* installation workers, 9,087 vendors for *johkasou* operation and maintenance, 63,104 *johkasou* operators, 4,806 vendors for *johkasou* de-sludging, 4,205 *johkasou* de-sludging technicians and 2,533 *johkasou* inspectors at the end of March 2005 (Ogawa, 2006). These kinds of structures are necessary to construct and maintain *johkasou* on-site municipal wastewater treatment systems in Japan.

3.3. Japanese Sanitation Sector Development with a New Initiative for Asian and Pacific Regions

Japan Sanitation Consortium (JSC) was established in October, 2009, which is aimed to work as a knowledge hub of sanitation sector in Japan (Japan Sanitation Consortium, 2009; Asia-Pacific Water Forum, 2009). The concept of JSC was approved at Asia Pacific Water Forum in June, 2009. JSC is consisted of Sewage Business Management Centre (Sewage Business Management Centre, 2008), Japan Environmental Sanitation Center, Japan Sewage Works Association, Japan Education Center for Environment Sanitation and so on and supported by Ministry of Land, Infrastructure, Transportation and Tourism and Ministry of the Environment. The purposes of JSC are to make a network of organizations in the field of sanitation, to accumulate, share and disseminate knowledge and information in the field of sanitation. JSC is financially assisted by Asian Development Bank (ADB). JSC will prepare a networking of the countries in the region, hold international seminars, conduct research activities, and advise and assist activities of ADB and Japan International Cooperation Agency (JICA) in the field of sanitation.

For cost, construction cost of combined *johkasou* is larger than on-site sanitation in the other Asian countries (Tsuzuki and Koottatep, 2010). Construction cost reduction by domestic industries in developing countries may be possible when recent development of low-price vehicles in a developing country is considered. For pollutant discharge reduction benefit among miscellaneous benefits, recent research and development of combined *johkasou* in Japan successfully achieved high removal efficiency of pollutants which can meet effluent concentration standards of 10 g-BOD m^{-3}, 10 g-TN m^{-3} and 1 g-TP m^{-3} (Kitao, 2006).

[11] Act No. 100 in 1998 established on 12 June 1998.

Japanese government and industry sectors have been tried to proliferate *johkasou* technology especially since the third water forum held at Kyoto in 2003 (Ministry of the Environments, 2006; Katayama 2006). Some amounts of combined *johkasou* have been introduced to Eastern Europe, Indonesia, Malaysia, Thailand, China, Middle East and the United States of America. Japanese industries seemed to leave on-site sanitation markets in China and Malaysia, however, domestic industries of these countries are dealing with *johkasou* type on-site treatment systems. There are about 500 private companies dealing with *johkasou* type on-site wastewater treatment systems in China. In Korea, official regulations and development systems of combined *johkasou* type on-site wastewater treatment systems have been developed, and there were about 90 thousands combined *johkasou* operated to treat both black and gray water.

In China, on-site treatment systems called *huafenchi* is installed in houses, which is mandated by the local government regulations regardless of existence of centralized WWTPs (Oh and Kitawaki, 2005). *Huafenchi* is often consisted of concrete structures and treats only black water whereas a modern *johkasou* is made of plastic package while an old one is with concrete structure and treats both black and gray water. *Huafenchi* is similar to a septic tank and simple *johkasou*. There are sometimes clogging problems during *huafenchi* operation. Residential people are not affordable to Japanese type *johkasou*. However, prediction of water resource shortage will increase possibility of technology transfer of Japanese combined *johkasou* because of effluent water quality which is suitable for reuse. Possible customers of combined *johkasou* will be collective housings, office buildings, hotels, recreational facilities and exclusive residential districts.

JSC activity will broaden activity of Japanese sanitation sector in Asian and Pacific countries especially in the fields of knowledge and information sharing not only sewage works and combined *johkasou* on-site treatment systems but also other sanitation alternatives.

4. Areal Deployment Strategies of Centralized and On-site Sanitation: Zoning or Intermixed

Areal deployment of sanitation facilities in Japan is zoned or geographically divided at each prefectural scale as described above, and some basin wide WWTPs include areas of more than one prefecture (Tsuzuki, 2010; Infrastructure Development Institute, Japan, 2004). On the contrary, in Thailand, a middle-class developing country with the MDGs sanitation indicator of 96% in 2006, on-site treatment systems such as septic tanks and cesspools are commonly applied in both urban and rural areas, even in the areas served with centralized WWTPs (Tsuzuki et al., 2009a). Besides large value of the MDG sanitation indicator, ambient water pollution is still a major environment problem in Thailand, especially in densely populated areas including the capital, Bangkok. Based on the results of pollutant discharge estimation of municipal wastewater, pollutant discharge per capita (PDC) in Pak Kret District, sub-urban of Bangkok, was estimated as 48.4 g-BOD person^{-1} day^{-1} with seepage and septage of septic tank and leachate from composting (SSL) and 29.2 g-BOD person^{-1} day^{-1} without SSL based on the material flow analysis (MFA) results (Tsuzuki et al., 2008). These PDC values were larger compared to those in Japan, which was found to be a reason for ambient water deterioration.

Systematic and strategic planning scheme of municipal wastewater treatment should be prepared for a mixture condition of on-site treatment systems and centralized WWTPs. WWTPs influent sometimes include certain industrial wastewater. The intermixed condition of the on-site and centralized systems is not commonly found in developed countries, however, may be not so unique in developing countries where on-site treatment system development has been preceded to centralized WWTPs development. Simple on-site treatment systems are required to prevent infectious diseases and to improve people's lifestyles when there is no centralized WWTP.

After the concept of step development from on-site treatment to centralized treatment in a draft guidelines of sanitation development in developing countries (Infrastructure Development Institute, Japan, 2004; Tsuzuki, 2009), scenario-based approaches including estimation of PDC have been proposed for improvement of the mixture condition of on-site and centralized municipal wastewater treatment systems (Tsuzuki et al., 2009a; Suharyanto and Matsushita, 2009). Pollutant discharge and pollutant load analyses including both on-site and centralized wastewater treatment systems in the project planning stage are important to improve ambient water quality as well as increasing percentage of people with appropriate sanitation.

5. Technological Alternatives in Consideration with Pollutant Discharge and Investment

UNDP data in 2004 (UNDP, 2007) showed that the MDGs water and sanitation indicators were related to Human Development Indicator Rank and Gross Domestic Product (GDP) per capita (Tsuzuki, 2008). Economic development in a country is considered to be necessary to increase water and sanitation indicators, percentage of people with access to safe drinking water and appropriate sanitation. There were many countries with GDP per capita of less than US$ 10,000 person^{-1} year^{-1}. In these countries, the MDGs water and sanitation indicators were small. On the contrary, most countries with GDP per capita with more than US$ 10,000 were with large values of the MDGs water and sanitation indicators. Therefore, US$ 10,000 was considered to be a rough indicator to improve water and sanitation conditions at the MDGs indicator levels.

Low-cost is a key word of sanitation in developing countries. Technical alternatives have been summarized as described above (Franceys et al., 1992; Serageldin, 1994; Mara, 1996a, 1996b; Infrastructure Development Institute, Japan, 2004; Tilley et al., 2008). For *johkasou* in Japan, a concrete or fiber reinforced plastics (FRP) structured on-site treatment system with anaerobic and aerobic chambers has been developed since 1980's (Ishii and Yamada, 1990). A lot of lactic acid beverage plastic bottles of about 20 ml each were used for biofilm attachment structures in the chambers to form biofilm on their surface and to enhance microorganism activities. The effluent wastewater quality of the so-called Ishii-type *johkasou* is usually as small as 1 g-BOD m^{-3}, and as clean as drinkable. There are currently also some products by major companies with BOD effluent concentration less than 5 g-BOD m^{-3}. Such kinds of on-site wastewater treatment system may be alternatives of on-site wastewater treatment systems in developing countries. *Huafenchi* in China would improve by some modification to resembling structure of those types of on-site wastewater treatment systems.

Wastewater treatment mechanisms of *johkasou* type on-site treatment systems are mainly biological treatment. Therefore, blowers operated by electricity for aeration of aerobic chamber are necessary. Operational cost will be a challenge in developing countries. In China, willingness to pay for wastewater treatment was not enough to afford Japanese combined *johkasou* especially for lower income people (Oh and Kitawaki, 2005). *Huafenchi* in China is included in appropriate sanitation in the MDGs sanitation indicator context. However, more improvement of people's lifestyles and ambient water quality will be necessary. How much should we improve people's life? Advanced wastewater treatment systems are being installed to improve people's lifestyles and ambient water quality in Japan. There may be invisible barriers on the country boarders.

Relationships between pollutant discharge indicators, BOD discharge per capita (PDC-BOD) and TP discharge per capita (PDC-TP), and income indicators of developing countries in several coastal regions were found to be inverted U-shaped curve, namely environment Kuzunets curve (EKC) (Tsuzuki, 2008,2009b). Moreover, the relathinship between an income indicator and PDC-BOD in Pacific, Asian and African countries was found to be the same as the origical Kuznets relationship between income and equity (Kuznets, 1955; Tsuzuki, 2007,2009). PDC-BOD was in wide range when GDP per capita was small and PDC-BOD was in a narrow range when GDP per capita was large. GDP per capita values at peak of the curves were US$ 5,050 for PDC-BOD and US$ 11,200 for PDC-TP. The value of PDC-BOD was consistent with the results of exiting literature, US$ 5,000-8,000, and that of PDC-TP was larger (Tsuzuki, 2009b). TN discharge per capita (PDC-TN) decreased linearly with GDP per capita increase. The difference of peak GDP per capita values between PDC-BOD and PDC-TP could be explained by wastewater treatment technological aspect that nutrient removals are targetted after some achievement of organic carbon removal. One of the reasons for the linear decreasing tendency of PDC-TN was that black water especially urine occupied much portion of PDC-TN. Black water is usually treated with basic on-site sanitation.

Construction cost and PDC of on-site wastewater treatment systems in Asian countries were summarized based on Water Environment Partnership in Asia (WEPA) database of municipal wastewater technologies (Tsuzuki and Koottatep, 2010). Construction cost and PDC were in a wide range. PDCs were estimated as 0.05-9.5 g-BOD person^{-1} day^{-1}, 0.4-5.7 g-TN person^{-1} day^{-1} and 0.02-0.82 g-TP person^{-1} day^{-1}, whereas construction cost per capita was US$ 4-2,942. Sanitation technologies should be suitably applied in consideration with socioeconomic conditions in the area to be developed.

For municipal wastewater source control of pollutant discharge, "sot interventions" in households were found to be effective. Estimated possible pollutant discharge reduction percentages were 38-53% for BOD, 26-40% for TN and 21-32% for TP in the Yamato-gawa River Basin, Japan (Tsuzuki et al., 2009b), and 39% for BOD, 21% for total Kjeldahl nitrogen and 34% for phosphate in the Bangkok Metropolitan Area, Thailand (Tsuzuki et al., 2010), depending on lifestyles, wastewater treatment methods, and combinations of "soft interventions". Some effective "soft interventions" are (1) to wipe plates, utensils and cookware with paper or rag before washing them in the kitchen, (2) to stop draining residual liquids and solids and put them in the trash can, and (3) not to drain rice washing water (Tsuzuki et al., 2009c; Yamato-gawa River Office, 2010; Chiba Prefecture, 2008).

Conclusions

For sanitation development, technological, institutional and regulation are important factors for comprehensive development in a country or in an area.

Posing first priority on the MDGs water and sanitation indicator achievements, pollutant discharge reductions should also be aimed in the sanitation sector to improve ambient water quality, which will benefit the society from miscellaneous aspects including decrease of ambient water pollution control costs, drinking water treatment cost reduction of surface water sources, and comfortableness increases around ambient water. Japanese technologies and experiences of sanitation development together with world experiences in the fields will contribute to sanitation development in the world especially in Asian and Pacific countries. "Soft interventions" in households are also measures to reduce municipal wastewater pollutant discharge as well as "hard interventions" such as wastewater treatment systems.

Global investment in sanitation sector, both the MDGs sanitation target perspectives and pollutant discharge reduction perspectives should be in good balance. Investment balance of the MDGs perspectives and pollutant perspectives is not so easy, however, should be determined in the actual sanitation sector investment by stakeholders including international governmental sectors, national and rural governments, policy makers, professionals, industries, communities and individuals.

There are several technologies for low-cost sanitation for developing countries. There are also technologies with large pollutant removal or small pollutant discharge. Japanese *johkasou* or their modifications could be alternatives in middle and large per capita income countries.

References

Asia-Pacific Water Forum (2009) Welcome to Regional Network for Water Knowledge Hubs! (http://www.apwf-knowledgehubs.net/, accessed on 10 April 2010)

Baroudi, A. (2006) *A Uniform Approach to HACCP*. Food Safety Institute (FSI), International, Newport Beach, CA. (http://www.foodprocessing.com/whitepapers/2006/014.html, accessed on 8 Apr 2010)

Bartram, J., Lewis, K., Lenton, R. & Wright, A. (2005) Focusing on improved water and sanitation for health. *Lancet, 365,* 810–812.

Basin Wide Sewage Works Comprehensive Planning System Design Council, Japan (2008) Survey on Basin Wide Sewage Works Comprehensive Planning: Guidelines and Explanation. Japan Sewage Works Association, 285p. (in Japanese)

Checkley, W., Gilman, R. H., Black, R. E., Epstein, L. D., Cabrera, L., Sterling, C. R., Moulton, L. H. (2004) Effect of water and sanitation on childhood health in a poor Peruvian peri-urban community. *Lancet, 363,* 112-118.

Chiba Prefecture (2008) Action Plan to Improve Environment of Tokyo Bay by Everyone. (http://pref.cihba.lg.jp/syozoku/e_suiho/1_sido/suishinkeikaku/plan/suiplanup.html, accessed on 10 April 2010) (in Japanese)

Ehrhardt,D., Groom, E., Halpern, J. & O'Connor, S. (2007) *Economic Regulation of Urban Water and Sanitation Services: Some Practical Lessons*. World Bank, Washington, DC. Water Sector Board Discussion Paper Series, Paper No. 9. 36p. (http://siteresources.worldbank.org/INTWSS/ResourcesWS9_EconReg_C.pdf, accessed on 7 Apr 2010)

Fewtrell, L., Kaufmann, R. B., Kay, D., Enanoria, W., Haller, L. & Colford Jr, J. M. (2005). Water, sanitation, and hygiene interventions to reduce diarrhoea in less developed countries: a systematic review and meta-analysis. *Lancet Infectious Diseases, 5 (1),* 42–52.

Franceys, R., Pickford, J. & Reed, R. (1992) *A Guide to the Development of On-site Sanitation.* WHO, 229p. (http://www.who.int/water_sanitation_health/hygiene/envsan/onsitesan/en/, accessed on 7 Apr 2010)

Hashimoto, K. (2009) Sanitation issues in the developing countries. *Journal of Japan Society on Water Environment, 32 (9)*, 2-6. (in Japanese)

Harada, H. (2010) Preface: Clear and present danger - Collapse of water environment in developing regions and its restoration by appropriate technology. *Japan Society on Water Environment, 32 (9)*, 1-1. (in Japanese)

Hoque, B. A., Juncker, T., Sack, R. B., Ali, M. & Aziz, K. M. (1996) Sustainability of a water, sanitation and hygiene education project in rural Bangladesh: a 5-year follow-up. *Bulletin of the World Health Organization, 74 (4)*, 431-437.

Infrastructure Development Institute, Japan (2004) *Guidelines for Low-cost Sewerage systems in developing countries (draft)*, 270p.

Ishii, K. & Yamada, K. (1990) *Revolution of Municipal Wastewater Treatment: Escape from River Environment Deterioration*. Fujiwara Publishing, Tokyo, 222p. (in Japanese)

Japan Education Center for Environment Sanitation (2010) Japan Education Center for Environment Sanitation homepage (http://www.jeces.or.jp/en/about/index.html, accessed on 7 Apr 2010)

Japan Environmental Sanitation Center (2009) Japan Environmental Sanitation Center homepage (http://www.jesc.or.jp/en/index.html, accessed on 7 Apr 2010)

Japan Sanitation Consortium (2009) *Reference material: Japan Sanitation Consortium (JSC) will be established.* (http://www.mlit.go.jp/common/000050452.pdf, accessed on 9 Apr 2010) (in Japanese)

Japan Sewage Works Agency (2010) Japan Sewage Works Agency homepage. (http://www.jswa.go.jp/english/index.html, accessed on 9 Apr 2010)

Japan Sewage Works Association (2007) Sewage Works Statistics in Japanese Fiscal Year 2005: Public administrative volume. Japan Sewage Works Association. (in Japanese)

Japan Sewage Works Association (2009) Japan Sewage Works Association homepage. (http://www.jswa.jp, accessed on 9 Apr 2010) (in Japanese)

Katayama, T. (2006) *Johkasou* in the world. Presentation material for 19th Specialist Meeting on *Johkasou*, Wastes and Recycle Section, Central Environment Council, 22 Sep 2006. (in Japanese)

Kitao, T. (2006) *Structure and maintenance of small johkasou.* Japan Education Center for Environment Sanitation, Tokyo, 196p. (in Japanese)

Kuznets, S. (1955) Economic growth and income inequality. *American Economic Review, 45 (1)*, 1–28.

Lens, P., Lettinga, G. & Zeeman, G. eds. (2001) *Decentralised Sanitation and Reuse: Concepts, systems and implementation.* International Water Association Publishing, London, 550p.

McGranahan, G, Jacobi, P., Songsore, J., Surjadi, C., & Kjellen, M. (2001) *The Citizens at Risk: From Urban Sanitation to Sustainable Cities*. Earthscan Pubns Ltd. (Risk, Society and Policy Series), 240p.

Mara, D. (1996a) *Low-cost Sewerage,* John Wiley & Sons, 238p.

Mara, D. (1996b) *Low-cost Urban Sanitation.* John Wiley & Sons, 240p.

Ministry of Agriculture, Forestry and Fisheries, Japan (2008a) Minutes of sixth meeting of Agriculture and Agriculture Village Promotion and Deployment Section Meeting, Food, Agriculture and Agriculture Village Council, Ministry of Agriculture, Forestry and Fisheries, Japan. 18 March 2008. (http://www.maff.go.jp/council/seisaku/ nousin/pdf/ report_2.pdf, accessed on 10 Apr 2010) (in Japanese)

Ministry of Agriculture, Forestry and Fisheries, Japan (2008b) Leaflet: Agriculture village WWTPs. (http://www.maff.go.jp/j/press/nousin/sousei/pdf/080528-03.pdf, accessed on 9 Apr 2010) (in Japanese)

Ministry of Agriculture, Forestry and Fisheries, Japan (2009a) Press release: Municipal wastewater treatment percentage in fishery village at the end of Japanese Fiscal Year 2008. (http://www.jfa.maff.go.jp/j/press/bousai/090930.html, accessed on 9 Apr 2010) (in Japanese)

Ministry of Agriculture, Forestry and Fisheries, Japan (2009b) Material to explain benefit-cost analysis results on agriculture area WWTPs projects. (http://www.maff.go.jp/nouson hyouka/h20/before/090331/ref_data02d.pdf, accessed on 10 Apr 2010) (in Japanese)

Ministry of Foreign Affairs, Japan (2006) Japan announces a new ODA Initiative: "Water and Sanitation Broad Partnership Initiative" (WASABI). (http://mofa.go.jp/announce/ announce/2006/3/0320.html, accessed on 11 Apr 2010)

Ministry of Land, Infrastructure, Transportation and Tourism, Japan (2009) Sewage works data room (http://www.mlit.go.jp/crd/crd_sewerage_tk_000104.html, accessed on 9 Apr 2010) (in Japanese)

Ministry of the Environments, Japan (2005) Waste & Recycling, Issues: Johkasou/ Wastewater Treatment. (http://www.env.go.jp/en/recycle/, accessed on 11 Apr 2010)

Ministry of the Environments, Japan (2006) Minutes of 19th Specialist Meeting on *Johkasou*, Wastes and Recycle Section, Central Environment Council, 22 Sep 2006. (in Japanese)

Nakanishi, J. & Okinoto, T. (1982) *Theory of municipal wastewater treatment planning: Environment impact assessment of municipal wastewater treatment planning at Komagane City, Japan*. Musashino-Shobo, Tokyo. 222p. (in Japanese)

Newman, J., Pradhan, M., Rawlings, L., Riddder, G., Coa, R. & Evia, J. L. (2002) An impact evaluation of education, health and water supply investments by the Bolivian Social Investment Fund. *The World Bank Economic Review 16(2)*, 241-274. Cited in Poulos et al. (2006)

Norman G. M. & Gravani, R. B. (2006) Principles of food sanitation, 5th ed. Springer, 413p.

Ogawa, H. (2006) Domestic wastewater treatment by *johkasou* systems in Japan. *Side Event: Potential for "Johkasoh" (Private Wastewater Treatment Tank) in Asia, Asia 3R Conference 30 October-1 Nov 2006, Tokyo*. (http://www.env.go.jp/recycle/3r/en/ asia.html, accessed on 10 April 2010)

Oh, H. & Kitawaki, H. (2005) Research on *johkasou* technology transfer in China: Case study of Harbin City. Master Thesis of Toyo University, Japan. (in Japanese)

Ohno, Y. (2005) Actual conditions of public project evaluation: Results brought from benefit-cost analyses by ministry and by project. *Financial Review, Ministry of Finance, Japan, Policy Research Institute, 77*, 66-89. (in Japanese)

Pattanayak, S.K., Yang, J.C., Patil, S., Poulos, C., Jones, K., Kleinau, E. Corey, C. & Kwok, R. (2005) *Environmental Health Impacts of Water Supply, Sanitation and Hygiene Interventions in Rural Maharashtra, India. Study Protocol.* Submitted to The World Bank. Cited in Poulos et al. (2006)

Poulos, C., Pattanayak, S. K. & Jones, K. (2006) *A Guide to Water and Sanitation Sector Impact Evaluations.* World Bank. 44p. (http://siteresources.worldbank.org/INTISPMA/Resources/383704-1146752240884/Doing_ie_series_04.pdf, accessed on 7 Apr 2010)

Prüss, A., Kay, D., Fewtrell, L. & Bartram, J. (2002) Estimating the burden of disease from water, sanitation, and hygiene at a global level. *Environ Health Perspect, 110*, 537–542.

Serageldin, I. (1994) Water supply, sanitation, and environmental sustainability: The financing challenge. The World Bank, Washington, D.C. 42p. (http:// www-wds.worldbank.org/external/default/WDSContentServer/WDSP/IB/1994/11/01/C0000926 5_3970716143355/Rendered/PDF/multi_page.pdf, accessed on 7 Apr 2010)

Sewage Business Management Centre, Japan (2008) Sewage Business Management Centre homepage (http://www.sbmc.or.jp/english/index.html, accessed on 9 Apr 2010)

Sewage Works Project Management Research Group (1997-2000*) Sewage Works Handbook,* 9th-12th editions. Gyousei, Tokyo. (in Japanese)

Suharyanto & Matsushita, J. (2009) Comparative study on integrated wastewater management system model for developing countries under rapid urbanization. *7th International Symposium on Southeast Asian Water Environment*, Bangkok, Thailand, 28-30 October 2009.

Tilley, E., Lüthi, C., Morel, A., Zurbrügg, C. & Schertenleib, R. (2008) *Compendium of sanitation systems and technologies.* Swiss Federal Institute of Aquatic Science and Technology (Eawag), Dübendorf, Switzerland. 158p.

Tsuzuki, Y. (2006) An index directly indicates land-based pollutant load contributions of domestic wastewater to the water pollution and its application, *Science of the Total Environment, 370,(2-3)*, 425-440.

Tsuzuki, Y. (2007) Relationships between pollutant discharges per capita (PDC) of domestic wastewater and the economic development indicators. *Journal of Environmental Systems and Engineering (Dobokugakkai Ronbunsyu, Category G), 63 (4)*, 224–232 (in Japanese with English abstract).

Tsuzuki, Y. (2008) Relationships between water pollutant discharges per capita (PDCs) and indicators of economic level, water supply and sanitation in developing countries, *Ecological Economics, 68*, 273–287.

Tsuzuki, Y. (2009a) Environmental kuznets curve (EKC) relationships between pollutant discharge per capita (PDC) of domestic wastewater and income indicators, *Journal of Global Environment Engineering, Japan Society of Civil Engineering, 14*, 37-46.

Tsuzuki, Y. (2009b) Comparison of pollutant discharges per capita (PDC) and its relationships with economic development: An indicator for ambient water quality improvement as well as the Millennium Development Goals (MDGs) sanitation indicator, *Ecological Indicators, 9*, 971-981.

Tsuzuki, Y. (2010) Chapter 5: Domestic wastewater pollutant discharge and pollutant load and water quality in the ambient water in developed and developing countries. In Kudret Ertuð and Ilker Mirza eds.: *Water Quality: Physical, Chemical and Biological Characteristics*. 125-164, Nova Science Publishers, 277p.

Tsuzuki, Y. & Koottatep, T. (2010) Municipal Wastewater Pollutant Discharge Indicator Estimation and Water Quality Prediction in Pak Kret District, Bangkok, Thailand, *Journal of Water and Environment Technology, Japan Society on Water Environment, 8 (1)*, 51-75. (http://www.jstage.jst.go.jp/browse/jwet/8/1/_contents/-char/ja/, accessed on 10 April 2010)

Tsuzuki, Y., Koottatep, T., Ahmed, F. & Rahman, M.M. (2008) Water quality and pollutant load in the ambient water and domestic wastewater pollutant discharges in the developing countries: Survey results in autumn and winter in 2006, *Journal of Global Environment Engineering, Japan Society of Civil Engineers, 13*, pp.121-133.

Tsuzuki, Y., Koottatep, T., Wattanachira, S., Sarathai, Y. & Wongburana, C. (2009a) On-site treatment systems in the wastewater treatment plants (WWTPs) service areas in Thailand: Scenario based pollutant loads estimation, *Journal of Global Environment Engineering, Japan Society of Civil Engineering, 14*, 57-65.

Tsuzuki, Y., Koottatep, T., Sarathai, Y., Fujii, M., Mochihara, Y., Matsuda, Y. & Yoneda, M. (2009b) Municipal wastewater pollutant discharge reduction: Possibilities of "soft interventions" in the households in Japan and in Thailand, *The 17th Global Environment Symposium, Japan Society of Civil Engineering*, 11-12 Sep 2009, Okinawa, Japan.

Tsuzuki, Y., Fujii, M., Mochihara, Y., Matsuda, Y. & Yoneda, M. (2009c) Domestic wastewater pollutant discharge reduction with the community participation program in the Yamato-gawa River drainage area, Japan. *The 12th International Riversymposium*, 21-24 Sep 2009, Brisbane, Australia.

Tsuzuki, Y., Koottatep, T., Jiawkok, S. & Saenpeng, S. (2010) Municipal wastewater characteristics in Thailand and effects of "soft intervention" measures in households on pollutant discharge reduction. *Water Science and Technology, 62 (2)*, 231-244.

UN Water (2008) Fact Sheet 2: Sanitation generates economic benefit. (http:// esa.un.org/iys/docs/IYS Advocacy kit ENGLISH/ Fact sheet 2.pdf, accessed on 10 April 2010)

UNDP (2007) *Statistics of the Human Development Report* (http://hdr.undp.org/en/ statistics/)

Water Environment Partnership in Asia (WEPA) (2007) *Database of technologies in operation*. (http://www.wepa-db.net/technologies/top.htm)

Water Supply and Sanitation Collaboration Council and World Health Organization (2005) *Sanitation and hygiene promotion : programming guidance*. 84p.

WHO/UNICEF Joint Monitoring Programme on Water Supply and Sanitation (2010) *Progress on Sanitation and Drinking-Water: 2010 Update*. 60p. (http://www.who.int /water_sanitation_health/publications/9789241563956/en/index.html, accessed on 9 April 2010)

World Health Organization (2004) Water, sanitation and hygiene links to health: Facts and figures updated November 2004. (http://www.who.it/water_sanitation_health/ publications/facts2004/en/index.html, accessed on 10 Apr 2010)

World Water Assessment Programme (2009) The United Nations World Water Development Report 3: Water in a Changing World. UNESCO and Earthscan, 349p. (http://www. unesco.org/water/wwdr/wwdr3/side_publications.shtml, accessed on 7 Apr 2010)

Yahya, A. S. (1994) Characteristics of chlorine tablet solubility and their disinfection efficiency of *johkasou* effluents. *Bull Inst Public Health, Japan, 43 (3)*, 316-319. (in English with Japanese abstract)

Yamato-gawa River Office, Ministry of Land, Infrastructure, *Transportation and Tourism (2010) Social Experiment Program on Municipal Wastewater Pollutant Discharge Reduction.* (http://www.kkr.mlit.go.jp/yamato/drainage/index.html, accessed on 10 April 2010) (in Japanese)

Yang, X. M., Yahashi, T., Kuniyasu, K. & Ohmori, H. (2001) On-site systems for domestic wastewater treatment (*johkasous*) in Japan. In: Lens, P., Lettinga, G. & Zeeman, G. eds. (2001) *Decentralised Sanitation and Reuse: Concepts, systems and implementation*. 256-280. International Water Association Publishing, London, 550p.

In: Potable Water and Sanitation
Editor: Joel M. McMann

ISBN 978-1-61122-319-4

Chapter 7

Industry Structure and Development in the Water and Wastewater Industry

Malcolm Abbott and Bruce Cohen
Swinburne University of Technology, Australia

1. Introduction

In recent years there has been a fair degree of analysis completed on the optimal size and structure of water supply and wastewater organisations, as well as reform undertaken of the structure of water supply and wastewater disposal organisations around the world. These reforms have included such things as: the commercialisation of state-owned water utilities, increased contracting out of functions to the private sector, some separation of regulatory and commercial responsibilities and in a few cases the privatisation of state-owned assets. In many cases, however, vertically integrated monopoly service providers have remained intact.

Complicating the task of reforming the structure and ownership of the water supply and wastewater industry has been the need to satisfy multiple industry and community objectives. These structural issues have arisen across all countries, regardless of the level of development, although circumstances can alter significantly. In the case of developed countries the issues of efficiency and environmental impact tend to have been given greater emphasis, whereas in developing countries improving access and reducing public health risks tend to receive greater attention.

The purpose of this chapter is to review the structural issues that have confronted the reform of the water supply and wastewater industries in the past, and examine them in light of those forces that can be expected to impact on the issue of structural reform in the future. These factors including such things as; changing climate patterns; rising demands for improved access to clean water, the desirability of diversifying water supplies and the reduction of environmental impacts.

To this end, the paper is structured as follows – Section 2 provides a discussion of some of the forces impacting on the water industry, with particular reference to issues affecting the structure of the water sector in developing countries. Section 3 then outlines various structural approaches that have been adopted with respect to the water and wastewater industry. In light of these arrangements, Section 4 introduces some of the findings of research on structural performance, and in Section 5 some concluding comments are made.

2. Forces for change in the water and wastewater sector

Across the world, there is a range of factors that can be expected to drive change in the water and wastewater sector in the future. Some are forces that are a constant in the sector, others arise due to changing circumstances or changing perceptions of importance, such as with respect to the environment.

Particularly in developing countries, the primary force driving the water industry is the ongoing need for sufficient investment in infrastructure and water sources so that people have access to clean supplies of water and have wastewater disposed of effectively. This is reflected in the Millennium Development Goals, which aims to halve by 2015 the proportion of the urban population who were without sustainable access to safe drinking water and basic sanitation in 1990. Progress towards achieving this goal has been mixed (Locussal et al, 2009). Nonetheless, service provision of this kind is generally regarded as a basic function of government in most countries, as there are important public health and environmental factors involved. These services not only involve the construction of delivery and disposal infrastructure, but also the development of multiple sources of water.

This public need for access to clean drinking water is one that is by no means new, but in recent times is becoming complicated by a range of other factors. These include the ongoing trend towards urbanisation, and the risks associated with climate change.

While accurate predictions of the impact of climate change remain problematic, there is a broad expectation that it will be associated with increasingly variable climatic conditions, and that most regions will experience an overall net negative impact on water resources and freshwater ecosystems (IPCC, 2007; 2008). In some regions, particularly those at high latitudes and in parts of the tropics, it is expected that climate change will result in increases in precipitation and higher annual runoff. The net benefits of this, however, are expected to be tempered by increased flood risks and threats to water quality. In other regions, particularly semi-arid and arid areas, the expectation is for water availability from traditional water sources to be reduced because of altered rainfall and river flow patterns, reduced run-off due to hotter weather and resultant dryer soil conditions, reductions in inland groundwater levels, and decreased flows fed by shrinking glaciers (IPCC 2008:70).

Such changes can be expected to reinforce the general trend for policy makers to place greater emphasis on environmental concerns, which will impact on both developed and developing countries. More directly, these changes will require new thinking about how water and wastewater services can be delivered at adequate quantity and quality.

For example, in regions where there are reduced inflows, the clear consequence will be an increased need for alternative water supply sources to be found. These alternatives will, in all likelihood, need to be from sources that are less vulnerable to climatic changes, and will, therefore, be expected to be higher cost options than that of more traditional, natural water sources.

One such alternative source is desalinated ocean water. This source, however, is problematic – particularly in developing countries – because of the high costs and energy intensity of the desalination process (Raluy et al 2006; Stokes and Horwath 2009). Desalinated water was first used extensively in countries such as the United Arab Emirates where energy costs are low and water scarce. In more recent times desalination plant has been built in other places such as Israel, Singapore and Western Australia, with the expressed purpose of reducing the dependence of supplying water from rainfall.

A second option is increased use of recycled water and stormwater. In a structural sense, recycled water gives rise to issues relating to the relationship between the retail function, wholesale water provision and disposal of wastewater. More broadly, the use of recycled water has implications for the physical design of wastewater infrastructure, as well as the maintenance of water quality. This is of particularly importance in developing countries where health risks associated with water provision are often already high, and which may be exacerbated by the development of alternate, fit-for-purpose water supplies of varying quality. Further, the technologies associated with treating effluent to suitable quality are themselves energy intensive and involve substantial infrastructure and operational costs. Nonetheless, in recent times the use of recycled water has become more common in places such as Israel, South Australia, California, and Japan.

Similarly questions exist with respect to the potential use of storm water, which also gives rise to various issues for industrial, commercial and domestic users, including the extent to which stormwater contributes waste into the water supply system and how regulatory regimes can be structured to control the nature and quantity of this waste.

Greater use of inter-regional trade in water is a third option, but it also has problems in terms of the high costs associated with the transportation of water over long distances. This can involve substantial pumping unless gravity-feed systems can be utilised. Inter-regional trade also involves environmental impacts, as well as social impacts of moving water across regions and from rural to urban use (see Chong and Sunding 2006). These social impacts are likely to be exacerbated where water is priced substantially differently across regions and historically urban users have not been subject to constraints or caps on usage in the same manner as rural users.

As indicated above, a related outcome of the search for greater diversity of water supply sources is an increased potential for water supply to be fit for purpose. There are a variety of structural and productivity issues associated with providing fit-for-purpose water – in particular, that such an approach may require the replication of infrastructure, i.e. third pipe systems, to supply lower quality water for industrial, agricultural or external domestic use. Such infrastructure comes at considerable cost (see CSU 2003; Tang et al, 2007). This is particularly the case for developed nations with existing potable water distribution networks; in many developing countries the major structural concern remains the delivery of sufficient potable supplies – whether that be through piped network infrastructure or distributed systems such as protected wells, boreholes and springs, or rainwater collection (see WHO/UNICEF 2005).

The development of alternative water sources is bound up with how the water and wastewater sector may be configured. This is because different structures can create different incentives for investment in alternative water sources. The biggest possible structural change is the greater reliance on market based structures, involving water supply options with different costs and reliability. Such market based structures have developed in other resource based industries such as natural gas and electricity, but only rarely in the water supply industry. A market-based approach to water supply would represent a substantial shift from the centrally planned and primarily monopoly structures, which currently dominate the water supply and wastewater disposal industry. While feasible, it should be recognised that the development of such market based structures is complex, particularly given the network elements associated with water and wastewater transmission, the task of appropriately defining rights to both the water and access to storage facilities and the need to maintain water quality standards where water is supplied from multiple sources (Cave 2009). As such, the appropriateness of such mechanisms for water provision in developing countries is far from clearly established.

The combination of these factors is expected to put pressure on the water and wastewater systems around the world. Furthermore, these factors are putting pressure on governments to reform the structure of water and wastewater sectors in a variety of ways. In saying this there is no general consensus of the optimal structure of the water supply and wastewater disposal sector. Indeed it may be possible that a range of options are available depending upon local circumstances. The nature of the options that have been adopted around the world are outlined in the section below, having regard to the underlying characteristics of the industry.

3. Industry Structure

Industry structures in the water supply and wastewater disposal sector vary across different countries around the world – in terms of the range of activities that individual entities undertake, geographical size, the composition of the customer base, the extent of state sector involvement, the amount of competition (if any), the character of regulation, and the bodies who are given these responsibilities.

In general, the range of activities that water businesses may undertake include such things as: bulk water collection and storage, bulk water transfer, water treatment, water distribution, reticulation and retail supply, sewage collection, distribution and treatment, drainage and irrigation. In many instances, water utilities are also responsible for such things as land and resource management, quality assurance, regulation and policy development. The range of activities undertaken by water businesses is illustrated in Figure 1.

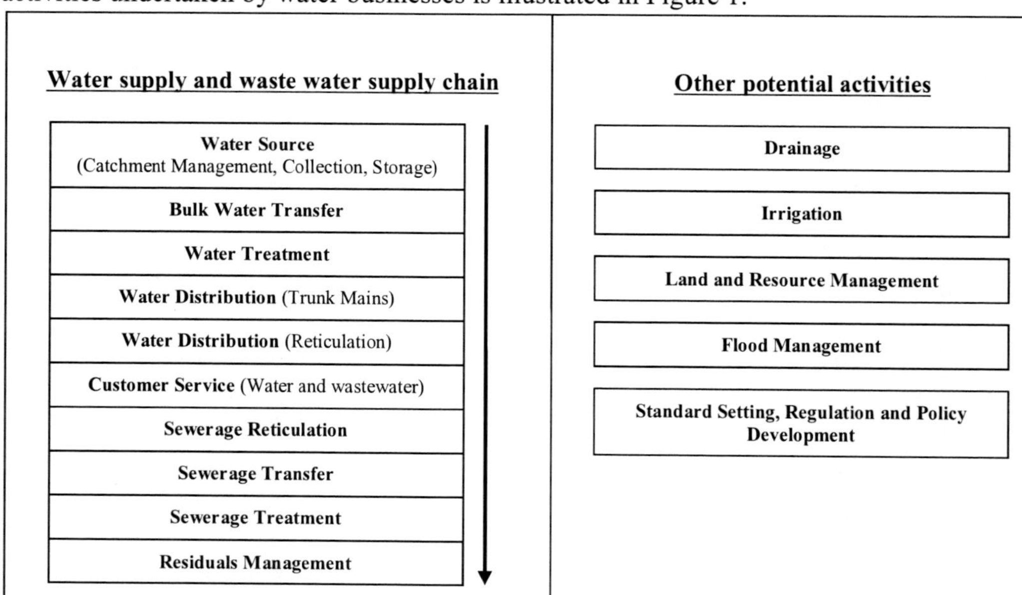

Figure 1. Water supply and waste water industry – key activities.

In most cases around the world the most common structure is that of a single, vertically integrated utility in each region of water supply. This is especially true in small and medium size markets, and is typical of both privately and government owned water supply agencies (Kessides 2003; Abbott and Cohen 2009). These vertically structured organisations can be government owned (as is common in many countries) or privately owned (as is the case in England for instance), and in the cases where they are government owned are sometimes integrated with other government functions. Generally examples of water utilities that are integrated with government functions, such as planning and environmental regulation, occur in smaller municipal centres. That said there are, however, cities such as Toronto and Hong Kong where the water utility is a division of government rather than a separated state owned company. Quasi government boards of control are also still popular such as the Public Utilities Board of Singapore and Chennai Metropolitan Water and Sewerage Board in India. In addition in recent years a number of jurisdictions have established commercialised, separate companies, while retaining government ownership (i.e. Johannesburg Water in South Africa and Aguas de Portugal in Portugal).

Sometimes in larger metropolitan markets horizontal separation occurs with a number of, separate vertically integrated entities operating in different parts of the city. This, for instance, is the case in cities such as Paris and Manila. Traditionally water in Paris was distributed by Lyonnaise des Eaux and Compagnie des Eaux de Paris. In Manila water is supplied by the Manila Water Company and Maynilad Water Services. Horizontal separation of this sort enables regulators, or other policy makers, to make comparisons of the performance of the respective companies.

As well as horizontal separation, vertical separation can also occur with a split between the wholesale and distribution/retail segments of the industry. This is the case in cities such as Auckland in New Zealand and Melbourne in Australia. In Auckland the government owned company, Watercare, supplies bulk water to a number of local councils and local council owned water companies. In Melbourne the government company, Melbourne Water supplies bulk water to three separate government owned water retailers. This sort of vertical separation is often undertaken in order to encourage the development of multiple sources of supply, although at times it has occurred simply to allow joint ownership of the bulk supply of water by smaller distribution companies. This is the case in New Zealand, for instance, in Auckland and Wellington. Although there are a few examples of this type of structural unbundling in the water supply industry it has not occurred to the same degree that it has in the case of industries such as electricity and natural gas supply.

The tendency toward the creation of single, vertically integrated monopolies is a function of the heavy up front capital costs, and engineering scale economies, which contribute to the creation of natural monopoly conditions. Further, as water has a low value added relative to its transport costs, centralised transmission over long distances through a large national or regional network, as occurs in the case of an electricity grid, is impractical. As a result, water supply systems tend to be highly decentralised. Competition in the industry is generally limited because of this difficulty in long distance transportation and because the large share of the cost of supplying water and collecting sewerage is tied up in distribution networks, which are costly to duplicate. Approximately two-thirds of the cost of water supply is related to the cost of the supply network for water. In contrast around 40 percent of the cost of supplying electricity is related to the transmission and distribution networks (Wallsten and Kosec 2005). The complexity of managing inflows and outflows within the network also impinges on the scope for competition, as do water quality and public health related externalities – the most serious being water contamination which can lead to disease and other health problems (see Garcia, Moreaux and Reynaud 2007). Management of a scarce natural resource (water) also plays a role, as does the planning of network extensions; both of which have broader public policy implications.

Even without competition in the industry a few policy issues surrounding the structure of the industry do arise. First and foremost is the optimal size of a water supply company. Over the years the trend has been toward merging companies to create larger sized units in order to reap greater economies of scale. A related issue is the degree to which economies of scope can be achieved through the joint provision of water supply and wastewater services. In most jurisdictions, but not all, water supply and wastewater services are jointly provided by single entities. The rationale for this is that there are economies of scope between the two functions. That said, water only companies are not unknown. In England, for example, there are both water and sewerage companies and water only companies (ten water and sewerage companies and fifteen water only companies; Stone and Webster 2004).

Another policy issue that has arisen has been the case of the privatisation of water assets. Although privatisation of the water industry has been less common than in the case of other utilities it has occurred in countries as diverse as Indonesia, Chile, England and the Philippines. As well some countries like France have a long history of private provision, and in many other instances (for example, in Latin America, Sub-Saharan Africa and Central Asia), private sector involvement has been instituted more recently through franchising arrangements and other forms of public-private partnerships (Marin 2008). In the case of the United States, however, around eighty-nine percent of people are supplied by government or cooperative owned utilities and there has been no substantial trend towards privatisation in recent years in that country (Wallsten and Kosec 2005).

Due to the extensive environmental and health externalities government ownership is regarded by many as the most appropriate structure for the industry and even in the case of private companies, they are generally heavily regulated (see Garcia, Moreaux and Reynaud 2007). This does raise the question of the type of regulation that is appropriate. Regulation generally seeks to guard against the extraction of monopoly rents and to ensure adequate water quality and environmental outcomes, while at the same time enabling investors to receive a necessary return on long lived assets. Generally, responsibility for such regulation vests either within or separate to the water businesses themselves. In more recent times government authorities in some jurisdictions have separated their commercial activities in water supply and wastewater management from policy and regulatory functions. Further public policy intervention takes place in the form of the subsidisation of favoured consumer groups. Often these groups are low income or farmer groups, and assistance comes in the form of implicit cross subsidies in tariffs or explicit rebates paid by government welfare authorities.

The choice of structure and size of water utilities has important implications not only for the development alternative water sources but also for the general efficiency performance of these organisations. Given the diversity of structural models around the world it seems that there is no clear consensus on the optimal organisational structure of water utilities.

4. The Industry's Productivity Performance

From the previous sections of this chapter it is quite clear that around the world a wide variety of water and wastewater industry structures, size of entities and ownership models. This diversity has encouraged a number of researchers to study the industry's productivity performance in order to come to some firm conclusions on the optimal size and structure of water utilities. In undertaking this research there has been a tendency to concentrate on the issues of the existence of economies of scale and scope, public versus private ownership and the effects of subjecting the industry to different forms of regulation.

Initially is was the possible existence and extent of economies of scale that attracted attention, the first such study being undertaken as far back as 1969 (Ford and Warford 1969). Numerous studies since then have found that significant scale economies do exist in the water supply industry, however, some studies raised doubts about the extent of these economies while others have suggested that although some aspects of the water supply chain are associated with economies of scale (e.g. distribution), while others are associated with diseconomies (e.g. production of water available for delivery).[1] The studies that looked at small scale water businesses generally found that economies of scale could be reaped if they became larger. This suggested that a degree of consolidation of water and wastewater utilities is a strategy that might reap benefits in the case where utilities are relatively small in size. The research, however, often found that at relatively large output levels the gains that could be achieved in terms of lower average costs from greater economies of scale were exhausted.[2]

After scale economies the issue of economies of scope was raised. Whereas economies of scale relate to efficiencies associated with the level of output of a single product type, economies of scope relate to efficiencies that are gained from combining multiple outputs together to achieve synergies. With respect to the integration of water supply and wastewater activities, there has been discovered evidence that economies of scope are enjoyed by organisations that operate water supply and wastewater disposal together, although it appears that this is more strongly the case for small organisation rather than large ones. This occurs because small organisations have greater capacity to gain from the more intense use of common costs such as maintenance and administrative staff.

In a number of countries the consolidation of utilities into larger entities has been pursued with some vigour. This would appear to be a reasonable approach where there are small utilities that might gain from the creation of both economies of scale and scope.

As well as the research that has been undertaken on the horizontal integration of water supply and wastewater disposal additional research has been undertaken on the impact of the separation of bulk water supply from distribution and retail. This research indicates that integration is efficient for small companies, but not necessarily for large ones. Further, there is some evidence that economies from vertical integration are most apparent for businesses that only undertake water supply activities (Stone and Webster 2004). Vertical disaggregation of the industry has not been common around the world, but the separation of regulatory and planning functions and establishment of stand alone, commercial water supply companies has been more frequent. This "corporatisation" has seen the separation of regulatory and policy

[1] Instead of make explicit references to all of these studies it is possible to see a review in Abbott and Cohen 2009.

[2] There is no absolute consensus around where this point of exhaustion is but there is little likelihood of additional economies of scale being enjoyed when utilities exceed 1,000, 000 connections (Abbott and Cohen 2009).

functions and their transfer to separate government agencies and the establishment of government owned, water supplying corporations. Generally, the water sector is subject to considerable environmental, planning, social and economic regulation. These regulations cover issues including public health (such as drinking water quality standards), pricing (such as rate of return and incentive-based mechanisms), trade waste, supply reliability and river and sea water quality. At times governments have given responsibility for regulation to government owned companies themselves, or they have subsumed these responsibilities into local government bodies. Steps have been taken in a number of jurisdictions to subject monopoly water suppliers to independent economic regulation of pricing.

There is some evidence that some economies can be achieved through combining in a single organisation environmental management and commercial water activities (Lynk 1993; Hunt and Lynk 1995). Amongst other things Lynk (1993) looked at the average level of inefficiencies of the regional water authorities in England and Wales prior to privatisation in 1991 to determine if there were economies of scope between water supply and environmental regulation. Lynk found that there was evidence that the regional Water Authorities enjoyed enhancements through joint production, which implied these were lost with privatisation and separation. In addition, a limited amount of research has found that environmental regulations can enhance the efficiency of water and sewerage businesses. Not a great deal of research, however, has so far been conducted on these issues, which is unfortunate because the separation of these activities from commercial water supply could potentially be a source of inefficiency. As such, areas for future research include the relationship between the responsibility for reducing undesirable water output and managing water losses (Garcia and Thomas 2001) and the opportunities to reduce costs associated with sewage treatment.

A final research focus has been the question of the impact of ownership (private versus government) on productivity levels. Initially, most of the work on this issue was undertaken in the United States. Later in the 1990s the issue of whether the privatisation of the English and Welsh water and wastewater businesses improved the productivity and efficiency of the industry was undertaken by a number of researchers. In the American literature, the research findings were inconclusive. A number of studies found that there was no discernable difference between government and privately owned companies. Other studies found that private operators are more efficient and still others that public operators rate better (Abbott and Cohen 2009). One possible explanation for these divergent findings, that has some support in the research, is that government firms operate more efficiently at high levels of output, while privately owned firms are more efficient at low output levels (Bhattacharyya et. al. 1995). More broadly, however, it appears that in driving efficiency, ownership may not be as important as the level of competition in the industry (Wallsten and Kosec 2005). In a related theme, a number of analysts have looked at the change in efficiency brought about by privatisation. In most instances, the evidence showed limited effects from privatisation or even a decline in overall efficiency. Further, any improvements following privatisation appear to be less than the improvements in productivity that occurred in the period immediately prior to privatisation. Given the ambiguous nature of the results of this research it is probably little wonder that the privatisation of assets has not been pursued in the water and wastewater industry to the same degree that it has in other utilities such as electricity and gas supply and telecommunications.

CONCLUDING COMMENTS

In the past the water supply and wastewater sector has enjoyed a more stable industry structure than most sectors, in large part due to its monopoly characteristics and the social importance and basic public good nature of water quality and wastewater treatment services. A range of factors, however, are requiring a rethinking of traditional water industry structures including population growth and movement, as well as the risks associated with climate change.

At present, however, there is no policy consensus on the optimal structure of the industry in terms of achieving economic efficiency and other water quality and environmental outcomes, innovation and social goals, and indeed the most appropriate structures may rest of a number of different factors. Further, it is important to determine which industry structure(s) allows for the achievement of the greatest level of productivity and efficiency, while at the same time enabling a diversity of water sources to be developed, innovation in service delivery as well as the maintenance of environmental and water quality standards.

The expectation that research of the kind detailed in this chapter has the capacity to provide a simple and singular template to guide future development of the water and wastewater sector must be tempered,by the knowledge that this industry is of a highly localised nature, and subject to distinctive regional economic, political, geographic, geological, meteorological and hydrological constraints.

Nonetheless, given the findings of the past research, a number of points can be made about the future industry structures for the water and wastewater sectors, with particular reference to developing countries.

First of all is that there is insufficient experience internationally to come to any definitive conclusions about the desirability, or otherwise, of fully privatising large scale operational water assets. Pending more examples of successful privatisations it is perhaps an issue about which developing countries not to get too distracted.

However, this position should be distinguished from the related but separate question of the role of the private sector in countries where the existing industry is not yet capable of delivering the quantity and/or quality of services that the community requires. This is particular so with respect to countries with low coverage levels, ineffective public utilities that provide inadequate or partial services and which have remote, difficult-to-access regions (see Kariuki and Schwartz, 2005).

While the potential forms of private sector participation are many and varied, and an assessment of their relative merits beyond the scope of this paper, private providers can potentially bring a range of skills and efficiencies to improve otherwise inadequate service provision, as well as placing pressure on remaining public sector operators.

In this context, public private partnerships – while not a universal panacea – have been associated with a range of improvements in the water sector in developing countries, including with respect to reducing water losses, improved labour productivity and enhanced billing practices and collections (see Marin 2008)

However, such arrangements require appropriate contract and risk management processes to be established to ensure that such providers do not take advantage of unreliable public services to gouge customers, or fail to deliver adequate levels of services at affordable prices where contractual arrangements are not sufficiently robust and/or regulatory oversight in ineffective due to a lack of skills or malfeasance. In this regard, it is noted that it has been argued that a country's institutional capacity and quality of governance are key factors determining efficiency in the water sector, and are more important than private participation in itself (see World Bank 2004:255).

Secondly, in many markets there has been some, though often relatively small, movements towards larger, more integrated entities. On balance, this is probably a prudent approach. In many instances, gains from economies of scale and scope are possible where the size of entities is increased. Indeed, there may even be a case for combining relatively small water and sewerage entities with regulatory functions as well, particular in regions with limited access to requisite skills. Creating more efficient entities in this process would mean it would be possible to release more resources to expand water access and increase quality (though in terms of the need to diversify sources of water in small water markets, this sort of activity in most cases would not be as urgent an issue). The gains from this process of aggregation, however, are not endless. Once water and sewerage entities reach a certain size there appears little to be gained combining organisations. Indeed in large urban areas the establishment of separate corporatised commercial businesses and regulatory agencies becomes an option. As well alternative structures such as those that embody the vertical separation of wholesale and retail and horizontal separation of water companies become an option. Indeed these options need to be examined in terms of whether they are likely to lead to additional investment in alternative water sources. Finally, in considering the scope for aggregation in the water sector, regard also needs to be given to its potentially adverse impact on local democracy and autonomy, as well as the requirement to ensure appropriate governance and oversight processes are developed and implemented in unison with such change.

REFERENCES

Abbott, M.J. and Cohen, B. (2009), Productivity and efficiency in the water industry, *Utilities Policy* 17, 233-244.

Bhattacharyya, A., Harris, T., Narayanan, R. and K. Raffiee (1995), Specification and estimation of the effect of ownership on the economic efficiency of the water utilities, *Regional Science and Urban Economics, 25*, 759-784.

Cave, M. (2009) *Independent Review of Competition and Innovation in Water Markets: Final Report*, Department for the Environment, Food and Rural Affairs, London.

Chong, H. and D. Sunding (2006), Water markets and trading, *Annual Review of Environmental Resources*, 31, pp. 239-264.

Colorado State University (CSU) (2003), *The Benefits and Costs of Pressurized Dual Water Systems in Colarado and the Potential Role of Canal Companies and Irrigation Districts in Providing the Pressured Irrigation Supply Portion of Dual Systems*, Sociology Water Lab and the Colorado Institute of Irrigation Management.

Ford, J. and J. Warford (1969), Cost functions for the water industry, *Journal of Industrial Economics, 18(1),* 53-63.

Garcia, S. and A. Thomas (2001), The structure of municipal water supply costs: application to a panel of French local communities, *Journal of Productivity Analysis, 16*, 5-29.

Hunt, L. and E. Lynk (1995), Privatization and efficiency in the UK water industry: an empirical analysis, *Oxford Review of Economics and Statistics, 57(3),* 371-388.

Intergovernmental Panel on Climate Change (IPCC) (2007) *Climate Change 2007: Impacts, Adaptation and Vulnerability: Contribution of Working Group II to the Fourth Assessment Report of the Intergovernmental Panel on Climate Change*, Cambridge University Press: Cambridge, UK.

Intergovernmental Panel on Climate Change (IPCC) (2008) *Climate Change and Water: IPCC Technical Paper VI*, Geneva.

Kingdom, W. (2005) *Models of Aggregation for Water and Sanitation Provision*; Water Supply & Sanitation Working Notes No. 1/2005, World Bank.

Lariuki, M. and Schwartz, J. (2005) *Small-Scale Private Service Providers of Water Supply and Electricity: A review of incidence, structure, pricing and operating characteristics*, World Bank Policy Research Working Paper No. 3727.

Locussol, A; Fall, M and Dickson, E. (2009) *Guiding Principles for Successful Reform of Urban Water Supply and Sanitation Sectors*, Water Working Notes No 19, World Bank.

Kessides, Ioannis (2003), *Infrastructure Regulation Promises, Perils and Principles,* AEI-Brookings Joint Center for Regulatory Studies, July.

Kim, H. (1987), Economies of scale in multi-product firms: an empirical analysis, Economica, 54(214), 185-206.

London Economics (1998), *Competition in Water*, London Economics, London.

Lynk, E. (1993), Privatisation, joint production and the comparative efficiencies of private and public ownership: the UK water industry case, *Fiscal Studies, 14*, 98-116.

Lynk, E. and Hunt, L. C (1995). Privatisation and efficiency in the UK water industry: an Empirical analysis, *Oxford Bulletin of Economics and Statistics*, 57(3), 371-88.

Marin, P (2008) *Public Private Partnerships in Urban Water Utilities – Findings from Developing Countries*, Public Private Infrastructure Advisory Facilities, World Bank.

Productivity Commission (PC) (2008) *Towards Urban Water Reform: A Discussion Paper*, Melbourne, Australia

Raluy, G., Serra, L., Uche, J. (2006), Life cycle assessment of MSF, MED, and RO desalination technologies. *Energy,* 31, 2361–2372.

Stokes, J. and Horwath, A. (2009) Energy and Air Emission Effects of Water Supply, *Envionrmental Science and Technology, 43*, 2680-2687.

Stone and Webster Consultants for OFWAT (2004), *Investigation into Evidence for Economies of Scale in the Water and Sewerage Industry in England and Wales: Final Report*, London.

Tang, S., Yue, D., and D Ku (2007) *Engineering and Costs of Dual Water Supply Systems*, IWA Publishing, London.

Victorian Competition and Efficiency Commission (VCEC) *Water ways: Inquiry into Reform of the Melbourne Retail Water Sector Final Report*, Melbourne.

Wallsten, S. and K. Kosec (2005), Public or private drinking water? The effects of ownership and benchmark competition on U.S. water system regulatory compliance and household water expenditures, *Working Paper 05-05, March, AEI-Brookings Joint Center for Regulatory Studies.*

World Bank (2004) *Reforming Infrastructure: Privatisation, Regulation and Competition, World Bank Policy Research Report*, Washington D.C.

World Health Organization and UNICEF Joint Monitoring Programme for Water Supply and Sanitation (WHO/UNICEF) (2005) *Water for Life: Making it Happen*, Geneva, Switzerland.

In: Potable Water and Sanitation
Editor: Joel M. McMann

ISBN 978-1-61122-319-4

Chapter 8

PLANT DERIVATIVES AS NATURAL SANITIZERS FOR THE CONTROL OF FUNGAL INVASION

Ahmed A. Tayel[*]
Genetic Engineering and Biotechnology Institute, Minufiya University, Sadat City, P. O. box 79/22857, Egypt

ABSTRACT

Plant kingdom always donate human with lots of precious compounds and materials. Most of plant derivatives are recognized as safe, eco-friendly and low-cost agents regarding synthetic and chemical substances.

The global trend has shifted towards the reduction of synthetic biocides application in general and in food-dealing fields in particular. Despite of the prevalence applications of semi-synthetic and synthetic compounds in sanitation for the control of microbial pathogens attack, natural products are still viewed as potential alternatives.

Two innovative systems were constructed for evaluating the sanitation potency of plant derivatives in closed and semi-closed places against many fungal pathogens; the first applied plant extracts aerosol as sanitizers, whereas the second employed fumigation with smoldered plant fumes. Both of the two techniques exhibited promising and effective sanitation competence toward many fungal pathogens that attack human and plants. *Candida albicans* and *Penicillium digitatum* strains could be completely inhibited using aerosolization with pomegranate peel extract, while growth of *Aspergillus flavus* and *Botrytis cinerea* could be successfully stopped after exposure to smoldered plant fumes either in vitro or on the infected crops using fumigation room. The biodegradable and inoffensive nature of sanitizers from plant origins recommends their application for the effective elimination of fungal pathogens.

It is an important step toward desirable substitution of synthetic and chemical sanitizers with other human-friendly agents from biological derivations.

[*] For correspondence: Dr Ahmed A. Tayel, PhD. Genetic Engineering and Biotechnology Institute, Minufiya University, Sadat City, P. O. box 79/22857, Egypt; E-mail: tayel_ahmad@yahoo.com

1. INTRODUCTION

Discovery of healing powers from plants is a prehistoric thought. Plants contain treasures of secondary metabolites' varieties, such as tannins, terpenoids, alkaloids, and flavonoids, which have been found in vitro to have anti-microbial properties. Searching for these potential drugs and dietary supplements derived from plants have accelerated in recent years. Researchers in ethnopharmacology, botany, microbiology, and natural-products chemistry are combing the Earth for innovative phytochemicals which could be applied for treatment of infectious diseases.

Clinical microbiologists have many reasons to be interested in the topic of antimicrobial plant materials; it is very likely that many of these phytochemicals have the potentialities as substitutes for antimicrobial drugs prescribed by physicians; several are already being experienced in humans

Antimicrobials from plants gain many advantages over chemical and synthetic biocidal agents since most of plants derivatives are generally recognized as safe. Furthermore, the biodegradable nature of plant derivatives, because of their biological origin, prevents human body from accumulation effects resulted from chemical and synthetic antimicrobials application. Globally, food safety is regularly compromised by the presence of mycotoxins produced from contaminating mycotoxigenic fungi either in growing crops in the field or after their harvesting and during storage. Mycotoxin contamination is prevalent in cereal grains, nuts, fruit and green coffee beans. If feeds are contaminated with mycotoxins, associated residues and metabolites may appear in animal products.

Because food ingredients are now often sold in areas of the world far remote from their production sites, the need for extended safe shelf-life for these products has also expanded to prevent consumers' health from any toxic substances produced by pathogenic spoilage organisms (Holley and Patel, 2005).

Fungi are omnipresent in the environment, and infection due to fungal pathogens has become more frequent [Walsh & Groll, 1999; Fleming *et al.,* 2002]. With the rise of HIV and immunocompromised patients, opportunistic fungal pathogens became a frequent cause of morbidity and mortality [Garbino *et al.,* 2001]. The increment of the number of immunocompromised hosts after organs transplantations has grown steadily since the first human kidney was transplanted in 1954, with the introduction of novel immunosuppressive agents and improved rates of survival. Thus the incidence of opportunistic infections, including fungal infections has also increased [Winston *et al.*, 1995; Fishman & Rubin, 1998].

There is a surplus of current and ethnobotanical literature listing plants with known pest control properties. Over 2000 plant species are known to have insecticidal properties [Grainge & Ahmed, 1988] and many of these plants are still used by natives in developing countries. It is estimated, however, that only 5 to 15% of the 250,000 to 500,000 known plant species have been assessed for biological activity, and there is petite information concerning biological properties in the infinite majority of tropical plant species [Balandrin *et al.,* 1993; Mendelsohn, & Balick, 1995].

Because of greater consumer awareness and concern regarding synthetic chemical additives, foods preserved with natural additives have become a popular need. This has led researchers and food processors to look for natural food additives with a broad spectrum of antimicrobial activity [Marino *et al.*, 2001].

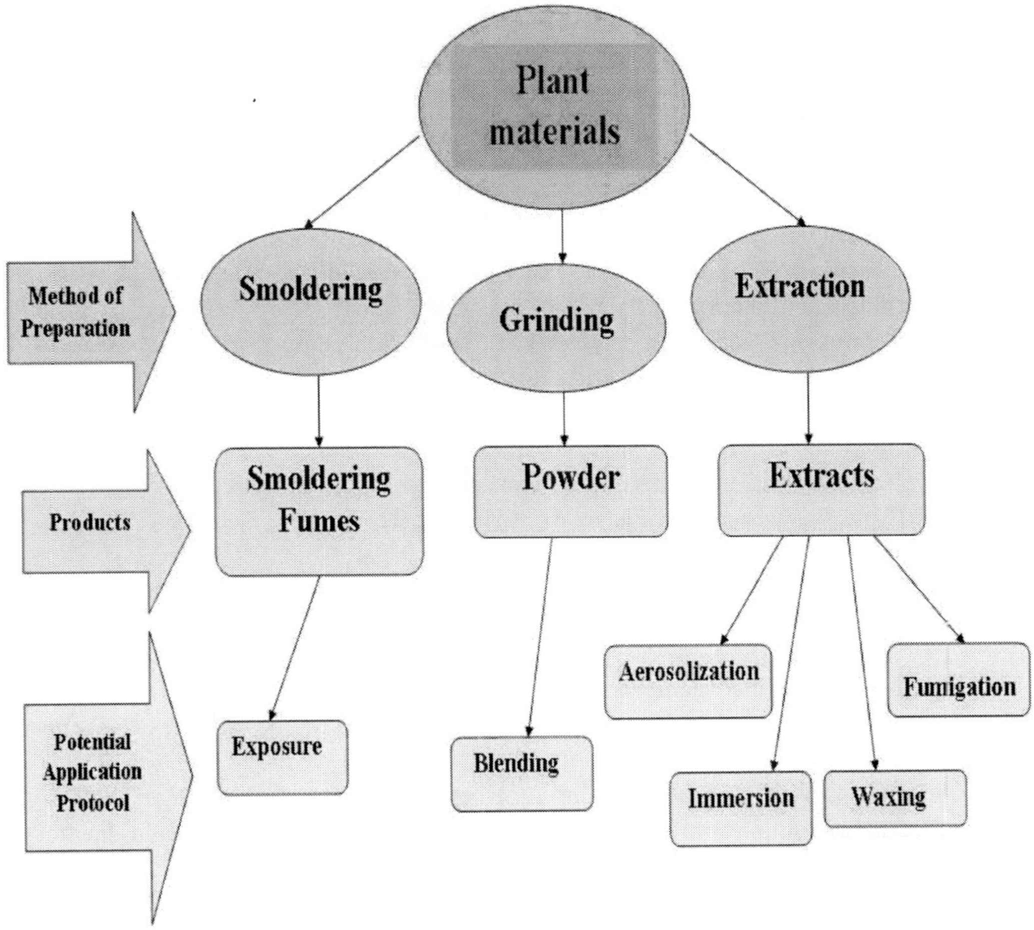

Figure 1. Schematic diagram for the possible handling procedure for dealing with plant materials to be used as antimicrobials.

Viewed globally, food safety is regularly compromised by the presence of mycotoxins occurring in cereal grains, nuts, fruit and green coffee beans. If feeds are contaminated with mycotoxins, associated residues and metabolites may appear in animal products.

Since many food ingredients are now often sold in areas of the world far remote from their production sites, the need for extended safe shelf-life for these products has also expanded [Holley & Patel, 2005]. A need has always been felt for some control on the quality of foodstuffs. Such control is intended to protect the health of the consumer and to prevent misrepresentation with respect to the composition of food. The abiding principle in food safety must be the prevention of contamination by the suppression of fungal pathogens responsible for mycotoxins production, as curative methods are of limited efficacy [D'Mello *et al.*, 1998].

Many cereal and other crops are susceptible to fungal attack either in the field or during storage. These fungi may produce, as secondary metabolites, a diverse group of chemical substances known as mycotoxins. There can be wide, year to year, variations in the levels of mycotoxins in foods, depending on many factors, such as adverse conditions favoring fungal invasion and growth. Since prevention is always better than cure, the former approach is now being utilized to reduce/ eliminate the chances of mycotoxins entering the food and feed production chains [DeVries *et al.*, 2002].

Encouraging results on the use of natural products to control postharvest fungal invasion indicate that we should be able to develop natural fungicides that would be as effective as synthetic fungicides, and actually safer for man and the environment [Mahmoud, 1999; Abbas, 2005].

There are many ways to deal with plant materials and derivatives for their use as fungicidal sanitary agents. Figure 1 illustrates some potential handling procedures for the preparation and usage of plant materials.

Two successful application procedures were conducted using plant derivatives for the control of fungal invasions in closed and semi-closed systems. The first used plant extracts aerosol for controlling the growth of *Candida albicans* [Tayel & El-Tras, 2009] and *Penicillium digitatum* [Tayel *et al.,* 2009] whereas the second investigated the exposure to smoldering plant fumes as a potent sanitation protocol to control growth of *Aspergillus flavus* [Tayel , 2010] and *Botrytis cinerea* [Tayel & El-Hadary, 2010].

2. Plant Extract Earosolization

Aerosolization is defined as dispersion in air of a liquid material or a solution in the form of fine mist, usually for therapeutic and sanitary purposes and room disinfection [Wanner & Rao, 1980; Hiom *et al.*, 2003]

Aerosolization of closed and semi-closed places with aerosols containing small droplets from plant extracts solution proved to be a successful practice for sanitizing and disinfecting places assumed to have such fungal contaminations.

A UV sterilized incubator with dimensions of 200 X 200 X 250 cm was designed to experiment the effect of plant extracts aerosolization (pomegranate peel extract was evaluated in these studies) on the growth of fungal pathogens in a semi-closed system (Figure 2).

The designed system is a suitable method to evaluate the antimicrobial, sanitizer and disinfectant capability of natural products against airborne pathogens. Plant extract should be examined first by qualitative and quantitative *in vitro* methods to assess its antimicrobial potentiality, e.g. Agar diffusion assay and determination of minimal inhibitory concentration, respectively.

For the usage of aerosolization room, sterilized Petri dishes, containing fresh uninoculated agar media, were placed on the steal network shelves at different placement heights in the room. Dishes covers were removed and then 50 ml of fungal cell suspension (10^3 cell/ml) were aerosolized using a nebulizer attached to air compressor to generate a uniformly droplets size. Control plates were recovered then a definite plant extract solution with a known concentration, to calculate the saturation per volume, was aerosolized and the first set of treated plates was recovered. Repeating the latest steps, final plant extract aerosol saturation in the incubator reached up to the maximum level to be tested. Treated plates were incubated at the appropriate temperature for pathogen growth for 24 h. then colony forming units (CFU) were counted and plotted for each treatment.

2.1. Factors Affecting Sanitation Using Plant Extracts Aerosol

2.1.1 Aerosol Carrier

The extraction solvent is mainly the preliminary choice to be used as carrier for plant extracts, but some of them could have toxic or allergenic effects, e.g. organic solvents, through inhalation or exposure. Sanitizers that are readily soluble in water can be easily delivered to target microorganisms by using a fine mist aerosol as a carrier [Oh *et al.,* 2005]. Although gaseous sanitizers are effective, only a limited number of sanitizers can convert into gas form. But aerosolization can deliver diverse antimicrobial agents [Fiser, 1978; Profe & Steiger, 1982].

2.1.2. Minimal Saturation Level (MSL)

The minimal saturation level (MSL, mg /m^3) expresses the lowest weight of plant extracts carried within aerosol per aerosolization room volume that could completely inhibit the growth of pathogenic microorganisms dispersed in room air. The MSL varies among plant extracts aerosols according to the antimicrobial activity of the applied extract.

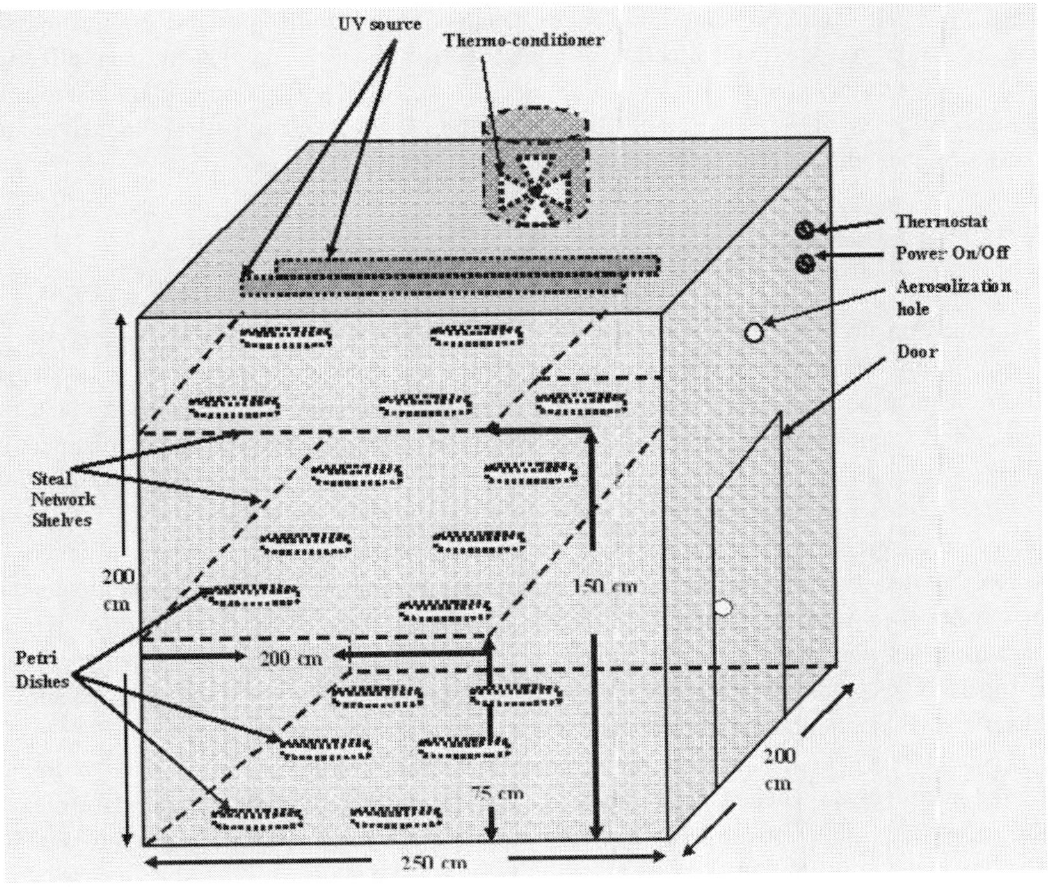

Figure 2. Schematic diagram of the aerosolization room design which used for evaluating the sanitation potency of plant extracts' aerosol.

2.1.3. Aerosol Droplet Size

Droplets size of the plant extract carrying aerosol is a critical factor for the efficiency of aerosolization as sanitizing method. This factor is highly correlated with the exposure time and the heights of places to be disinfected.

The settling time is decreased with the increase of droplets size [Oh *et al.,* 2005]. The strong sanitizing effect will be in the middle and lower positions could be with the application of bigger aerosol droplets' sizes, whereas by using an aerosol or fog generator adjusted to produce very small particle size, 3-4 μm or less, a higher diffusibility and efficiency could be expected all through the room space.

2.2. Method Applicability

With its diffusible activity and broad applicable antimicrobial spectrum, aerosolization may be introduced as an alternative antimicrobial delivery system. It could be recommended the application of plant extracts' aerosol as an effective method for sterilizing semi- closed places which are suspected with *C. albicans* contamination, e.g., operation rooms, hospital wards, jail cells, poultry farms. Safe room disinfection is another possible application for aerosolization. Aerogenic disinfection of poultry houses has been thought to be an effective tool for increasing poultry production [Steiger *et al.*, 1982]. Regarding plant pathogens, aerosolization is an effective tool for disinfecting crops packinghouse, shipment silos, transporting cabinet as well as food industry equipments.

3. Smoldering Plant Fumes

Based on the historical application of modified atmosphere created by Joseph; son of Jacob, in his mentioned story in Qur'an and bible, for protecting stored food grains up to seven years from spoilage, the construction of a fumigation system was carried out using fumes and vapors released from smoldered plants as sanitizing, protecting and disinfecting agents [Tayel, 2010].

The expression "Modified atmosphere" is used primarily for the technology applied to protect food products from quality loss and extend the period during which products of acceptable quality can be marketed. The modification process often tries to lower the amount of oxygen (O_2), moving it from 20% to 0%, In order to slow down the growth of aerobic organisms and the speed of oxidation reactions. The removed oxygen can be replaced with nitrogen (N_2), commonly acknowledged as an inert gas, or carbon dioxide (CO_2), which can lower the pH and inhibit the growth of microorganisms [Phillips, 1996].

Smoldering is defined as slow, low-temperature, flameless form of combustion, sustained by the heat evolved when oxygen directly attacks the surface of a condensed-phase fuel. In this proposed sanitation system, plant materials smoldering normally generate notable amounts of CO_2 in addition to many aromatic volatile compounds, so that; this feature is one of suggested fungicidal factors of the presented system.

Release of volatile oils accompanied with phenolic compounds and CO_2 during plant smoldering provided potent fungicidal factors which were capable of preventing pathogenic fungi contamination and growth.

Figure (3) depicts the schematic design of smoldering room. The system was experimentally designed for evaluating the efficacy of fumes and vapors from smoldered plants on the inhibition of pathogenic fungi growth. Plant smoldering was applied using electric gradable hot plates at approximate surface temperature of 500 °C. Smoldering time was considered from the beginning of each treatment up to smoke suction and plates recovering. The smoldering experiment included weighing of powdered plant materials in correlation to the cabinet volume, e.g. g/m^3, and putting them on hot plates surface (maximum 10 g/ hot plat).

Inoculated Petri dishes with fungal spores were placed uncovered on the network shelves and exposed to smoldering fumes for 45 min.

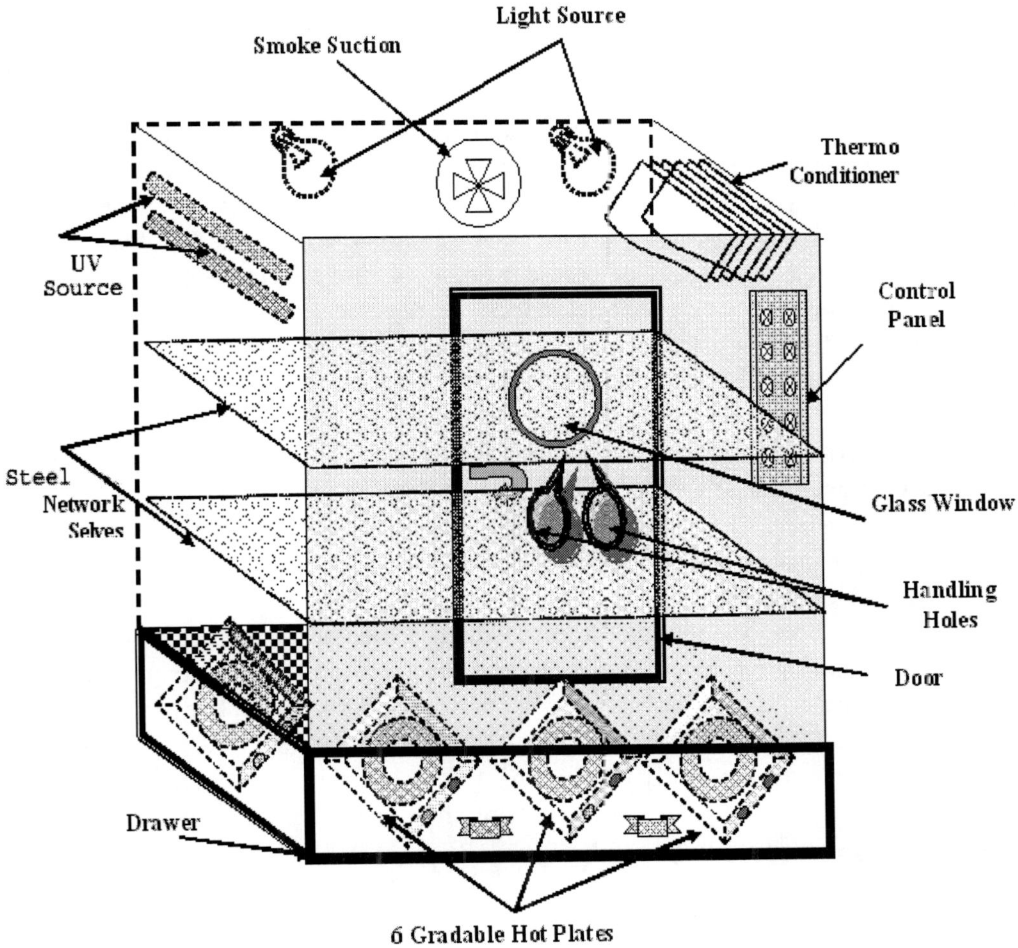

* Dimensions: 125 cm (height) X 100 cm (length) X 60 cm (width).

Figure 3. Schematic diagram of the smoldering room design which used for evaluating the fungicidal effect of smoldered plants.

Fume was then sucked and the plates were recovered through handling holes, which attach with rubber expandable gloves, and incubated for 7 days. The inside room temperature was automatically adjusted to 40-45 °C during experiments using thermo conditioner.

3.1. Factors Affecting the Efficiency of Plant Smoldering Process

Considering the novelty of plant smoldering fumes' exposure as a sanitation and biopreservation protocol, the following factors were proved to influence the efficiency of the process:

3.1.1. Smoldered Plants Type

The evaluation of antimicrobial activity for many smoldered plant types revealed that sanitation potency is highly dependent on the species of plant to be smoldered. Most of examined smoldered plants exhibited notable antifungal activity against pathogenic fungi, e.g. *A. flavus* and *B. cinerea*, but with varied potentialities. Many compounds are formed and released during plant materials smoldering due to heat exposure and oxygen attacking, this released compounds include CO_2, CO, essential oils and phenolic compounds. The composition, structure, active principals as well as functional groups of the plant play an important role in determining their antimicrobial activity. Usually compounds with phenolic groups are most effective [Deans & Ritchie, 1987; Dorman & Deans, 2000]. Plant materials, in this protocol, have to be powdered before their smoldering to maximize their surface area, which facilitate the O_2 attack of the particles. The released amounts from each individual compounds varied from a plant to another due to their constituents and active components; this could be the explanation of the in antifungal disparity among examined smoldered plants.

3.1.2. Minimal Inhibitory Relative Amount (MIRA)

The sanitation and biopreservation effect of smoldering plant fumes' exposure could be attributed the combined involvement of many antimicrobial agents released from plants smoldering. The enhancement of fungitoxic activity of mixtures of the oils was attributed to the joint action of two or more substances present in these oils [Scardavi, 1966]. The synergism between many antifungal agents would be beneficial in postharvest protection because the pathogen would not easily generate resistance against several fungitoxicants [Tripathi & Dubey, 2004]. However, the minimal inhibitory relative amount (MIRA) calculation for each plant, lowest amount from smoldered plant per volume unit (g/m^3) that released fumes able to completely inhibit fungal growth within this space, could emphasize the actual required amount from the plant, as a whole, including its combined constituents releasing during smoldering process.

3.1.3. Target Crops

The delivery system and selection of an antimicrobial should be based on the sensory and chemical compatibility of the antimicrobial with the target material, its stability given the type of primary preservation system used, its effectiveness against the expected undesirable microorganisms, its safety and its cost effectiveness [Wanger & Moberg, 1989; Holley &Patel, 2005]. The nature of targeted surface to be exposed to smoldering fumes is an important factor for adjusting process conditions; walls and large solid surfaces are generally easier to treat with minimal limitations. Regarding food materials and crops, solid hard food, e.g. cereal grains, dried foods, nuts and egg shells, are favored to be treated with smoldering fumes without harmful or undesirable effects on their appearance and sensory characteristics (Figure 4)

On the other hand, perishable and soft foods, e.g. fresh fruits and meat products, need more accurate considerations for their treatment by smoldering fumes as biopreservatives to avoid the possible interactions between their tissues and smoldering fumes' constituents. The choice of a plant with less aromatic scent, decreasing exposure period, reduction of interior room temperature, complete fume suction after treatment and its replacement with fresh sterile air are potential precautions to avoid alteration in sensory and physical characteristics of treated foods.

3.2. Method Applicability

The diffusibility of gaseous sanitizers is the best among commonly applied agents. Fume has the ability to penetrate and reach to every minute and hidden pore through over the target surface, as well as flat surfaces themselves, to perform its sanitation and disinfection action. The proposed technique is highly suitable for sanitizing any food containers, refrigerators, processing rooms, transporting shipments and cereal storage silos. Exposure to smoldered plant fumes is also an effective method for the elimination of fungal pathogens attacking food and feed materials, preventing of foodstuffs from fungal spoilage and prolonging storability period and shelf life of both hard and soft foods.

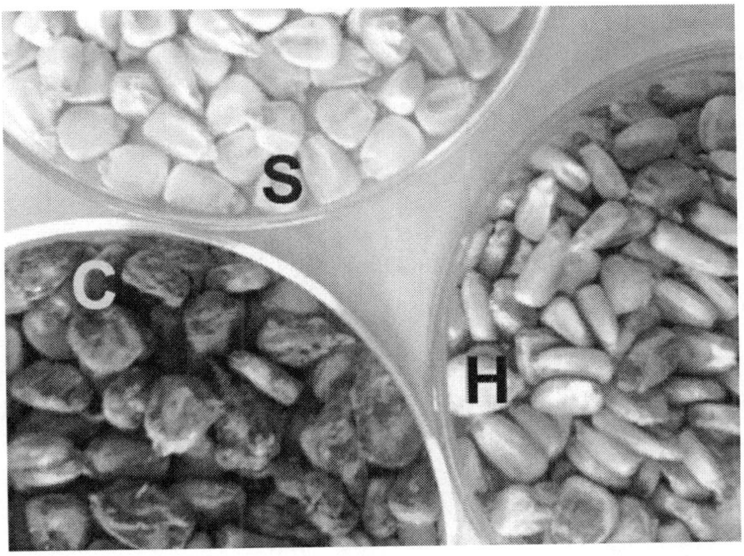

Figure 4. Effect of exposure to smoldering fumes, from different relative amounts of cinnamon and storage for 15 days at 30 °C, on the infection progress with *Aspergillus flavus* on experimentally infected maize grains: S: Exposure to 20 g/m^3; H: Exposure to 10 g/m^3; C: Control group (non-exposed grains).

Conclusion

Many plants include potent antimicrobial constituents not only against plant and insect pathogens but also against human fungal pathogens, these biological agents, because of their natural origin, are highly biodegradable and most of them have almost no-residues in the environment. Encouraging results on the use of natural products to control fungal invasion indicate that it is greatly recommended to explore and develop novel natural fungicides that would be as effective as synthetic fungicides, and presumably safer for man and the environment. The investigation of plant derivatives, i.e. extract carrying aerosols and smoldering plant powders, verified their potentialities as powerful sanitizer and disinfectant agents against many fungal pathogenic strains invading both human and plants. Designed systems using these plant derivatives, i.e. aerosolization room and smoldering room, are efficient tools to be applied for the prevention and elimination of fungal pathogen within closed and semi-closed places. Aerosolization room using plant extracts carrying aerosols is suitable for sanitizing medical operation rooms, hospital wards, jail cells, poultry farms, transporting ships and food containers after carrying out further studies for optimizing all operation conditions. Whereas plant smoldering process could be effectively applied for preserving food and feed stuffs, especially hard foods like grains, nuts and dried fruits, with minimal subsequent effects on sensory characteristics. The exposure to smoldered plant fumes is also a successful treatment for sanitizing the different types of food holders and containers with controlled volumes, crops packinghouse, food and dairy industries' equipments.

Acknowledgment

Always and foremost, Thanks are indebted to our god "ALLAH" forever for his mercy guiding and helping.

References

Abbas, H. K. (2005). *Aflatoxin and food safety.* Boca Raton, CRC Press, Taylor & Francis Group.

Balandrin, M. F., Kinghom, A. D. & Famsworth, N. R (1993). Plant derived natural products in drug discovery and development. *in Human Medicinal Agenfs fiom Plants. ACS* Symposium Series 534, Washington, American Chemical Society, D.C. 2-12.

D'Mello, J. P .F.; Macdonald, A. M. C.; Postel, D.; Dijksma, W. P. T. & Dujardin, A.(1998). Pesticide use and mycotoxin production in Fusarium and Aspergillus phytopathogens. *European Journal of Plant Pathology*, 104 (8), 741–751.

DeVries, J. W.; Trucksess, M. W. & Jackson L. S. (2002). *Mycotoxins and Food Safety.* New York, NY, Kluver Academic/Plenum Publishers.

Deans, S.G. & Ritchie, G. (1987). Antimicrobial properties of plant essential oils. *International Journal of Food Microbiology*, 5, 165–180.

Dorman, H. J. D. & Deans, S. G. (2000). Antimicrobial agents from plants: antibacterial activity of plant volatile oils. *Journal of Applied Microbiology,* 88 (2), 308–316.

Fiser, A. (1978). Disinfection of air and dust in fattening houses for chickens by lactic acid aerosol. *Acta Veterinaria,* 47, 173-83.

Fishman, J. A. & Rubin, R. H. (1998). Infection in organ transplant recipients. *New England Journal of Medicine*, 338, 1741.

Fleming, R.V.; Walsh, T. J. & Anaissie, E. J. (2002). Emerging and less common fungal pathogens. *Infectious Disease Clinics of North America* 16, 915– 933.

Garbino, J.; Kolarova, L.; Lew, D.; Hirschel, B. & Rchner, P. (2001). Fungemia in HIV-infected patients: a 12-year study in a tertiary care hospital. *AIDS Patient Care STDs,* 15, 407–410.

Grainge, M. & Ahmed, S. (1988) *Handbook of Plants with Pest-Control Properties,* New York, John Wiley & Sons.

Hiom, S. J.; Lowe, C. & Oldcorne, M. (2003). Assessment of surface bioburden during hospital aseptic processing. *The International journal of pharmacy practice,* Sept: R62.

Holley, R. A. & Patel, D. (2005). Improvement in shelf-life and safety of perishable foods by plant essential oils and smoke antimicrobials. *Food Microbiology,* 22(4), 273–292.

Mahmoud, A. L. (1999). Inhibition of growth and aflatoxin biosynthesis of Aspergillus flavus by extracts of some Egyptian plants, *Letters of Applied Microbiology,* 29, 334–336.

Marino, M., Bersani, C. & Comi, G. (2001). Impedance measurement to study antimicrobial activity of essential oils from Lamiaceae and Compositae. *International Journal of Food Microbiology,* 67, 187–195.

Mendelsohn, R. & Balick, M. J. (1995).The value of undiscovered pharmaceuticals in tropical forests, *Economic botany, 49*, 223-228.

Oh, S.W.; Gray, P. M.; Dougherty, R. H. & Kang, D. H. (2005) Aerosolization as novel sanitizer delivery system to reduce food-borne pathogens. *Letters in applied microbiology,* 41, 56–60.

Phillips, C. A. (1996). Review: Modified atmosphere packaging and its effects on the microbial quality and safety of produce. *International Journal of Food Science and Technology,* 31, 463-479.

Profe D. & Steiger A. (1982). *Aerosol distribution in the air on surfaces of the animal house. Taguingsbericht,*197, 113–120.

Scardavi, A. (1966). Synergism among fungicides. *Annual Review of Phytopathology,* 4, 335–348.

Steiger, A.; Trenner, P. & Profe, D. (1982) *Technique of aerogenic disinfection during the service period in large animal houses. Taguingsbericht,* 197, 93–97.

Tayel A. A.; El-Hadary M. H. (2010). Grey mould control and shelf life extending in strawberry through exposure to smouldered plant fumes as natural fungicides. *Proceedings of World Academy of Science, Engineering and Technology,* International Conference on Bioscience, Biotechnology, and Biochemistry , Rome, Italy, April 28-30.

Tayel A. A.; El-Baz A. F.; Salem M. F. & El-Hadary M. H. (2009). Potential Applications of Pomegranate Peel Extract for the Control of Citrus Green Mold. *Journal of Plant Diseases and Protection*, 116 (6), 252-256.

Tayel, A. A. (2010). Innovative system using smoke from smoldered plant materials to control *Aspergillus flavus* on stored grain. *International Biodeterioration & Biodegradation.* 64 (2) , 114-118

Tayel, A. A. & El-Tras, W. F. (2009). Anticandidal activity of pomegranate peel extract aerosol as an applicable sanitizing method. *Mycoses.* 53, 117–122.

Tripathi, P. & Dubey, N. K. (2004). Exploration of natural products as an alternative strategy to control postharvest fungal rotting of fruit and vegetables. *Postharvest Biology and Technology,* 32 (3), 235–245.

Walsh, T. J. & Groll, A. H. (1999). Emerging fungal pathogens: evolving challenges to immunocompromised patients for the twenty-first century. *Transplant Infectious Disease* 1, 247–261.

Wanger, M. K. & Moberg, L. J.(1989). Present and future use of traditional antimicrobials. *Food Technology,* 43 (1), 143–147.

Wanner, A. & Rao, A. (1980). Clinical indications for and effects of bland, mucolytic, and antimicrobial aerosols. *The American review of respiratory disease,* 122, 79–87.

Winston, D. J.; Emmanouilides, C. & Busuttil, R.W. (1995). Infection in liver transplant recipients. *Clinical Infectious Diseases* 21, 1077-1091.

INDEX

A

absorption, 8, 13, 19
abstraction, viii, 78, 109
accessibility, 107, 160, 161
accountability, 183
acetic acid, 4
acid, 3, 4, 5, 6, 7, 8, 11, 15, 16, 19, 20, 23, 24, 28, 29, 34, 35, 89, 93, 94, 129, 130, 191
acidity, 9, 93, 99
adaptation, 22
additives, 2, 10, 216
administrators, 181
adsorption, 104, 130, 143
advantages, 2, 11, 159, 216
advocacy, 70
aerobic bacteria, 27
aerosols, 218, 219, 224, 226
aflatoxin, 225
Africa, 107, 113, 116, 152, 169, 175, 176, 208
agar, 34, 218
agencies, 43, 44, 109, 151, 157, 158, 159, 168, 169, 170, 172, 184, 192, 207, 210, 212
agglomeration, 125
aggregation, 212
agricultural sector, 155, 156
agriculture, ix, 78, 107, 115, 144, 150, 159, 169, 184, 185, 186, 187, 188, 198
AIDS, 225
airplanes, 180
alcohols, 10
alfalfa, 9, 13, 28, 33, 34
alkaloids, 216
aluminium, 169
amino acids, 5, 11
ammonia, 3
anaerobic bacteria, 16
ANC, 93
anger, 222
antibacterial properties, 25
antioxidant, 2, 4, 10, 12, 13, 19, 22, 23, 30, 31
apples, 10, 11, 15, 22, 29, 30, 33
appropriate technology, 197
aquifers, viii, ix, 77, 78, 80, 85, 86, 87, 88, 89, 90, 93, 95, 96, 99, 103, 104, 105, 108, 109, 110, 112, 113, 114, 115, 116, 117, 122, 123, 127, 129, 130, 132, 135, 138, 139, 140, 141, 142, 143, 144, 145, 146, 149, 150
Argentina, 39
argon, 27, 28, 30, 32, 36
arrest, 170
arsenic, 169
Arunachal Pradesh, 46
asbestos, 169
ascorbic acid, 4, 7, 10, 13, 20
aseptic, 225
Asia, 75, 192, 195, 196, 198, 200
Asian countries, 192, 195
assessment, viii, 31, 37, 41, 48, 74, 80, 84, 86, 110, 111, 116, 117, 122, 146, 147, 148, 156, 163, 174, 177, 211, 213
assets, 42, 203, 208, 210, 211
attachment, 7, 31, 194
authorities, 158, 166, 208, 210
autonomy, 212
average costs, 209
avoidance, 42

B

Bacillus subtilis, 29
bacteria, 5, 7, 8, 9, 10, 12, 14, 27, 38, 154
Bangladesh, 39, 186, 197
banks, 169
barriers, 195
basic services, 182
bauxite, 78

beef, 180
beer, 180
behaviors, 157, 170, 183
Belgium, 3, 167, 173
benzene, 180
bible, 220
bicarbonate, 89, 124, 125, 133, 143, 144
biological activity, 216
biopreservation, 222
biosynthesis, 13, 225
birth weight, 39
blame, 166
Bolivia, 183
boreholes, 86, 87, 88, 107, 108, 122, 123, 139, 167, 205
Botswana, 152, 175, 176, 183
Brazil, 117, 149, 181
breakdown, 15, 154, 160
breathing, 156
breeding, 157
bridges, 17
brittle deformation, 86
bromine, 5
buildings, 193
Burkina Faso, 78, 155, 183
by-products, vii, 1, 3, 5, 6, 9, 31

C

cabbage, 4, 5, 7, 14, 20, 28
cadmium, 169
calcium, 4, 7, 20, 22, 30, 32, 94, 104, 105, 124, 133, 143
calibration, 149
Cambrian, 83
Cameroon, 166, 174
campaigns, 169, 170, 172
capacity building, 43
carbohydrate, 13
carbohydrates, 11
carbon, ix, 23, 26, 35, 94, 111, 116, 139, 146, 148, 195, 220
carbon dioxide, 23, 26, 35, 94, 111, 148, 220
carotenoids, 10, 13
case study, 64, 173, 174, 176
categorization, 105
category a, 105, 144
cation, viii, 77, 88, 93, 96, 103, 110, 130, 133
cattle, 128, 139
Census, 155, 174
Central Asia, 208
child mortality, 38, 39, 74, 153
childhood, 196
Chile, 174, 208
China, vii, 37, 193, 194, 195, 198
chlorination, 3, 31, 34, 191
chlorine, vii, 1, 3, 4, 5, 9, 16, 19, 20, 25, 26, 28, 29, 30, 33, 34, 35, 191, 200
chloroform, 3
chlorophyll, 12, 19
cholera, 38, 46, 180
chromium, 169
churches, 109
citizenship, 176
City, 37, 198, 215
class, 65, 89, 144, 193
clay minerals, 87
cleaning, 2, 4, 186
cleavage, 87
climate, x, 79, 108, 113, 117, 118, 120, 149, 203, 204, 211
climate change, 108, 118, 204, 211
climatic factors, 107
cluster analysis, viii, ix, 78, 89, 112, 113, 115, 117, 125, 128, 130, 149
clustering, 99, 125, 126
clusters, 99, 117, 125, 126, 128, 129, 131, 133
CO2, 4, 12, 16, 17, 18, 19, 23, 26, 30, 33, 159, 220, 221, 222
coatings, 11
cocoa, 128, 158
coffee, 216, 217
Colombia, 183
combined effect, 16, 93, 99
combustion, 220
commodity, 3, 17, 41
community, vii, viii, 37, 40, 42, 44, 45, 75, 151, 153, 157, 165, 166, 168, 170, 172, 173, 181, 184, 185, 186, 188, 189, 190, 196, 200, 203, 211
comparative advantage, 44
compatibility, 222
competition, 206, 208, 210, 213
complement, 86
complexity, 44, 207
compliance, 43, 154, 213
composition, 2, 15, 81, 82, 83, 84, 87, 106, 113, 121, 123, 206, 217, 222
compost, 43
composting, ix, 151, 157, 159, 169, 172, 181, 193
compounds, x, 4, 5, 10, 11, 13, 14, 21, 23, 25, 28, 29, 31, 157, 169, 215, 220, 221, 222
compression, 27
conceptual model, 86
conditional mean, 56
conditioning, 2
conductivity, 86, 124
confinement, 117

conformity, 170
confounding variables, 38
connectivity, 139
consensus, 206, 208, 209, 211
conservation, 169, 191
consolidation, 87, 122, 209
constant prices, 46, 63
Constitution, 42
consumer protection, 154
consumption, 2, 19, 25, 31, 42, 93, 153, 167, 168
consumption rates, 22
contact time, 5, 15
contamination, viii, ix, 6, 10, 22, 29, 35, 40, 46, 77, 89, 90, 93, 99, 100, 107, 108, 116, 139, 156, 180, 208, 216, 217, 220, 221
conviction, 171
cooling, 86
cooling process, 86
coordination, viii, 37, 43, 62, 64, 71, 167, 184
correlation, 93, 221
corrosion, 3, 9
corruption, 166
cortex, 8
cost effectiveness, 222
cost-benefit analysis, 185
cotton, 144
covering, 78, 117, 139
creativity, 165
crops, x, 22, 25, 79, 81, 108, 121, 139, 143, 144, 154, 215, 216, 217, 220, 223, 224
crystalline, 19, 110, 122, 147
cultivation, 12
culture, 34, 160
Cyprus, 175

D

danger, 157, 159, 197
database, 195
datasets, 99
deaths, 152, 156, 180, 182, 184
decay, 12, 17, 25, 32, 33, 35, 36
decentralization, 42, 44, 70, 71, 158
decision makers, 183
decomposition, 14, 160
defecation, 160, 162, 170, 172, 181
defence, 11, 12
deficiency, 52, 61, 70, 187
deformation, 82, 83, 114
degradation, 12, 99, 153, 158
democracy, 212
dengue fever, 153
Denmark, 3, 174
Department of Health and Human Services, 26
dependent variable, 38, 46, 52, 59, 69
deposition, 152, 169
deposits, 78, 81, 82, 85, 120, 158
depreciation, 184
derivatives, x, 215, 216, 218, 224
destruction, 154, 158
detection, 22, 43, 89
developed countries, 166, 194, 197, 203
developed nations, 205
developing countries, 75, 107, 152, 153, 154, 166, 170, 176, 180, 181, 185, 187, 192, 194, 195, 196, 197, 199, 200, 203, 204, 205, 206, 211, 216
development assistance, 184
deviation, 51, 65, 89, 128, 185, 186
diarrhea, 39, 75, 180, 182
diffusion, 19, 218
dignity, 154
disability, 180
discharges, 156, 199, 200
discriminant analysis, 113, 149
disinfection, vii, 1, 3, 5, 6, 12, 22, 27, 31, 33, 34, 191, 200, 218, 220, 223, 225
dispersion, 48, 52, 59, 62, 69, 143, 218
distribution of income, 62, 64, 71
diversity, 205, 208, 209, 211
dizziness, 16
DNA, 11, 13
doctors, 50, 65, 69, 72
dominance, 133, 168
donors, 153, 185
draft, 194, 197
drainage, 40, 46, 47, 93, 129, 144, 153, 154, 157, 158, 170, 200, 201, 206
drinking water, 5, 32, 35, 39, 40, 41, 42, 43, 44, 46, 52, 58, 64, 69, 71, 89, 90, 107, 114, 150, 180, 181, 187, 194, 196, 204, 210, 213
drought, 44, 173
drug discovery, 224
drugs, 182, 216
drying, 2, 108
dumping, 157, 158, 169, 170, 172

E

Eastern Europe, 193
ecology, 107, 169
economic activity, 169
economic development, 107, 154, 164, 172, 187, 199
economic efficiency, 211, 212
economic growth, 154, 158
economic status, 46
economies of scale, 208, 209, 212
economy, vii, 107, 116, 156, 169, 176
EEA, 167, 174

Efficiency, 38, 72, 73, 74, 75, 213, 222
effluent, 156, 191, 192, 193, 194, 205
egg, 29, 223
Egypt, 215
elders, 160
electrical conductivity, 89, 104, 143, 144
electricity, 58, 64, 71, 170, 189, 195, 205, 207, 210
electrodes, 8
electrolysis, 8, 9
electron, 30
emission, vii, 1, 2, 9, 157, 160
employment, 170
empowerment, 43
endogeneity, 56, 58, 60
enforcement, ix, 108, 110, 151, 166, 170, 171, 172
engineering, 24, 41, 43, 44, 180, 191, 207
England, 174, 207, 208, 210, 213
environmental conditions, 80, 159, 168
environmental degradation, 152, 154, 156
environmental effects, 109
environmental factors, 204
environmental impact, x, 22, 31, 203, 205
Environmental Protection Agency, 26, 175
environmental quality, ix, 151
environmental regulations, 110, 210
environmental sustainability, 199
enzymatic activity, 12
enzymes, 8, 10, 14, 17, 19
EPA, 5, 26
epidemiology, 73, 75
equality, 71
equilibrium, 35
equipment, 3, 9, 11, 43, 167
equity, 71, 185, 195
erosion, 82, 86
ESR, 104
ethylene, 27
European Union, 154, 166
exercise, 52, 59, 69
expenditures, 182, 213
experiences, x, 179, 181, 196
expertise, 43
exploitation, ix, 80, 107, 108, 115
exploration, 110, 111
exposure, x, 3, 6, 10, 13, 15, 17, 22, 31, 33, 158, 163, 182, 215, 218, 219, 220, 222, 223, 224, 225
externalities, 208
extraction, ix, 80, 99, 109, 115, 116, 126, 208, 219

F

facies, 126
factor analysis, ix, 88, 89, 93, 94, 96, 113, 115, 117, 126, 132, 149
factories, 156
fairness, 185
farmers, 99
farms, 95, 108, 128, 139, 220, 224
fast food, 180
fat, 15
faults, 82, 85, 121
FDA, 4, 5, 6, 14, 26
feldspars, 87, 103
fermentation, 16, 17, 18
fertility, 159, 169
fertilization, 33
fertilizers, ix, 95, 96, 115, 128, 135, 139, 146, 159, 169
fever, 39
fiber, 194
films, 5
filters, 175
fish, 8, 108, 169
fitness, 127
fixed effect model, 58
flavonoids, 10, 12, 216
flavour, 7, 18
flexibility, 13, 107
flocculation, 43, 143
flooding, 118, 154, 158
flora, 11, 12, 33
food industry, 8, 9, 11, 19, 25, 27, 28, 220
food products, 220
food safety, 22, 216, 217, 224
Ford, 209, 212
foreign exchange, 154, 158
fractures, 85
fragments, 81
France, 164, 181, 208
free radicals, 14
freshwater, 94, 99, 129, 130, 146, 204
fructose, 15
fruits, 2, 6, 9, 10, 19, 23, 24, 25, 27, 28, 29, 30, 32, 33, 34, 35, 36, 154, 223, 224
funding, 41, 43, 109
fungal infection, 216
fungi, 5, 32, 216, 217, 221, 222

G

garbage, 153, 154, 157, 164, 167, 172
GDP per capita, 194, 195
gender differences, 160, 164
geography, 107
geological history, 117
geology, 107, 112, 113, 114, 117, 149
Germany, 3, 112, 164, 181
glucose, 15

glutathione, 9
GNP, 181, 182
google, 179
governance, 174, 212
government intervention, 172
government policy, 168
grades, 79, 82, 120, 121
graduate students, 180
granite, 81, 82, 83, 84, 85, 92, 121, 138, 144
grass, 43
gravity, 205
gross domestic product, 59
groundwater, viii, ix, 41, 42, 77, 78, 79, 80, 85, 86, 87, 88, 89, 90, 93, 96, 99, 100, 102, 103, 104, 105, 106, 107, 108, 109, 110, 111, 112, 113, 114, 115, 116, 117, 119, 122, 123, 124, 125, 126, 127, 129, 132, 133, 134, 135, 138, 139, 140, 141, 142, 144, 145, 146,뫰147, 148, 149, 150, 204
grouping, 85
growth rate, 10
guidance, 43, 200
guidelines, 89, 90, 173, 188, 194
guiding principles, 169
Guinea, 39, 72, 78, 141, 155
Guinea, Gulf of, 78, 141, 155
Guyana, 183

H

habitats, 46
harvesting, 3, 216
hazards, 22, 166, 172
health care system, 38, 64
health education, 44
health effects, 40
health expenditure, 50, 53, 64, 70, 71
health information, 45
health problems, 99, 208
health services, vii, 37
health status, 38, 41
health systems, vii, 37, 41, 48, 70
heat treatment, 30
heavy metals, 90, 93, 100, 154
height, 221
helium, 28
hemisphere, 118
hepatitis a, 47
HIV/AIDS, 154, 176
Hong Kong, 207
hookworm, 38, 182
horizontal integration, 209
host, 39, 81, 120
hotels, 193
household income, 39
hub, 48, 192
human development, 46, 52, 69
Human Development Report, 45, 48, 74, 200
human dignity, 170
human resources, 43
hydrogen peroxide, 22, 27, 28, 32, 35
hydrolysis, 5
hydroxyl, 6, 14
hygiene, 39, 40, 168, 180, 182, 186, 197, 199, 200
hypothesis, 8, 40

I

illumination, 13
illusion, 176
immersion, 5
immunization, 51
immunocompromised, 216, 226
immunosuppressive agent, 216
impact assessment, 198
impacts, 128, 135, 153, 154, 157, 159, 166, 205
incidence, viii, 37, 38, 41, 46, 48, 156, 213, 216
income inequality, 197
independence, 155
independent variable, 51, 54
India, v, vii, 37, 38, 39, 41, 42, 43, 44, 45, 46, 48, 50, 56, 58, 60, 64, 69, 70, 71, 73, 74, 75, 147, 173, 181, 199, 207
Indonesia, 181, 193, 208
inefficiency, 56, 72, 73, 210
inequality, 70
infant mortality, 38, 46, 181
infectious hepatitis, 38
information sharing, 193
infrastructure, ix, 42, 48, 57, 58, 64, 71, 151, 158, 161, 169, 172, 181, 204, 205
inhibition, 12, 18, 26, 27, 221
inhibitor, 19
injections, 4
inoculation, 31
insects, 39, 160, 187
inspectors, 192
integration, 155, 209
internal consistency, 125
international financial institutions, 159
international trade, 154
Internet, 180
intervention, 52, 54, 62, 64, 70, 71, 113, 182, 183, 186, 200, 208
intrusions, 81
investors, 208
ionic strength, 104

ions, 3, 89, 104, 105, 117, 125, 128, 130, 132, 133, 143
Iran, 148
irradiation, 11, 34
isotope, 110, 111, 112, 114, 124, 140, 141, 147, 148
Israel, 149, 205
Italy, 30, 34, 225
Ivory Coast, 78

J

Japan, v, x, 8, 179, 180, 184, 185, 186, 187, 188, 189, 190, 191, 192, 193, 194, 195, 196, 197, 198, 199, 200, 201, 205
joints, 86
jurisdiction, 158, 168, 171

K

Kenya, 176
kidney, 3, 216
kinetics, 18, 30
Korea, 193
Kuwait, 149

L

lack of control, 38
lactic acid, 10, 12, 194, 225
lakes, 156
landscape, viii, 77, 153, 157
Latin America, 152, 181, 208
law enforcement, 170
leaching, 93, 99, 100, 129, 139
leakage, 11
legislation, 158
leisure, 154
life expectancy, 38, 46, 48, 52, 54, 58, 62, 64, 70, 71, 72, 153, 182
limestone, 82, 83, 88, 112, 121
lipids, 5
lipoproteins, 4
liquids, 191, 195
Listeria monocytogenes, 23, 24, 25, 26, 28, 29, 30, 31, 33, 34, 35
literacy, 48, 52, 59, 60, 61, 62, 69
liver, 3, 226
liver transplant, 226
livestock, 78, 108
local community, 41
local government, 44, 158, 159, 166, 168, 188, 189, 191, 193, 210
logging, 109
Luo, 23, 28, 35, 36
lycopene, 13
lying, 78, 81, 120

M

magnesium, 94, 104, 105, 124, 143
majority, 44, 107, 130, 132, 155, 158, 159, 163, 165, 169, 216
malaria, 46, 47, 156
Malaysia, 111, 148, 193
malfeasance, 212
management, vii, viii, 12, 37, 41, 42, 77, 78, 109, 110, 114, 150, 152, 153, 154, 166, 169, 170, 171, 172, 173, 175, 176, 181, 182, 199, 208, 210
Mandarin, 21
manganese, 78, 156
MANOVA, 113, 149
manpower, 52, 64, 69, 70, 71
manufacture, 169
manure, ix, 99, 115, 128, 146, 169
mapping, 41, 82, 84
markers, 27
mass media, 170
mass spectrometry, 112
matrix, 44, 84, 117
meat, 9, 223
media, 218
median, 38, 87, 89, 93, 99, 123, 128
medicines, 46
melon, 4, 5, 6, 7, 11, 15, 16, 17, 22, 31, 32, 33, 34
membership, 99
membranes, 4, 9
meta-analysis, 197
metabolism, 13, 17, 18, 26, 143
metabolites, 216, 217
metals, 100, 129
meter, 89, 129
methodology, 69, 73
mice, 153
microbial cells, 5
microorganism, 11, 29, 32, 194
Middle East, 193
migration, 51, 168
mineral water, 107
mineralogy, 117
mining, 93, 107, 155, 156, 168, 171
modeling, 88, 111, 133, 147
modelling, 33
modification, 19, 34, 194, 220
moisture, 16, 159
molds, 11, 29, 34
monitoring, 41, 42, 43, 90, 109
monopoly, 203, 205, 207, 208, 210, 211
morbidity, ix, 38, 70, 71, 73, 75, 151, 153, 216
mortality rate, 38, 39, 183

mosquitoes, 153, 157
multivariate statistics, 112
mycotoxins, 216, 217, 218

N

NaCl, 8
national policy, 41
natural food, 217
natural gas, 205, 207
nebulizer, 218
necrosis, 9
negative relation, 39
net migration, 168
Netherlands, 3, 25, 28, 111, 113, 149, 183
network elements, 206
networking, 192
New England, 75, 225
New Zealand, 183, 207
NGOs, 44, 109
Nigeria, 77, 112, 148, 173
nitrate, ix, 99, 115, 124, 127, 136, 139, 146
nitrogen, 28, 30, 192, 195, 220
nitrous oxide, 25, 31, 32
noble gases, vii, 1, 3
noise, 72, 132
normal distribution, 73
North America, 225
nursing home, 50
nutraceutical, 13, 25
nutrients, 143, 186, 192

O

oceans, 73
offenders, 170
oil, 11, 34, 78, 88
opportunities, viii, 37, 62, 64, 70, 71, 154, 168, 170, 175, 210
optimization, 109
organ, 225
organic compounds, 28
organic matter, vii, 1, 3, 15, 136, 146
organic solvents, 219
osmotic pressure, 143
oversight, 212
ownership, 203, 207, 208, 209, 210, 212, 213
oxidation, ix, 8, 9, 14, 15, 17, 29, 93, 100, 116, 129, 135, 136, 146, 220
oxidative reaction, 3
oxygen, 14, 21, 23, 26, 27, 28, 31, 35, 36, 105, 111, 140, 141, 142, 148, 220, 222
ozonation, 191
ozone, 15, 19, 22, 24, 27, 28, 29, 31, 32, 33, 34, 35, 191

P

Pacific, 75, 192, 193, 195, 196
packaging, vii, 1, 2, 6, 13, 23, 24, 25, 26, 28, 34, 35, 225
parallel, 120
parentage, 82, 121
pathogens, x, 3, 5, 7, 8, 9, 11, 15, 25, 29, 31, 35, 36, 38, 215, 216, 217, 218, 220, 223, 224, 225, 226
PCA, 126
PCR, 22
penalties, 170
penetrability, 5
per capita income, 46, 48, 52, 59, 69, 196
performance, 43, 51, 52, 64, 65, 69, 70, 71, 164, 191, 204, 207, 208, 209
peri-urban, 196
permeability, 11, 18, 86, 94, 105, 122, 143, 144, 145, 147
permission, iv
peroxide, 20, 33
personal hygiene, 40
pests, 153, 157
pharmaceuticals, 225
Philippines, 38, 73, 74, 208
phosphorus, 192
physiology, 28, 33
pilot study, 40
plants, x, 43, 143, 184, 187, 188, 189, 200, 215, 216, 220, 221, 222, 224, 225
plaque, 46
plastics, 194
policy initiative, 41
policy makers, 196, 204, 207
policy reform, 73
polio, 38, 152
politics, 183
pollution, 42, 94, 110, 147, 156, 157, 158, 159, 169, 171, 173, 182, 193, 196, 199
polymeric films, 2
polyphenols, 9
polythene, 89
population density, 59, 62, 78, 188, 189
population growth, 51, 107, 155, 160, 211
population size, 161
porosity, 87, 122
Portugal, 27, 207
positive correlation, 93, 96
potassium, 24, 94, 124
potato, 8, 12, 20, 24
poultry, 9, 34, 108, 220, 224

poverty, 39, 40, 48, 50, 52, 62, 69, 79, 158, 163, 168, 173, 175, 176, 177
poverty line, 48, 50, 52, 69
poverty reduction, 79, 163, 173, 177
precipitation, 105, 110, 140, 204
preservative, 30
prevention, 5, 191, 217, 218, 224
primacy, 41
principal component analysis, 117
private ownership, 209
privatization, 39, 159, 173
productivity, vii, 37, 80, 87, 143, 153, 205, 209, 210, 211
profit, 9, 159, 184
profitability, 158
programming, 200
project, 42, 43, 44, 79, 111, 124, 148, 171, 173, 175, 181, 183, 186, 190, 194, 197, 198
proliferation, 109, 139, 154
proteins, 5
Pseudomonas aeruginosa, 11, 22
public access, 184
public education, 166, 169, 170, 172
public expenditures, 185
public health, vii, 37, 41, 43, 44, 48, 52, 64, 70, 71, 73, 74, 157, 203, 204, 208, 210
public investment, 50, 52
public policy, 208
public sector, 51, 62, 64, 71, 211
public service, 211
public-private partnerships, 208
pumps, 43, 86, 107, 108, 186
purchasing power, 168
pure water, 8
purification, 11
purity, 15
pyrimidine, 11
pyrite, 81, 93, 100, 120

Q

quality assurance, 206
quality of life, 107
quality of service, 211
quality standards, 210
quartz, 82, 84, 86, 87, 121, 123

R

radiation, vii, 1, 11, 12, 13, 21, 22, 23, 24, 26, 27, 30, 32
Radiation, 11, 147
radiation damage, 30
radicals, 6
rain forest, 79, 120
rainfall, ix, 78, 94, 105, 115, 117, 118, 120, 135, 141, 153, 154, 158, 167, 169, 170, 204, 205
rate of return, 210
raw materials, 169
reactants, 5
reaction rate, 18
reactions, 6, 133, 141, 220
reactive oxygen, 2
realism, 175
reality, 160
recall, 50
recommendations, iv, 27, 70, 71
recreation, 154
recurrence, 155
recycling, 159, 170, 171
reforms, 42, 70, 71, 203
regression analysis, 46
regression model, 39
regulatory framework, 108
regulatory oversight, 211
rehabilitation, 175
relevance, 117, 153
reliability, 159, 205, 210
remediation, 116
remote sensing, 111
removals, 195
repair, 11
replacement, 185, 223
replication, 11, 205
research and development, 191, 192
residues, 6, 14, 216, 217, 224
resistance, 2, 11, 15, 26, 31, 34, 222
resource allocation, 168
resource management, 41, 166, 206
resources, vii, viii, ix, x, 43, 71, 77, 78, 108, 109, 110, 114, 116, 172, 173, 179, 181, 184, 212
respiration, 2, 16, 25, 26, 31, 94
responsiveness, 42
restructuring, 155
retail, 205, 206, 207, 209, 212
risk management, 211
RNA, 5
rocks, viii, 77, 81, 82, 83, 84, 85, 86, 87, 88, 90, 93, 94, 110, 111, 112, 120, 121, 122, 123, 133, 136, 142, 144, 147, 148
room temperature, 14, 222, 223
rubber, 222
runoff, 139, 154, 158, 204
rural areas, vii, ix, 37, 44, 50, 58, 64, 69, 71, 79, 115, 158, 159, 161, 162, 165, 185, 188, 189, 193
rural population, 44, 54, 56, 57, 58, 65, 70, 71, 78, 107, 152, 155, 168

rural poverty, 62, 64, 71

S

salinity, ix, 93, 94, 96, 101, 104, 105, 106, 110, 115, 116, 129, 132, 143, 144, 146
salts, 8, 22
sample variance, 125
sanctions, ix, 151, 172
SAP, 159
saturation, 218, 219
savannah, 79, 120
savings, 184
scale economies, 207, 209
scarcity, 69, 175
scatter plot, 95
schistosomiasis, 182
sea level, 78
seafood, 9, 38
secondary data, 45
secondary education, 52, 69
sedimentation, 156, 158
sediments, 81, 82, 83, 84, 85, 88, 120, 121, 128, 173
seed, 10
seminars, 192
senescence, 2, 12, 22, 25, 26
sensitivity, 11, 12, 17, 22
sentencing, 170
septic tank, 139, 181, 185, 186, 193
service provider, 42, 203
service quality, 42
settlements, 118, 156, 161
sewage, ix, 34, 38, 43, 89, 93, 96, 108, 115, 156, 165, 191, 193, 206, 210
ships, 224
shock, 8, 30, 32
short supply, 41
shortage, 65, 193
SIGMA, 56
signs, 51, 54, 65
silica, 94
silver, 6
simulation, 114, 150
Singapore, 205, 207
skin, 34
sludge, 185, 186, 187, 189
sludge disposal, 186
social development, 48
social responsibility, 176
social services, 161
social structure, 156
sodium, 3, 5, 7, 23, 24, 29, 34, 35, 94, 104, 105, 124, 143, 144
solid surfaces, 223
solid waste, 42, 152, 153, 156, 157, 171, 173, 175, 176, 191
solubility, 5, 15, 17, 19, 191, 200
South Africa, 152, 174, 175, 207
South Asia, 74
Southeast Asia, 199
Spain, 1, 31, 32, 167
species, 2, 6, 17, 216, 222
spore, 15, 26
sprouting, 7
stable isotopes, 89, 105, 142
staffing, 52, 65, 70, 71
stakeholders, 196
standard deviation, 99, 101, 102, 125, 127, 185
standardization, 89, 125
statistics, 200
steel, 9
sterile, 223
stomach, 173
storage, 2, 7, 9, 10, 12, 13, 14, 15, 16, 17, 19, 22, 25, 28, 29, 30, 31, 32, 35, 86, 110, 206, 216, 217, 223
stormwater, 157, 205
strategic planning, 194
streams, 85, 108, 167, 170
structural adjustment, 159
structural changes, 13
subsidy, viii, 37, 40, 159, 172
substitutes, 216
substitution, x, 215
success rate, 87, 122
succession, 81
sucrose, 19
Sudan, vii, 79, 120
Sun, 23
supervision, 90, 108
supply chain, 209
support services, 154
suppression, 16, 217
surface area, 222
surfactant, 33
surplus, 216
survey, 39, 155, 186
survival, 25, 30, 170, 216
susceptibility, 18
suspensions, 15, 22, 26
sustainability, viii, 40, 42, 78, 109, 168, 173, 175, 181, 186
sustainable development, 181
Switzerland, 3, 114, 150, 177, 199, 214
synthesis, 13, 15

T

tanks, 139

tannins, 216
Tanzania, 175
tariff, 42, 189
technical assistance, 43, 168, 171, 172
technical efficiency, 48, 73
technology transfer, 193, 198
telecommunications, 210
temperature, 2, 3, 4, 5, 11, 14, 15, 17, 25, 28, 29, 33, 78, 87, 89, 103, 117, 120, 140, 218, 220, 221
tertiary sector, 62
test data, 87, 123
testing, 22, 89
textbooks, 181
texture, 7, 8, 9, 14, 16, 30, 32, 84, 87, 123
Thailand, 75, 193, 195, 199, 200
Third World, 153, 154
threats, 170, 204
tissue, 8, 9, 12, 26
tobacco, 19
Togo, viii, 77, 78, 81, 82, 85, 87, 91, 96, 99, 103, 105, 106, 112, 117, 121, 123, 130, 135, 138, 141, 143, 144, 147, 148, 155
total energy, 78
tourism, 154, 156, 158, 170
toxic effect, 154
toxic products, 15
toxic substances, 216
toxicity, 3, 6, 17
trachoma, 38, 182
tracks, 173
training, 43, 191
transformation, 133
transgression, 81
transmission, 3, 39, 163, 206, 207
transparency, 166
transpiration, 79
transplant recipients, 225
transport costs, 207
transportation, 3, 187, 205, 207
treatment methods, 189, 190, 195
trial, 170
trichuriasis, 182
tropical forests, 225
turbulence, 15
turgor, 8
typhoid fever, 38

U

U.S. Geological Survey, 148
UNESCO, 113, 200
uniform, 82, 118, 121
United Arab Emirates, 204
United Kingdom, 183
United Nations Development Programme, 164
urban area, 42, 45, 46, 48, 51, 58, 158, 161, 169, 177, 188, 189, 212
urban areas, 42, 45, 46, 48, 51, 58, 158, 161, 169, 177, 188, 212
urban population, 42, 44, 78, 153, 158, 168, 204
urbanisation, 204
urbanization, 48, 51, 52, 70, 199
urine, 195
USDA, 114, 149, 150
UV irradiation, 25
UV light, 11, 13, 28
UV radiation, 29

V

vacancies, 65, 69
vacuum, 15, 16, 18, 24
valleys, 85, 87
valuation, 41, 185
Vanuatu, 183
variations, 111, 116, 117, 148, 160, 161, 162, 163, 218
varimax rotation, 93, 126
vector, 38, 39, 72, 187
vegetables, 2, 4, 6, 7, 8, 9, 10, 11, 12, 13, 14, 16, 17, 18, 20, 22, 23, 24, 25, 27, 28, 29, 30, 31, 33, 34, 35, 36, 107, 110, 154, 170, 226
vegetation, 79, 120, 139
vehicles, 2, 192
vein, 83, 86, 170
vertical integration, 209
viruses, 5
vision, 44
vitamin C, 4
vulnerability, 99

W

wages, 40
Wales, 210, 213
walking, 42
washing procedures, 22, 24
waste, viii, ix, 28, 77, 135, 139, 151, 152, 153, 154, 156, 157, 158, 159, 160, 161, 164, 165, 166, 167, 168, 169, 170, 171, 172, 173, 205, 206, 210
waste disposal, ix, 151, 152, 153, 154, 156, 157, 158, 159, 160, 164, 165, 166, 167, 169, 170, 171
waste management, ix, 151, 153, 154, 156, 165, 166, 167, 168, 169, 171
wastewater, vii, x, 39, 179, 184, 186, 187, 188, 189, 190, 191, 192, 193, 194, 195, 196, 198, 199, 200, 201, 203, 204, 205, 206, 208, 209, 210, 211

water quality, 38, 40, 41, 43, 99, 114, 116, 146, 149, 150, 180, 182, 183, 185, 186, 187, 193, 194, 195, 196, 199, 204, 205, 206, 208, 210, 211
water resources, vii, 38, 41, 42, 107, 116, 129, 150, 173, 204
water rights, 108
water supplies, x, 73, 87, 203, 205
watershed, 43, 174
wavelengths, 13
wealth, 160, 162, 168
websites, 46
welfare, 41, 70, 71, 107, 208
wells, viii, 44, 77, 79, 85, 86, 87, 88, 105, 107, 108, 110, 122, 123, 135, 139, 145, 205
West Africa, 78, 110, 111, 113, 147, 148, 155, 173
Western Europe, 167
wholesale, 205, 207, 212
withdrawal, 42, 159
workers, 4, 16, 81, 83, 120, 192
working population, 153
World Bank, 44, 74, 152, 153, 164, 168, 175, 177, 181, 182, 183, 196, 198, 199, 212, 213, 214
World War I, 187
worms, 156

X

xenon, 36

Y

yeast, 7, 8, 12, 16, 17, 29, 34
yellow fever, 153

Z

Zimbabwe, 175
zinc, 89